THEORETICAL [illegible]GY

PART TWO THEORETICAL IMMUNOLOGY

THE PROCEEDINGS OF THE THEORETICAL IMMUNOLOGY WORKSHOP, HELD JUNE, 1987 IN SANTA FE, NEW MEXICO

Alan S. Perelson, *Editor*
Theoretical Division
Los Alamos National Laboratory

Volume III

SANTA FE INSTITUTE
STUDIES IN THE SCIENCES OF COMPLEXITY

Addison-Wesley Publishing Company, Inc.
The Advanced Book Program
Redwood City, California • Menlo Park, California
Reading, Massachusetts • New York
Don Mills, Ontario • Wokingham, U.K. • Amsterdam • Bonn
Sydney • Singapore • Tokyo • Madrid • San Juan

Publisher: *Allan M. Wylde*
Production Administrator: *Karen L. Garrison*
Editorial Coordinator: *Pearline Randall*
Electronic Production Consultant: *Mona Zeftel*
Promotions Manager: *Celina Gonzales*

Director of Publications, Santa Fe Institute: *Ronda K. Butler-Villa*

Library of Congress Cataloging-in-Publication Data

Theoretical Immunology Workshop (1987 : Santa Fe, N.M.)
Theoretical immunology.

(Santa Fe Institute studies in the sciences of complexity ; v. 2-3)
Includes index.
1. Immunology--Mathematical models--Congresses.
2. Immunology--Philosophy--Congresses. 3. Immunologic diseases--
Mathematical models--Congresses. I. Perelson, Alan S., 1947-
II. Title. III. Series. [DNLM: 1. Allergy and Immunology--congresses.
2. Models, Theoretical--congresses.
QW 504 T3955t 1987]
QR182.2.M36T485 1987 616.07'9 88-6169
ISBN 0-201-15682-2 (pt. 1)
ISBN 0-201-15683-0 (pbk. : pt. 1)
ISBN 0-201-15687-3 (pt. 2)
ISBN 0-201-15688-1 (pbk. : pt. 2)

This volume was typeset using TEXtures on a Macintosh computer. Camera-ready output from an Apple LaserWriter Plus Printer.

ABCDEFGHIJ-AL-898

About the Santa Fe Institute

The *Santa Fe Institute* (SFI) is a multidisciplinary graduate research and teaching institution formed to nuture research on complex systems and their simpler elements. A private, independent institution, SFI was founded in 1984. Its primary concern is to focus the tools of traditional scientific disciplines and emerging new computer resources on the problems and opportunities that are involved in the multidisciplinary study of complex systems -- those fundamental processes that shape almost every aspect of human life. Understanding complex systems is critical to realizing the full potential of science, and may be expected to yield enormous intellectual and practical benefits.

All titles from the *Santa Fe Institute Studies in the Sciences of Complexity* series will carry this imprint which is based on a Mimbres pottery design (circa A.D. 950-1150), drawn by Betsy Jones.

Santa Fe Institute Studies in the Sciences of Complexity

VOLUME	EDITOR	TITLE
I	David Pines	Emerging Syntheses in Science, 1987
II	Alan S. Perelson	Theoretical Immunology, Part One, 1988
III	Alan S. Perelson	Theoretical Immunology, Part Two, 1988
IV	Gary D. Doolen, et al.	Lattice Gas Methods of Partial Differential Equations
V	Philip W. Anderson, et al.	The Economy as an Evolving Complex System
VI	Christopher G. Langton	Artificial Life: Proceedings of an Interdisciplinary Workshop on the Synthesis and Simulation of Living Systems

ALAN S. PERELSON
Los Alamos National Laboratory, Los Alamos, NM, December 1987

Foreword

Immunology is one of the most exciting branches of biology. Advances are being made almost daily in a broad range of experimental areas. The current notoriety of AIDS has made us all keenly aware, even in this age of modern medicine and miracle drugs, of the absolute necessity of a functioning immune system. Our inability to cure HIV infection or even to design therapies that eliminate its symptoms points out our lack of detailed understanding about the operation of the immune system.

The immune system is a complex system of cells and molecules distributed throughout the body. Analogies have been drawn between the immune system and the nervous system. Like the nervous system, the immune system performs pattern recognition tasks, learns and retains a memory of the antigens that it has fought. Many of the "players" in the immune system, the specific cells and molecules whose coordinate activity produce the phenomena of immunology, have been identified. The interactions between these cells and molecules are slowly being elucidated. The mechanisms that actually regulate the immune system are by and large still unknown.

Unlike many areas of biology, theoretical ideas have played a major role in the development of the field. Controversies such as instructive vs. selective theories of antibody formation, germ-line vs. somatic mutation models for the generation of antibody diversity, and regulatory circuits vs. idiotypic networks have dominated both the intellectual development of the field and determined the direction of much experimental effort. Quantitative theories, while playing a role in the design and interpretation of various assay systems (e.g., complement fixation, the precipitin reaction, the hemolytic plaque assay) have to date not been significant in the intellectual development of the field. This may be changing as the field addresses more quantitative issues such as the role of somatic hypermutation in the generation of antibody diversity, the role of receptor clusters in cell stimulation and desensitization, the effects of competition between solution phase antibodies and cell surface receptors of the same specificity for binding various ligands including anti-idiotypic antibodies, the effects of different concentrations of growth factor, the effects of changes in receptor affinity and receptor number on cell stimulation, cell proliferation and cell differentiation, etc. Quantitative theories will also be essential for the development of global models of the immune system in which both the spatial and temporal aspects of humoral and cell-mediated responses are integrated into a single model. They also may well be needed in order to understand the role of idiotypic networks in immune regulation.

In order to foster further development of the field of theoretical immunology the Santa Fe Institute and the Theoretical Biology and Biophysics Group, Los Alamos National Laboratory, sponsored a three-day workshop in Santa Fe, June 10-12, 1987. The workshop was made possible by a generous grant from the Office of Health and Environmental Research, U. S. Department of Energy. These two volumes represent the proceedings of that workshop.

Both the workshop itself and the production of these volumes could not have been done without the help of many people. Geoffrey Hoffmann, Ronald Mohler and George Bell were involved in the initial conception of the meeting, helped choose participants, and provided assistance in writing a grant proposal. The staff of the Santa Institute, and in particular Ginger Richardson, and Ronald A. Zee assumed the heavy burden of making local arrangements for the participants and handling the financial affairs of the meeting. Ronda K. Butler-Villa with great efficiency and proficiency changed our rather dull-looking manuscripts into the beautiful camera-ready form that appears here. Finally, I wish to thank both George A. Cowan, President, L. M. Simmons, Vice-President for Academic Affairs, and the Science Board for leading the Santa Fe Institute in many intellectually exciting areas including Theoretical Immunology.

One of the participants at the meeting, Petr Klein, whose work appears in this volume met a tragic death, along with his wife, in a head-on automobile collision some months after the meeting. His colleagues and I helped in the proofreading and final production of his contribution. Charles DeLisi, Chairman, Biomathematics Sciences, Mt. Sinai School of Medicine, New York, whom Petr worked with for two years, has provided the following remarks:

> One of the interesting facts which emerges on scanning the references in George Bell's pioneering papers on immune response dynamics, is early activity in mathematical immunology by Czechoslovokian scientists. A small group including Sterzl, Jílek and Hraba were breaking new ground theoretically and experimentally, but their research was relatively unknown in

the West. It was in this isolated but unusually innovative atmosphere that Petr Klein obtained his doctoral degrees in immunology and mathematics.

I first learned of Petr during one of Tomáš Hraba's visits to the National Institutes of Health (N.I.H.). Petr was soon thereafter awarded a Fogarty International Fellowship. When the Czechoslovokian government refused to allow him to accept it, he and his wife moved to Rome, and from there to Canada. His wife obtained a position with Jack Dainty at the University of Toronto and Petr joined my laboratory at the N.I.H.

Petr broke new scientific ground during his two years in Bethesda. He developed a new approach to predicting protein function from sequence properties, and also published the most reliable methods available for classifying membrane proteins and predicting secondary structure. by early 1985, his accomplishments were sufficiently impressive to enable him to secure a position in Canada near his wife.

I was hoping Petr would rejoin us in the United States during the next year or two, and I know many of our colleagues here had similar hopes. His gentle disposition and his scientific accomplishments made him a valued friend and colleague. He will be greatly missed.

Contents

PART ONE

PART TWO

THEORETICAL IMMUNOLOGY

Immune Surveillance

DOUGLAS LAUFFENBURGER, DANIEL HAMMER, ROBERT TRANQUILLO, HELEN BUETTNER, and ELIZABETH FISHER
Department of Chemical Engineering, University of Pennsylvania, Philadelphia, PA 19104

How Immune Cells Find Their Targets: Quantitative Studies of Cell Adhesion, Migration, and Chemotaxis

SUMMARY

Many of the critical functions of immune cells require direct contact between these cells and target cells. Rates of these functions may often be limited by the rate of encounter between the cells and targets. Since these rates are key variables in most theoretical analyses of the immune response, fundamental understanding of the underlying processes governing them is of great importance. This paper summarizes our recent work directed toward the quantitative analysis of the processes of cell adhesion, migration, and chemotaxis, and their effects on the rates of encounter between immune cells and their targets. Our results should prove to be helpful in improving theoretical understanding of the dynamics of the immune response.

INTRODUCTION

It is becoming increasingly clear that the the immune response is essentially based on cell-cell interactions requiring direct contact between different cells.[1,2] These interactions may be between cells of the host response and those perceived as foreign, or between different types of host cells. Examples of the former class include phagocytosis by neutrophils and macrophages, and cytotoxic killing by macrophages, lymphocytes, and killer cells. Examples of the latter include antigen presentation by macrophages, and activation of lymphocytes. In general, all these interactions can be seen to involve an immune cell coming into direct contact with a "target" cell, whether that target is another immune cell or a "foreign" cell. Further, this contact typically must occur somewhere within the host tissue space, either in connective tissue of the various host organs or in lymphoid tissue. These simple considerations give rise to the realization that many of the key interactions comprising the immune response conform to a scheme something like the following outline, with five fundamental processes underlying cell/target contact:

a. adhesion of circulating immune cells to the microvessel endothelium in a particular organ tissue;

b. emigration of immune cells through the microvessel endothelium into the tissue;

c. migration of immune cells through the tissue toward the target cells;

d. encounter of an individual immune cell with an individual target cell;

e. adhesion of the immune cell to the target cell.

Only after these steps are successfully accomplished can the function of the immune cell be carried out, whether phagocytosis, cytotoxic killing, antigen presentation, or activation. (In some situations, when the immune cell and target cell are already located within the same tissue, steps a and b are not necessary.) Thus, theoretical models of the immune response, such as those found elsewhere within this volume, implicitly involve all these processes whenever a rate constant for a cell/target interaction is included. It is quite likely, moreover, that the rate of the particular function under consideration is often governed by the rate of one of these processes. That is, the final functional step (i.e., killing, activation, etc.) may actually take place quite rapidly compared to the time needed for the immune cell to encounter the target cell in the first place.[3] So, theoretical understanding of the dynamics of the immune response will require corresponding understanding of the cellular processes underlying cell/target encounter. Further, manipulation of the immune response will require the capability to manipulate these processes in a reliable manner. It seems apparent, then, that the phenomena of cell adhesion and migration deserve increased attention, in order that theoretical analysis of the immune response move beyond abstract characterization to biophysical and biochemical science. Step e—adhesion between two cells that have already come into static contact—has been the subject of a number of excellent previous studies.[4–8] We will, therefore, not treat it further here. However, step a—adhesion of a cell to a surface in a dynamic fluid flow situation—is a very different matter, and we will present some results of our initial work on this phenomenon.[9] Steps b, c, and d all involve migration of immune cells, and probably chemotactic guidance. (Chemotaxis is the biased locomotion of cells in the direction of a concentration gradient of a chemical stimulus.) Cell migration and chemotaxis have long been presumed to be crucial phenomena in the immune response, but only recently have truly quantitative, theoretically-based, studies emerged. We will, therefore, describe some of our recent work in this area.[3,10,11]

DYNAMICAL CELL ADHESION

Figure 1 illustrates the general situation under consideration. An immune cell in a microcirculatory vessel approaches the vessel endothelium under conditions of low Reynolds number (i.e., negligible inertial forces) fluid shear flow. The fluid thus exerts both a translational shear force and a rotational shear torque on the cell near the endothelial surface; these forces will attempt in concert to deny adhesion of the cell to the surface. Adhesion requires formation of a number of bonds between specific cell "receptor" macromolecules and endothelial surface "ligand" macromolecules sufficient to resist the distractive forces on the cell.[12] Such bonds can form only in the cell/surface "contact area"—the region over which cell and surface are in close enough proximity to permit receptor/ligand bond formation. The size of the contact area is initially determined by "nonspecific" physical forces between the cell and surface, and by cell mechanical properties; as adhesion progresses the size of the contact area may also reflect receptor/ligand bond forces.[13–15] Our model for dynamical cell adhesion in the presence of fluid shear forces is based on the following view of the chronology of events:

1. Upon approach of the cell to the surface due to fluid forces, the contact area forms quickly (i.e., fraction of seconds), its size governed by mechanical equilibrium between van der Waals, electrostatic, and steric stabilization forces, and cell deformability properties.
2. During a very short "contact time," receptor/ligand bonds begin to form within the contact area.
3. As the fluid mechanical forces begin to pull the cell away from the surface, the receptor/ligand bonds are stressed; this stress acts to enhance the bond dissociation rate constant.[12]
4. If the dynamic balance between bond formation and dissociation, even in the presence of bond stress, attains a stable asymptotic state with a finite number of bonds, the cell becomes adherent. If the stable asymptotic state possesses no bonds, the cell becomes detached.

Assuming that the bonds within the contact area are stressed equally by the distractive forces on the cell, theoretical prediction of the flow conditions and cell and surface properties permitting adhesion can be reduced to a set of two nonlinear dynamical equations governing the numbers of bonds and free receptors within the contact area as a function of time. Conditions and properties leading to a stable

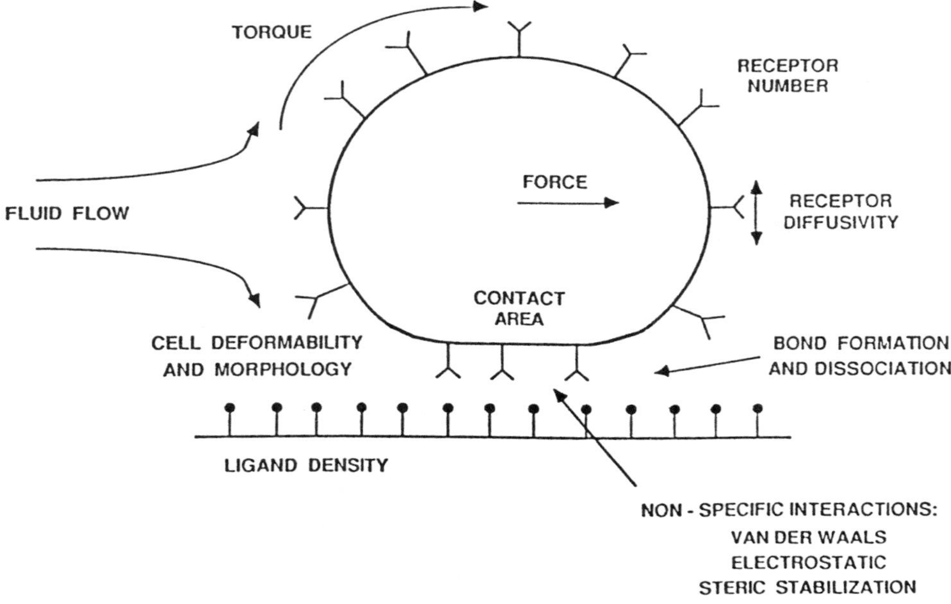

FIGURE 1 Schematic illustration of parameters involved in dynamical cell adhesion.

asymptotic state with a finite number of bonds can thus be elucidated using phase plane analysis.[16] A set of example calculations that illustrate the most germane theoretical predictions is shown in Figure 2. This figure plots the limiting fluid shear rate permitting adhesion as a function of the number of cell receptors for a corresponding surface ligand, at a given contact area size, bond formation rate constant, and receptor diffusivity. The curves are parameterized in terms of a dimensionless quantity characterizing the receptor/ligand bond affinity: κ is the ratio of the receptor/ligand equilibrium dissociation constant to the surface ligand density. Hence, κ^{-1} represents the affinity. Note that an increase in surface ligand density provides an effective affinity increase. Each curve represents the maximum shear rate for which adhesion can occur for the corresponding receptor number and affinity. Shear rates above the particular curve would not permit adhesion for that receptor number and affinity. Clearly, as receptor number increases at a constant affinity, the permissible shear rate increases; this increase is linear with receptor number when the affinity is sufficiently low. As affinity increases at a constant receptor number, the permissible shear rate also increases.

This type of plot enables investigation of a number of questions regarding immune cell adhesion to microvessel endothelium in target organs. For instance, it allows analysis of the role of specific cell receptors for selective adhesion. That is, one can determine what number and affinity of receptors for an endothelial ligand would be necessary to obtain adhesion of a given cell type in a particular tissue, without adhesion of other cell types in that tissue or of that cell type in other tissues. Also, it allows evaluation of the comparative roles of reduced blood flow rate and enhanced endothelial surface ligand expression for increased immune cell adhesion in a given tissue during infection. Other features of this model and its predictions are discussed in more detail elsewhere.[9,16] For the present, it represents the first theoretical analysis explicitly attempting to elucidate the parameters governing the adhesion of immune cells to endothelial surfaces under dynamic fluid flow conditions. As this process is the first step in eventual encounter between immune cells and their targets, it certainly merits increased attention. In this paper, however, we now turn to the succeeding processes in cell/target encounter: the migration of immune cells into and within tissue containing target cells.

CELL MIGRATION AND CHEMOTAXIS

The various classes of white blood cells have been shown to be capable of migration on two-dimensional substrates and through three-dimensional matrices, with cell paths resembling random walk behavior.[17] In its most basic mode, this locomotion is termed random motility. Certain chemical stimuli, including a number of immune

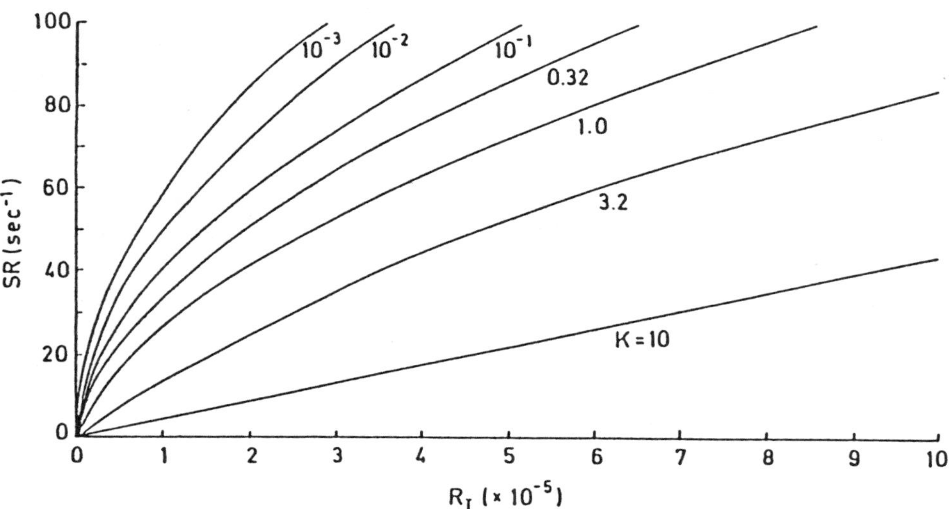

FIGURE 2 Plot of example results for dynamical cell adhesion model, from [9] showing maximal permissible fluid shear rate for adhesion as a function of cell receptor density. Curves are parameterized in terms of dimensionless inverse receptor/ligand affinity (see text).

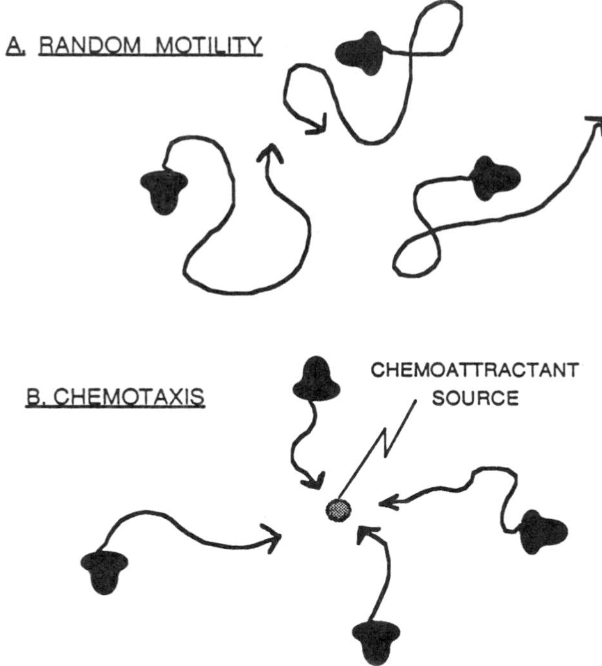

FIGURE 3 Illustration of typical paths of white blood cells exhibiting (a) random motility and (b) chemotaxis.

and inflammatory response mediators, influence the speed of this locomotion, in a phenomenon referred to as chemokinesis. Many of these same mediators have been demonstrated *in vitro* to cause a bias in the direction of cell locomotion toward higher stimulus concentration; this is the phenomenon known as chemotaxis.

Figures 3a and 3b illustrate typical paths of white blood cells exhibiting random motility and chemotaxis, respectively. Although it is obvious that cell migration is absolutely essential for encounter between immune cells and target cells within tissue, it is not self-evident whether the migration process is rate-limiting for any particular overall cell function. Further, it is not yet clear how critical chemotactic guidance actually is in any *in vivo* situation. Answers to both of these questions require measurement of appropriate parameters quantitatively characterizing cell random motility and chemotaxis, along with use of these parameters in mathematical models for the process of migration and encounter. We have been pursuing both of these avenues of investigation over the past few years, and will summarize recent advances here.

PARAMETER MEASUREMENT

There are two basic levels of approach to analysis of cell migration: individual cell and cell population. Correspondingly, each level possesses its own appropriate set of parameters characterizing cell movement; these sets must, of course, be related to each other. At the individual cell level, two parameters describe random motility: the *cell speed*, S, and the *cell persistence time*, P.[18] These can be defined rigorously in terms of cell path quantities (displacements and turn-angle distributions), but it will suffice here to explain their essential meaning. Cell speed is the center-of-mass displacement with time, and persistence time is the time interval between significant direction changes. For neutrophils, speed has been measured to range between about 5 to 30 μm/minute[17] (remember, speed can vary with chemical stimulus concentration) and persistence time to be roughly 5 minutes.[19] For alveolar macrophages, speed ranges between 1 to 5 μm/minute[17] and persistence time is approximately 15 minutes.[20] These are *in vitro* measurements on protein-coated plastic or glass surfaces, so extrapolation to *in vivo* conditions is uncertain.

An additional parameter is required in order to characterize chemotaxis. One common parameter is the *chemotactic index*: the ratio of net cell displacement toward the source of the chemoattractant to the total cell path length.[21] This parameter ranges between 0 and 1; 0 represents purely random movement (on the average) and 1 represents perfectly directed movement. An alternative parameter is the fraction of cells moving toward the chemoattractant source at any given time; in one dimension, this parameter should range between 0.5 (purely random movement) and 1 (perfectly directed movement). These two parameters should bear some relationship; a crude approximation gives $CI = 2f - 1$, where CI is the chemotactic index and f is the fraction cells moving up the gradient, or *fractional bias*.[22] The values depend on the magnitude of the chemoattractant concentration gradient and on the absolute value of the chemoattractant concentration. Under what seem to be optimal conditions, f has been measured to be about 0.9 for neutrophil leukocytes[23] and 0.6 for alveolar macrophages.[20] The corresponding values for CI would, thus, be 0.8 and 0.3, respectively. The best current methods for measurement of these individual cell parameters have been presented by Dunn[18] for speed and persistence time, by Allan and Wilkinson[24] for chemotactic index, and by Zigmond[23] for fractional bias. The essence of these methods is the tracking of movement of individual cells, in the presence of either uniform chemoattractant concentrations or established chemoattractant concentration gradients.

For cell populations, random motility is characterized by a single parameter, the *random motility coefficient*, μ. This coefficient is analogous to the diffusion coefficient for molecular Brownian motion. The relationship between the value of μ for a cell population, and S and P for individual cells can be derived from theoretical arguments: $\mu = PS^2/n$, where n is the number of dimensions in which cell locomotion is occuring.[22] Because S, and possibly P, vary with chemoattractant concentration (which we will denote by a), μ will also. Thus, the variation of μ with a quantitatively characterizes chemokinesis for a cell population.

A population-level analysis of chemotaxis again requires an additional parameter, the chemotaxis coefficient, χ. For sufficiently small chemoattractant concentration gradients, the product of χ and the attractant gradient, Δa yields a net *"directed cell velocity,"* $V_d = \chi\Delta a$, analogous to a convective fluid flow velocity for molecular transport.[25] Theoretical arguments similar to those mentioned above yield a relationship between χ for a cell population, and CI and S for individual cells: $\chi\Delta a = CI \cdot S$.[26] Again, because CI and S depend on the chemoattractant concentration, so will χ.

There presently exist two assays for experimental determination of values for the parameters μ and χ: the millipore filter assay and the linear under-agarose assay. In both of these assays, chemoattractant gradients can be established by means of diffusion across the cell movement substratum. In the under-agarose assay, the cells locomote on a protein-coated microscope slide, underneath a layer of agarose gel through which the attractant can diffuse from a source well. In the filter assay, the cells locomote within a porous membrane through which the attractant can diffuse from a source chamber. In both assays, cell migration is permitted to occur for a period of time (on the order of a few hours, typically), after which the cells are fixed and stained so that cell number density profiles can be counted. Values for μ and χ are then determined by matching these experimentally-measured cell number density profiles to theoretical profiles obtained by applying a mathematical model for the cell number density within the movement substratum:[27,28,10,11]

$$\frac{\delta c}{\delta t} = \frac{\delta}{\delta X}\left[\mu\frac{\delta c}{\delta X} - \left(X - \frac{1}{2}\frac{\delta\mu}{\delta a}\right)\frac{\delta a}{\delta X}c\right] \tag{1}$$

where $c(x,t)$ is the cell number density and $a(x,t)$ is the chemoattractant concentration, both functions of position and time within the movement substratum. The values of μ corresponding to various attractant concentrations should be obtained first, using uniform attractant concentrations within the assay. Then, χ can be evaluated in an experiment in which an attractant gradient is established. In the filter assay, the absolute concentration of attractant does not vary significantly throughout the relatively thin filter,[29] so that a single value of χ can be determined corresponding to the mean attractant concentration within the filter. Doing this over a range of concentrations allows determination of χ as a function of attractant concentration. In the under-agarose assay, on the other hand, the absolute concentration of attractant does vary widely with respect to both spatial postion and time,[29,30] so that a functional form for $\chi(a)$ must be specified *a priori*. We have, indeed, developed an appropriate *a priori* functional form based on empirical data correlating individual cell directional orientation bias with a spatial difference in cell receptor occupancy by attractant :[23,30]

$$\chi(a) = \chi_o\, N_T(a)\left[\frac{K_d}{(K_d + a)^2}\right] S(a) \tag{2}$$

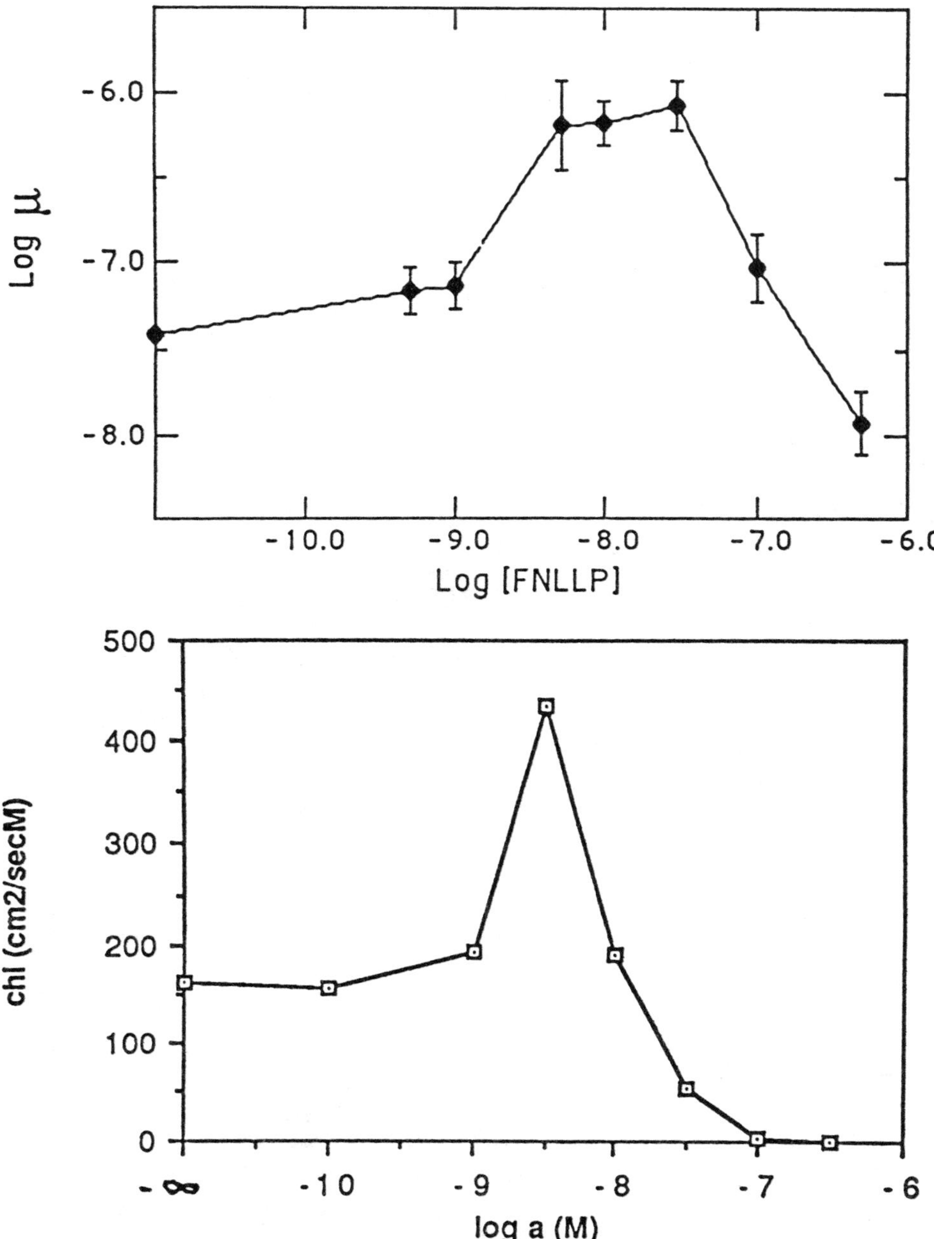

FIGURE 4 Plots of experimentally-measured values for $\mu(a)$ and $\chi(a)$ for human neutrophils responding to FNLLP, using the linear under agarose assay (from [30]).

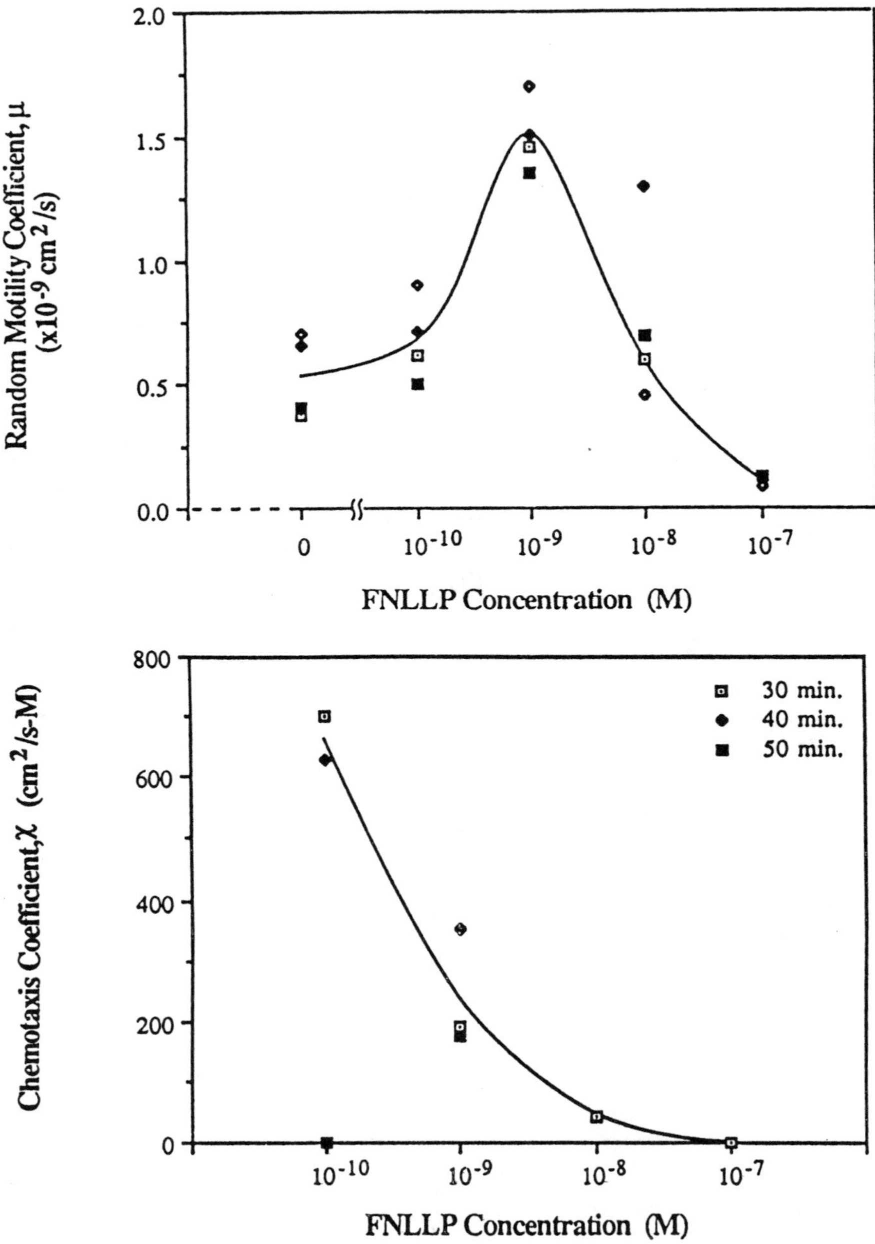

FIGURE 5 Plots of experimentally-measured values for $\mu(a)$ and $\chi(a)$ for rabbit neutrophils responding to FNLLP, using the millipore filter assay (from [31]).

where K_d is the receptor/attractant equilibrium dissociation constant, $N_T(a)$ is the number of cell surface receptors after downregulation, and χ_0 is the chemotactic sensitivity. $S(a)$ is the cell speed, as before, and can be determined from previous measurement of $\mu(a)$, using the relationship $\mu(a) = P[S(a)]^2/n$, if P can be estimated reliably. We have recently determined values for $\mu(a)$ and $\chi(a)$ for neutrophil leukocytes in response to the chemoattactant peptide FNLLP, using rabbit peritoneal exudate cells in the filter assay and human peripheral blood cells in the under-agarose assay. Figure 4 presents our results for $\mu(a)$ and $\chi(a)$ for human neutrophils in the under-agarose assay,[10,30] while Figure 5 presents the corresponding results for rabbit neutrophils in the filter assay.[11,31] Remember that $\chi(a)$ in the under-agarose assay was obtained by assuming the form of Equation (2), and determining a best-fit value for χ_0; in this case this value is 5×10^{-6} cm. At the same time, remember that $\chi(a)$ in the filter assay was obtained "point-by-point" without assuming the form of Equation (2). Interestingly, when the measured values of $\chi(a)$ for the filter assay are compared to the *a priori* form of Equation (2), a best-fit value $\chi_0 = 5 \times 10^{-6}$ cm also results. This comparison implies that these two assays are, in fact, reflecting similar chemotactic responses. The chief difference in the overall migration behavior between the two assays actually resides in the relative values for $\mu(a)$, as can be seen by comparison of Figures 4a and 5a. So, it appears to be the cell speed that is very different in the two assays, while the directional orientation bias is about the same.

MODELS FOR IMMUNE CELL/TARGET ENCOUNTER

It should be emphasized that Figures 4 and 5 represent the first truly quantitative measurements of fundamental leukocyte population migration parameters. These parameters, furthermore, are essential for adequate theoretical treatment of immune cell functions requiring direct cell-target encounter; they are necessary for mathematical modeling of steps b and c of the overall scheme outlined in the Introduction. A small number of theoretical models for the inflammatory host defense against bacterial infection have been published;[32–34] all of these demonstrate that μ and χ are key parameters for prediction of the outcome of the inflammatory response to a bacterial tissue challenge. Thus, experimental measurements of these quantities will be essential for evaluation of such models. On the other hand, the individual cell parameters S, P, and CI are the appropriate quantities for theoretical analysis of step d of the overall scheme. A brief summary of our recent work along this line will now be presented.

Once the immune cells have successfully accomplished steps a, b, and c of the overall cell/target contact scheme—that is, they have moved from the microcirculation into the tissue and have migrated to the locus of their target cells—the next important step is the actual encounter between an immune cell and its target. We have recently presented the first theoretical treatment of this process, for cell/target encounter in two dimensions.[3] Our model assumes that immune cells and target cells are distributed uniformly over a two-dimensional substratum (Figure 6a) at given number densities. Assuming the targets to be present in numbers greater than the immune cells, the average distance between targets, R, is inversely proportional to the square root of the target density. An immune cell is then considered to locomote within a unit space of radius R around a target, with specified speed S and persistence time P. The length of a "step" made by a locomoting cell in a given direction is defined to be $\delta = SP$. The probabilities of cell movement in the four major coordinate directions are specified as in Figure 6b: p, q, and m are the probabilities of movement toward, away from, and neutrally with respect to the target. If the target generates a chemoattractant for the immune cell, the resulting directional bias is characterized by the chemotactic index, $CI = p - q$.

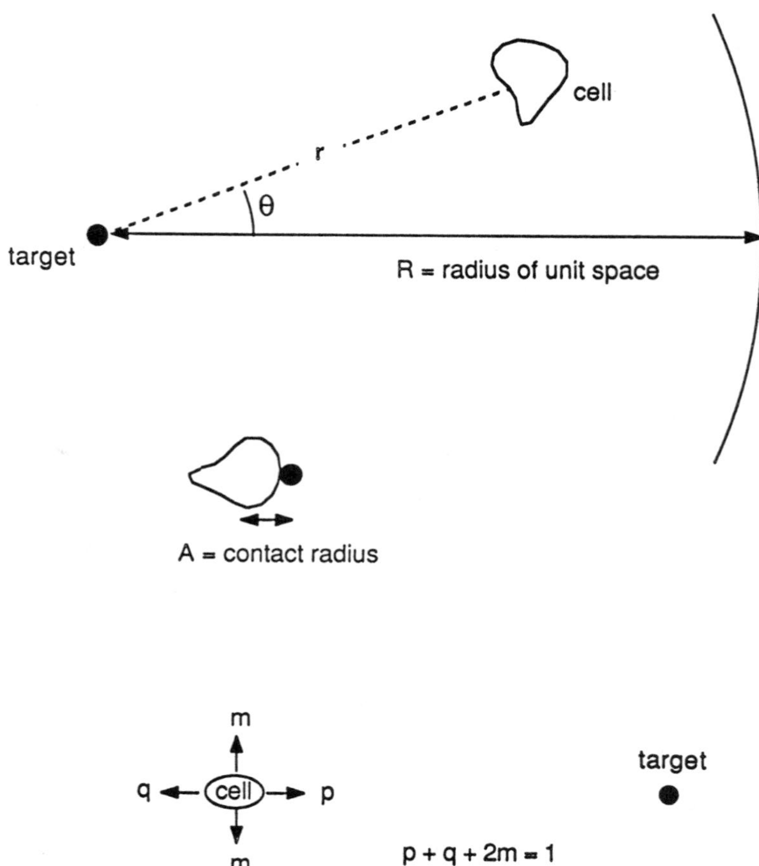

FIGURE 6 Illustration of model system for theoretical analysis of mean cell/target encounter times.

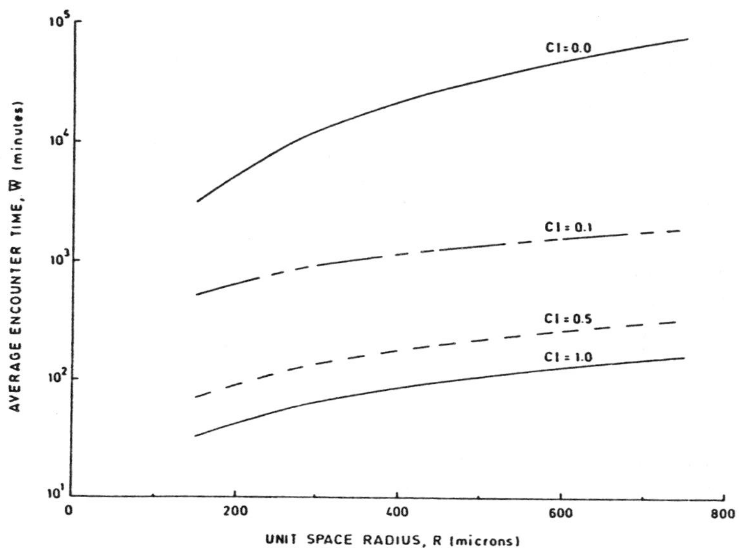

FIGURE 7 Plot of example results for mean cell/target encounter time as a function of radius of a unit space surrounding a target, from [3]. Curves are parameterized in terms of the chemotactic index (see text).

Encounter is said to occur when the immune cell approaches within a distance A of the target cell, where A is equal to the sum of the effective cell and target radii. Given these assumptions, a differential equation for the mean "encounter time" for contact between the cell and the target can be derived.[3] This equation can then be solved to obtain the encounter time as a function of the various system parameters. Some of the key results of this analysis are shown in Figure 7. The parameter values are specified for the particular application of lung macrophages encountering inhaled bacteria on the lung surface.[35] Thus, A, R, S, and P can be reasonably estimated from experimental reports. The only remaining parameter is the macrophage chemotactic index, CI. Figure 7 presents a plot of the mean encounter time between an individual macrophage and an individual bacterium on the alveolar surface as a function of the bacterial density present on the surface, for a series of values of CI between 0 and 1. Notice that the encounter time increases mildly as the target density decreases, but that it decreases dramatically for a small amount of chemotactic bias above purely random motility. An increase in CI from 0 to 0.1 leads to a decrease in the encounter time of roughly an order of magnitude. Further increases in CI, to 0.5 and to 1, lead to more reduction in the encounter time but with diminishing returns. Thus, even a small amount of chemotactic bias can be of tremendous benefit in increasing the rate of encounter between an immune cell and a target cell; nearly perfectly directed movement may not be justified. It is interesting to note that experimentally-measured values of CI for alveolar macrophages to small peptides (which are likely to be present in the lung as bacterial products) are in the range of 0.3.[20] It is also important to note that the encounter times predicted here are surprisingly long; the times shown in Figure 7 are quite long relative to typical times observed for ingestion of bacteria by macrophages. This suggests that the cell/target encounter step is likely to be rate-limiting in the overall process of phagocytic clearance of bacteria by macrophages in the lung. The reciprocal of the mean encounter time can be used as a rate constant for immune cell/target cell encounter in mathematical models for the dynamics of immune cell function, with the overall rate of this step also dependent on the local densities of immune cells and target cells within the tissue.[36]

CONCLUSIONS

In this paper, we have summarized recent work directed toward theoretical analysis of a number of key steps necessary for direct contact between immune cells and target cells with which they must interact in tissue spaces, including cell adhesion to blood vessel endothelium, migration into and within tissue, and encounter with a target. Since these steps depend on the cellular behavioral processes of adhesion, migration, and chemotaxis, we have described key parameters quantitatively characterizing these processes, along with theoretical analyses elucidating the dependence of these parameters on cell and tissue properties. Further, we have indicated how these parameters should find use in mathematical models for the dynamics of various aspects of immune cell function. We hope that this work will aid in the development of physiologically realistic theoretical analyses of the immune reponse.

ACKNOWLEDGMENTS

This work has been supported by an NSF Presidential Young Investigator Award and an NIH Research Career Development Award, along with NIH Grant AI-21538, to D. Lauffenburger. Much of this work has been done in collaboration with Prof. Sally Zigmond and Prof. Ron Daniele of the University of Pennsylvania Departments of Biology and Medicine.

REFERENCES

1. Paul, W., Editor (1984), *Fundamental Immunology* (New York: Raven Press).
2. Alberts, B., D. Bray, J. Lewis, M. Raff, K. Roberts, and J.D. Watson (1983), *Molecular Biology of the Cell* (New York: Garland Publishing), chapter 17.
3. Fisher, E., and D. Lauffenburger (1987), "Mathematical Analysis of Cell-Target Encounter Rates in Two Dimensions: The Effect of Chemotaxis," *Biophysical J.* **51**, 705.
4. Bell, G., M. Dembo, and P. Bongrand (1984), "Cell Adhesion: Competition Between Nonspecific Repulsion and Specific Bonding," *Biophysical J.* **45**, 1051.
5. Bell, G. (1979), "A Theoretical Model for Adhesion Between Cells Mediated by Multivalent Ligands," *Cell Biophysics* **1**, 133.
6. Capo, C., F. Garrouste, A.-M. Benoliel, P. Bongrand, A. Ryter, and G. Bell (1982), "Concanavalin-A-Mediated Thymocyte Agglutination: A Model Study of Cell Adhesion," *J. Cell Science* **56**, 21.
7. Evans, E. (1985), "Detailed Mechanics of Membrane-Membrane Adhesion and Separation: I. Continuum of Molecular Cross-Bridges." *Biophysical J.* **48**, 175.
8. Evans, E. (1985), "Detailed Mechanics of Membrane-Membrane Adhesion and Separation: II. Discrete Kinetically-Trapped Molecular Cross-Bridges," *Biophysical J.* **48**, 184.
9. Hammer, D., and D. Lauffenburger (1987), "A Dynamical Model for Receptor-Mediated Cell Adhesion to Surfaces," *Biophysical J.* **52**, 475.
10. Tranquillo, R., S. Zigmond, and D. Lauffenburger, "Measurement of Neutrophil Population Motility and Chemotaxis Parameters Using the Under-Agarose Migration Assay," *Cell Motility and the Cytoskeleton* (accepted for publication).
11. Buettner, H., S. Zigmond, and D. Lauffenburger, "Measurement of Neutrophil Population Motility and Chemotaxis Parameters Using the Millipore Filter Migration Assay" (submitted for publication).
12. Bell, G. (1978), "Models for the Specific Adhesion of Cells to Cells," *Science* **200**, 618.
13. Bongrand, P., and G. Bell (1984), "Cell Adhesion: Parameters and Possible Mechanisms," *Cell Surface Dynamics: Concepts and Models* , Eds. A. S. Perelson, C. DeLisi, and F. W. Wiegel (New York: Marcel Dekker), 459.
14. Bongrand, P., C. Capo, and R. Depieds (1982), "Physics of Cell Adhesion," *Progress in Surface Science* **12**, 217.
15. Evans, E., and R. Skalak (1981), *Mechanics and Thermodynamics of Biomembranes* (Boca Raton, FL: CRC Press).
16. Hammer, D. (1987), Ph.D. Thesis, University of Pennsylvania.
17. Wilkinson, P. (1982), *Chemotaxis and Inflammation* (New York: Churchill-Livingston), 2nd edition.

18. Dunn, G. (1982), "Characterizing a Kinesis Response: Time-Averaged Measures of Cell Speed and Directional Persistence," *Agents and Actions Supplements* **12**, 13.
19. Zigmond, S., R. Klausner, R. Tranquillo, and D. Lauffenburger (1985), "Analysis of the Requirements for Time-Averaging of Receptor Occupancy for
Gradient Detection by Polymorphonuclear Leukocytes," *Membrane Receptors and Cellular Regulation* , Eds. M. Czech and R. Kahn, 347.
20. Fisher, E., Ph.D. Thesis, University of Pennsylvania (expected 1988).
21. McCutcheon, M. (1946), "Chemotaxis in Leukocytes," *Phys. Rev.* **26**, 319.
22. Alt, W. (1980), "Biased Random Walk Models for Chemotaxis and Related Diffusion Approximations," *J. Math. Biol.* **9**, 147.
23. Zigmond, S. (1977), "Ability of Polymorphonuclear Leukocytes to Orient in Gradients of Chemotactic Factors," *J. Cell Biol.* **75**, 606.
24. Allan, R., and P. Wilkinson (1978), "A Visual Analysis of Chemotactic and Chemokinetic Locomotion of Human Neutrophil Leukocytes," *Exp. Cell Res.* **111**, 191.
25. Keller, E., and L. Segel (1971), "Models for Chemotaxis," *J. Theor. Biol.* **30**, 225.
26. Lauffenburger, D. (1982), "Measurement of Phenomenological Parameters for Leukocyte Motility and Chemotaxis," *Agents and Actions Supplements* **12**, 34.
27. Lauffenburger, D., C. Rothman, and S. Zigmond (1983), "Measurement of Leukocyte Motility and Chemotaxis Parameters with a Linear Under-Agarose Migration Assay," *J. Immunol.* **131**, 940.
28. Stickle, D., D. Lauffenburger, and R. Daniele (1985), "Measurement of Chemokinesis of Alveolar Macrophages Using the Linear Under-Agarose Assay," *J. Leukocyte Biol.* **38**, 383.
29. Lauffenburger D., and S. Zigmond (1981), "Chemotactic Factor Concentration Gradients in Chemotaxis Assay Systems," *J. Immunol. Methods* **40**, 45.
30. Tranquillo, R. (1986), Ph.D. Thesis, University of Pennsylvania.
31. Buettner, H. (1987), Ph.D. Thesis, University of Pennsylvania.
32. Lauffenburger, D., and K. Keller (1979), "Effects of Leukocyte Random Motility and Chemotaxis in the Tissue Inflammatory Response," *J. Theor. Biol.* **81**, 475.
33. Lauffenburger, D., and C. Kennedy (1983), "Localized Bacterial Infection in a Distributed Model for Tissue Inflammation," *J. Math. Biol.* **16**, 141.
34. Alt, W., and D. Lauffenburger (1987), "Transient Behavior of a Chemotaxis System Modeling Certain Types of Tissue Inflammation," *J. Math. Biol.* **24**, 691.
35. Green, G., and E. Kass (1964), "The Role of the Alveolar Macrophage in the Clearance of Bacteria from the Lung," *J. Exp. Med.* **119**, 167.
36. Fisher, E., D. Lauffenburger, and R. Daniele, "The Effect of Alveolar Macrophage Chemotaxis on Bacterial Clearance from the Lung" (submitted for publication).

J. R. HIERNAUX*† and R. LEFEVER†
*Laboratory of Microbial Immunity, NIH
†Service de Chimie - Physique II, Université Libre de Bruxelles

Population Dynamics of Tumors Attacked by Immunocompetent Killer Cells

INTRODUCTION

Usually, tumors *in vivo* present a complex nonmonotonic dynamical behavior. Notably:

1. the phases of progression during which the tumoral size increases may be interrupted by regression phases during which the tumoral size decreases;
2. in some tumors, after phases of rapid growth and regression, a dormant state[1] occurs such that the number of cancer cells remains rather low and nearly constant;
3. the establishment of a progressive tumoral state after the injection of an initial inoculum of cancer cells may involve intuitively hard-to-grasp threshold phenomena:[2-4] small-sized inoculum progressively grow (the so-called "sneaking-through phenomenon"), medium-sized inoculum are rejected, and large inoculum quickly proliferate; and
4. inoculum of identical size can grow differently in distinct animals: total regression, progressive growth, or dormancy have been observed in different animals subjected to an identical treatment.

Figure 1 illustrates this last point.

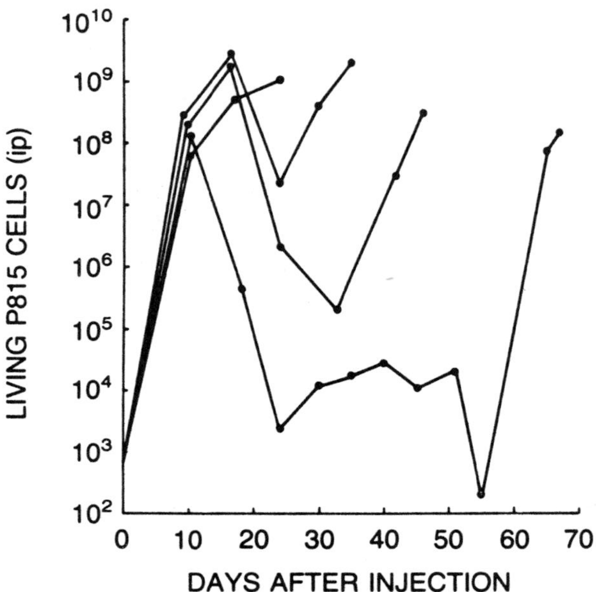

FIGURE 1 Patterns of tumor growth. Four DBA/2 mice were injected i.p. with 600 living P815 cells (P815 is a mastocytoma of DBA/2 origin). At various times thereafter, samples of the i.p. fluids were collected and diluted in culture medium that contained 0.4% Bactoagar. After a 3-to-4-day incubation at 37° C, the P815 colonies that contained more than six cells were counted. The number of living i.p. P815 was estimated, taking into account the cloning efficiency of *in vitro* growing P815, which was ± 0.3. Peritoneal exudate cells of control animals do not form colonies under these conditions. Each curve represents an individual mouse and ends on the last sampling before the death of the animal.

There is no doubt that at least part of these phenomena originate from the interactions between the neoplasm and the host's defense mechanisms. These interactions can be decomposed into four main steps: (1) neoplastic proliferation independent from the immune system; (2) recognition of immunogenic cancer cells by the immune system; (3) triggering of an anti-tumor immune response; and (4) development of a regulated anti-tumor response. Once triggered, the immune response leads to the production of immune effector cells. The latter proliferate, differentiate, and migrate to the tumoral site. Different types of anti-tumor effectors have been identified:[5–8] cytotoxic T lymphocytes (CTL), activated macrophages, natural killer (NK) cells, and antibody-dependent K cells. They are involved in distinct killing pathways which have the same ultimate goal: the cytolysis of cancer cells. Hence, tumor growth *in vivo* results from a competition between the proliferative capacity of the neoplastic cells and the efficiency of the anti-tumor immune response.

The proliferative capacity of cancerous cells depends on growth factors activating or inhibiting mitosis. The immune system is a highly regulated network of specialized cells and molecules: any immune response is controlled by positive and negative feedbacks which either amplify or suppress the production and the activity of immunocompetent effector cells. In the case of an anti-tumor immune response, the regulatory processes determine the quantity of killer cells. Moreover, in the case of a specific response (i.e., mediated by CTL), it might be assumed that immunocompetent effectors are selected on the basis of their affinity for tumor-associated antigens. There is no experimental evidence for this assumption. Nevertheless, the maturation of alloreactive CTL has been demonstrated in one case.[9] The regulation of the anti-tumor immune response could control the proliferation or regression of the tumor. This seems reasonable, because tumor regression is related to the production of killer cells in numerous experimental systems. In some systems, a switch from regression to progression is related to immunosuppression.[10–12] Interestingly, in one experimental system, the occurrence of immunosuppression has been correlated with the size of the tumor.[13] Tumor escape has also been linked to antigenic modulation, which induces the loss of key tumor-associated antigens involved in the stimulation of the immune system.[14] These facts illustrate the interdependence of neoplastic growth and of the immune system.

Nevertheless, paradoxical situations where tumor cells escape an active immune response have been described. Such situations challenge the idea that the goal of the anti-tumor immune response is the destruction of malignant cells. We distinguish at least three types of such paradoxical situations:

1. immunostimulation of tumor growth[15] has been observed, i.e., a situation where immunocompetent cells stimulate tumoral proliferation rather than destruction;
2. a dormant state arises after nearly complete rejection of a progressive tumor, when rejection is mediated by cytotoxic T cells recognizing tumor-associated antigens still displayed on the surface of dormant cells;[1,16] and
3. concomitant immunity is characterized by the rejection of a second implant of the original tumor, whereas the progressive growth of the initial implant persists.[17] At the time of rejection, the size of the second implant is smaller than the size of the initial one.

Carcinogenesis or the interaction between a killer and its target can be studied at the single cell level. By contrast, the dynamics of tumor growth must be examined at the population level, because the onset of the various growth patterns appears as a multicellular property. Indeed, the sensitivity to the initial conditions reflected by the dependence of tumor progression or regression on the size of the initial inoculum, as well as the multiplicity of growth behaviors following identical treatment, typically illustrate population dynamics properties. In the case of cancer growth, the latter essentially rely on the competition between two cellular populations: the malignant cells, and the cytotoxic cells able to kill them. The outcome of this competition is sometimes hard to predict on the basis of intuitive or logical reasoning. This is particularly true in the case of the paradoxical situations described previously. It is the fact that the dynamics of interacting cellular populations can be rather complex which leads to nonmonotonic kinetic patterns. Mathematical modelling allows us to have a better understanding of complex dynamics. In the sequel, we describe how a mathematical model of the predator-prey type, adapted to the problems of tumor growth, can offer a better grasp of the dynamics of interacting cellular populations.

MODELLING OF TUMOR GROWTH

We have developed a model describing the interactions between tumor cells and immunocompetent killer cells.[18,19] We focus on the tumor dynamical properties resulting from the kinetic interactions by which effector cells destroy cancer cells, rather than on the multiple immunoregulatory processes of the anti-tumor immune response. The purpose of this approach is to investigate the complex dynamical behaviors resulting from the effector-target kinetic competition, independently from the immunoregulation of killer cell production. The role of the latter, in our view, is to account for the switching between the various dynamical behaviors allowed by the kinetics of effector-target interactions. This switching is determined by the quantity (i.e., the actual number) and the quality (i.e., the binding affinity and the lytic capacity) of the immunocompetent effectors.

Broadly speaking, the interaction between a cytolytic cell and a cancerous cell can be schematized as follows:

$$E_0 + T \xrightarrow{b} E_1 \xrightarrow{\ell} E_0 + D$$

A free effector E_0 binds a target T, leading to the formation of a complex E_1. This binding step is characterized by a binding constant b. The bound effector lyses the target; this lytic step, characterized by a lytic constant ℓ, produces a free effector E_0 and a dead target D. According to this scheme, cytolysis is treated as an enzymatic reaction where the effector is equivalent to the enzyme. This analogy is based on the fact that a single effector can kill several targets.[20,21] It has been discussed by various authors.[18,22–24]

Let us now consider a volume element of tissue containing E_T effector cells ($E_T = E_0 + E_1$). We suppose that this volume element is sufficiently large to contain a great number of cells ($E_T \gg 1$), but, on the other hand, small enough for the target and effector cellular populations to be homogeneous, i.e., the scale of dimensions of interest here is much smaller than that over which the heterogeneity of the tumors manifests itself. On this local scale, we consider the outcome of the competition between the replication of target cells and their cytolysis. We expect that a necessary condition for the overall progression (or regression) of the tumor is the balance of this competition *locally* in favor (or disfavor) of the target cell population.

The local dynamics involves four kinetic parameters: E_T, the binding and lytic constants b and ℓ, and the replication constant λ describing the proliferation of the target cells. Formally, λ corresponds to the rate of a step which can be schematized as:

$$T \xrightarrow{\lambda} 2T$$

The kinetics of this reaction is obviously more complicated than that of a simple autocatalytic reaction. In particular, it is essential to account for the saturation exhibited by the growth of the T population in any finite volume element. Indeed, there is an upper limit N to the number of cells which can occupy a certain volume. Therefore, the proliferation of target cells T obeys a logistic growth curve and its kinetics can be described by the following equation:

$$\frac{d\rho_T}{dt} = \lambda\rho_T(1-\rho_T), \tag{1}$$

where we have introduced the cellular density

$$\rho_T = \frac{T}{N}.$$

If we assume that the total number of effector cells E_T remains constant within the considered volume element, the interaction between the targets and the effectors is described by the term[18]

$$F(\rho_T) = -\frac{b\ell E_T\rho_T}{\ell + b\rho_T}, \tag{2}$$

with

$$E_T = E_0 + E_1.$$

Hence, the kinetic equation accounting for the proliferation of T cells and for their interaction with immunocompetent cells has the following form:

$$\frac{d\rho_T}{dt} = \lambda\rho_T(1-\rho_T) - \frac{b\ell E_T\rho_T}{\ell + b\rho_T} \tag{3}$$

The steady state solution of Eq. (3) give us some information about the possible attractors and, therefore, about the potential kinetic patterns of tumor growth. The nature of the solution is influenced by the values of the four parameters E_T, b, ℓ, and λ. The value of the latter depends on regulatory processes controlling the dynamics of the competition between the tumor and the immune system. Growth-promoting or growth-inhibiting substances controlling the mitosis of cancer cells modify the value of λ. Immunosuppression can be reflected in decreasing the value of E_T or in modifying b and/or ℓ. Antigenic modulation of the tumor reflected by a decreasing density (or a lack) of tumor-associated antigens lead to smaller values of b.

In order to grasp how such variations affect the local dynamics of tumors, it is useful to formulate this question as a stability problem and to introduce a nondimensional switching parameter, controlling the exchanges of stability taking place in the system between the normal and tumoral states of the tissue. More precisely, we say that the normal state of a tissue is *stable* if the level of surveillance which the immune system exerts is sufficient to eliminate the neoplastic cells. On the contrary, if the appearance of a single cancerous cell is the starting point of a tumor, we say that the normal tissue is *unstable*. This notion of stability can be quantified in term of the switching parameter:

$$\beta = \frac{bE_T}{\lambda}.$$

This parameter compares the efficiency of the immune response and the proliferative capacity of the malignant cells. Clearly, if the value of β is small, the immune response is weak and the tumor can proliferate; the tumorless normal state is unstable. By contrast, in presence of a strong immune response, the neoplasm is rejected; the normal state is stable. It can be shown, as depicted on Figure 2, that the exchange of stability occurs when the value $\beta = 1$ is crossed.

Those intuitive stability considerations in term of the switching parameter β are, however, not sufficient to predict the final outcome of the balance between neoplastic proliferation and cytolysis, i.e., to predict if the tumor is in a dynamical state of progression or regression. Indeed, the analysis of our model[18] shows that tumor rejection is not automatically insured once the normal state is stable, i.e., β larger than 1, which implies the stability of tissue's normal state is not automatically sufficient to insure the rejection of an established tumoral state. The latter may be stable as well. This happens when the conditions are such that, for a given set of parameters values, Eq. (3) admits, besides a stable normal state of the tissue, another stable solution corresponding to a cancerous state. In other words, a bistability phenomenon occurs: the tumoral and the normal state are stable simultaneously.

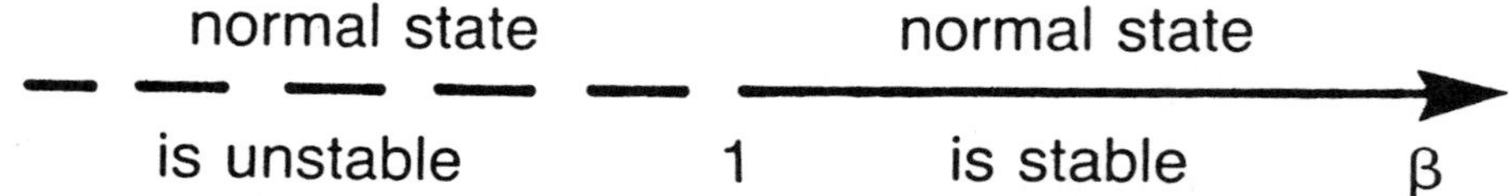

FIGURE 2 Presentation of the switching parameter β.

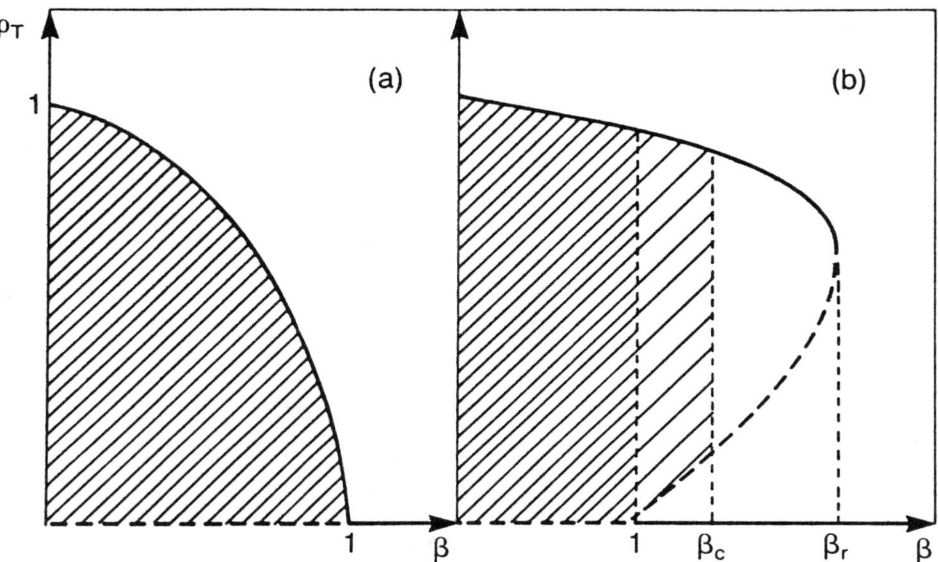

FIGURE 3 Variation of the density ρ_T of tumoral cells as a function of the switching parameter β. (a) For $\beta < 1$, the neoplasm proliferates, and for $\beta > 1$, it is rejected. (b) For $\beta > \beta_c$, the tumor is rejected. For $\beta < \beta_c$, tumor progression is observed.

In fact, during the unfolding of the immune response, two situations with markedly different properties may be encountered when the switching parameter β increases, e.g., as a result of the production of immunocompetent killer cells which increases the values of E_T and b. These situations are represented in Figures 3a and b, where the density of proliferative target cells in the tissue is plotted against β. Those curves present the solutions of Eq. (3) for two different sets of parameters values. Obviously, in both cases, when $\beta = 0$, i.e., in the absence of immune attack, the tumoral cells proliferate freely and, hence, their density is maximal (i.e., $\rho_T = 1$. The normal state of the tissue corresponding to $\rho_T = 0$ is then unstable: one cancer cell which appears somewhere in the tissue will proliferate and develop into a stable tumoral state corresponding to $\rho_T = 1$. The distinction between the case 3a and 3b, then, concerns the mechanism of exchange of stability between the normal and the tumoral state. Indeed, in case 3a, the density ρ_T is a monotonously decreasing function of β which, as expected intuitively, crosses the axis $\rho_T = 0$ (normal state) at the point $\beta = 1$. In the interval $0 < \beta < 1$, the tumor is progressive; rejection occurs as soon as $\beta > 1$. On the contrary, in the case 3b, the tumoral branch $\rho_T \neq 0$ extends and remains stable for $1 < \beta < \beta_r$. Notably, the transition

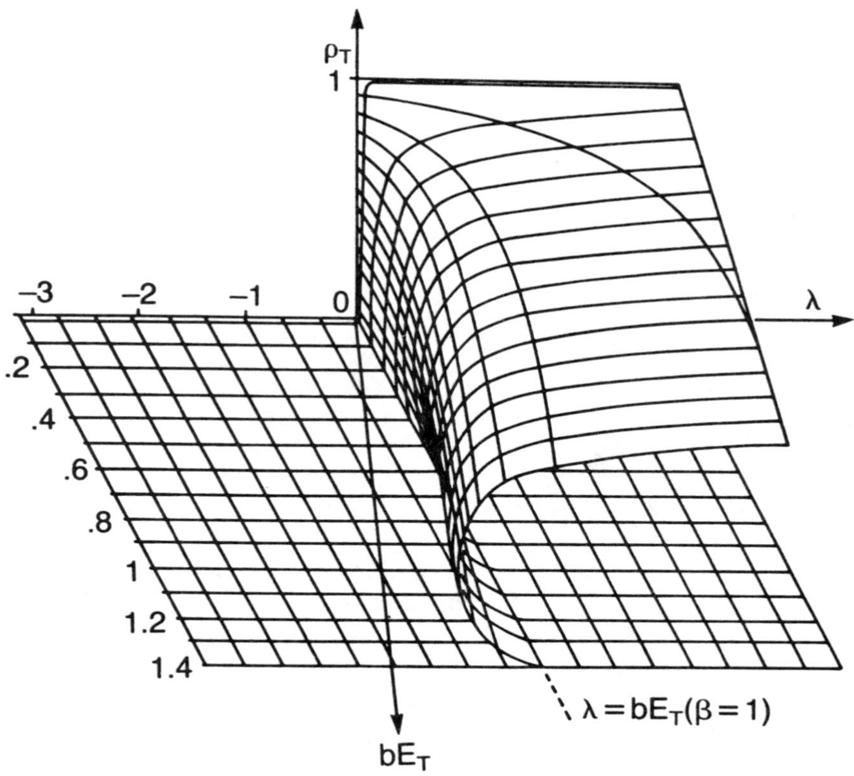

FIGURE 4 Variation of the density of tumoral cells as a function of λ and the product bE_T (ℓ being kept constant). As the value of bE_T increases, one observes a transition from the situation depicted in Figure 2a to the one presented in Figure 2b.

from progression to regression does not occur at the point $\beta = 1$ where the normal state $\rho_T = 0$ becomes stable, but at some larger value β_c as indicated on the figure.

It is interesting from the point of view of the signification of the immune response to examine how the transition between these two situations takes place in terms of the parameters of the cytolytic process. This is summarized in Figure 4. It presents the variation of the density of tumor cells as a function of the value of λ and bE_T, ℓ being kept constant. If there is no immune response (i.e., if $bE_T = 0$), tissue stability only depends on the value of λ. If λ is smaller than 0, the tumorless state is stable. By contrast, if λ is larger than 0, the tumoral state is the only stable attractor. This means that the tumor will progressively invade the normal tissue through local growth. Now, as soon as there is an immune response (i.e., bE_T larger than 0), the transition between the normal and tumoral state depends on the value of the switching parameter β (Figure 3a). If we progressively increase the intensity of the anti-tumor immune response (i.e., we consider larger values of bE_T), we observe a progressive transition from this simple situation to the case presenting multiple stationary states which is illustrated in Figure 3b.

The possible existence of bistables states is interesting. It allows the maintenance of a tumoral state in presence of a strong immune response, i.e., when the switching parameter β is larger than 1. Such a situation is compatible with paradoxical immunostimulation of tumor growth. Indeed, once a tumoral tissue is in a state corresponding to the upper branch, no regression will be observed as long as β is smaller than β_c.

Our data also have some broader implications concerning the cell-mediated anti-tumor immune response. In order to be rejected, a tumor must stimulate the immune system. This results in increasing the value of β, because the number of killer cells increases. Now, it is well established that lysis and binding are distinct steps. We can assume that immune selection will choose killer cells binding more strongly the target (i.e., b and β increase) and presenting comparable lytic capacity (i.e., comparable ℓ). In the course of this process, b might become larger than ℓ, and this would induce a switch from the situation of Figure 3a to the situation of Figure 3b. So improving the immune response by selecting clones of higher affinity might finally favor tumor growth. In the sequel, we describe dynamical situations compatible with tumor dormancy.

It has been shown that an effector can bind several targets.[25] The various stages of the cytotoxic reaction occurring within multicellular conjugates have been analyzed in detail in a mathematical model developed by Machen and Perelson.[26,27] Let us examine the new dynamical situations observed as a result of multibinding. If a killer cell can bind one or two targets, the cytotoxic reaction can be schematized as follows:

$$\begin{array}{ccccc} E_0 + T & \xrightarrow{2b_0} & E_1 + T & \xrightarrow{b_1} & E_2 \\ & & \Big\downarrow \ell_1 & & \Big\downarrow 2\ell_2 \\ & & E_0 + D & & E_1 + D \end{array}$$

A free effector cell binds a target T. This step is characterized by a binding constant b_0; 2 is a combinatorial factor indicating that E_0 can bind two targets. The bound effector E_1 can either bind a second target T (this step is characterized by a binding constant b_1) or lyse the bound target (lysis is characterized by the lytic constant ℓ_1. When a killer cell has bound two targets (E_2), it lyses them sequentially.[25] The first step of this lytic sequence is characterized by a lytic constant ℓ_2 (the factor 2 indicates that two targets can be potentially lysed). It leads to E_1 and dead target D.

In this case, the kinetic equation describing the proliferation of the target cells as well as their interactions with immune effectors is the following:

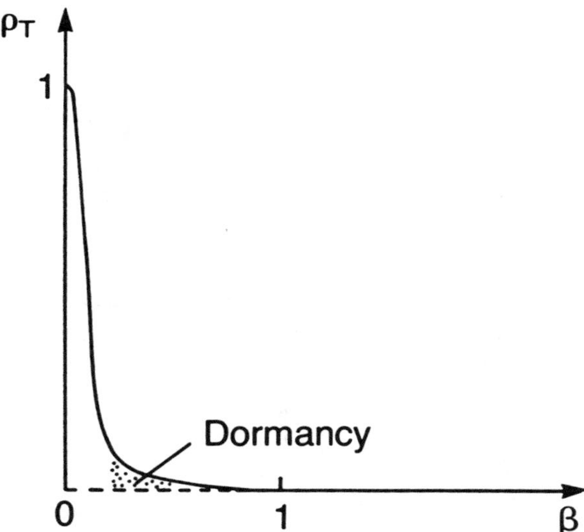

FIGURE 5 Variation of the density of tumoral cells as a function of β. The stippled area corresponds to a "microcancer state" (i.e., to the maintenance of a small number of tumor cells, which can be assimilated to dormancy); for $\beta > 1$, the tumor regresses.

$$\frac{d\rho_T}{dt} = +\lambda\rho_T(1-\rho_T) - \frac{2b_0\ell_1\ell_2E_T\rho_T + 2b_0b_1\ell_2E_T{\rho_T}^2}{\ell_1\ell_2 + 2b_0\rho_2\rho_T + b_0b_1{\rho_T}^2} \tag{4}$$

We now have size kinetic parameters: b_0, ℓ_1, b_1, ℓ_2, and E_T ($E_T = E_0 + E_1 + E_2$). Four types of solutions can be obtained. The first two solutions correspond to the situation illustrated on Figures 3a and 3b. Figures 5 and 6 present new solutions which can also occur. Each situation corresponds to distinct relative values of the parameters. If the ratio ℓ_1/b_0 is smaller than 1, the solution of the kinetic equation describing the time evolution of the cellular density of tumor cells is presented in Figure 3b. If this ratio is greater than 1, either one of the three other situations can occur. The curves presented in Figures 5 and 6 reflect interesting dynamical properties. In Figure 5, for values of β smaller than 1 and fairly different from 0, the system evolves to a "microcancer" state characterized by a low density of tumoral cells. We propose to establish an analogy between this state observed for values of β corresponding to a weak immune response and the tumor dormant state. Figure 6 presents a more complex type of solution where multiple steady states again are observed for β bigger than β_p and smaller than β_r. Once more, the cellular densities corresponding to the intermediate branch represent unstable attractors, which means that the system does not evolve to a state corresponding to those values for the cellular density. When β is smaller than β_p, the immune response is weak and the tumor progresses. When β is greater than β_r, a very strong immune response induces tumor regression. If β is smaller than 1 and larger than β_p, the tumor will proliferate either strongly (upper branch) or weakly (lower branch). The cellular densities of the lower branch are compatible with the establishment of dormancy. Now if, in the course of the dynamics of tumor growth, a state characterized by a value of β between β_p and 1 and by a tumor cell density within the stippled area is reached, the system evolves to a "microcancer" state corresponding to the lower branch. Such a state will be maintained as long as β is smaller than 1 and larger than β_p. When β is greater than 1 and smaller than β_r, the tumor either regresses (lower branch corresponding to $\rho_T = 0$) or progresses (upper branch). If the system reaches a dynamical state corresponding to the area located above the unstable branch (dashed line), neoplastic proliferation is observed. Once more, a tumor can grow in the presence of a strong immune response. Broadly speaking, the situation presented in Figure 5 will be obtained when ℓ_1 is greater than b_0 and when ℓ_2 is much smaller than b_1.

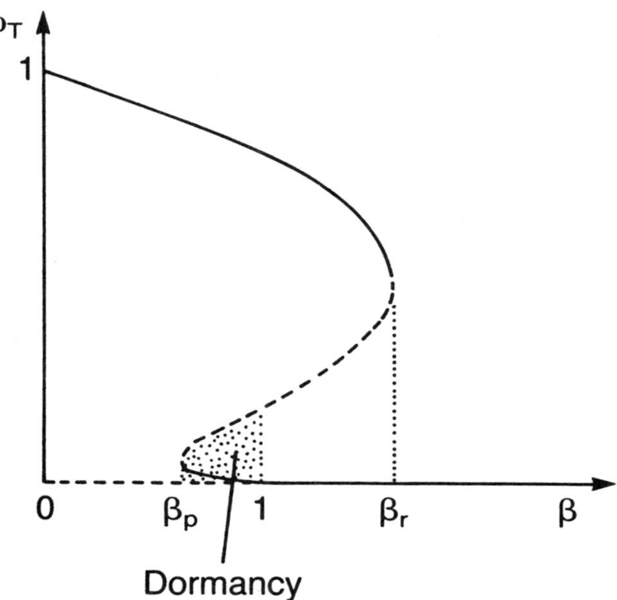

FIGURE 6 Multiple steady states are observed for $\beta_p < \beta < \beta_r$. The stippled area corresponds to the establishment of a dormant state.

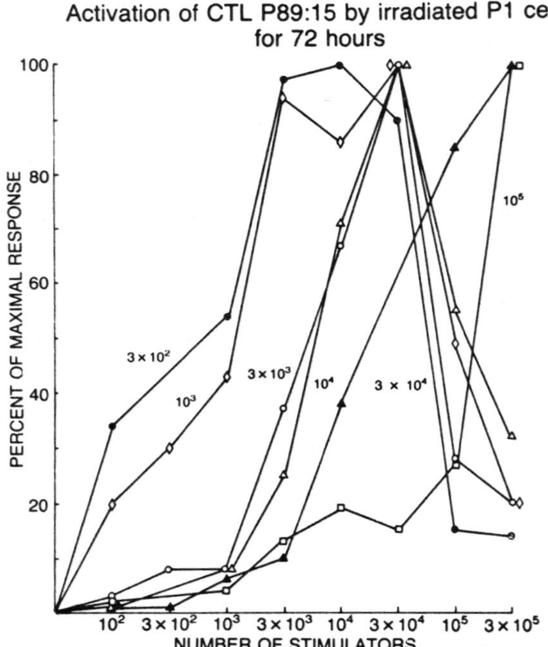

FIGURE 7 Proliferative response of varying numbers of the CTL clone CTL-P89:15 to various numbers of stimulating P815 cells. Proliferation was evaluated by pulsing the cultures with 1 μCi/well of [^{3}H]thymidine. In order to compare the various curves, the data are expressed as percentage of the maximal response obtained for each curve.

DISCUSSION

The results of our theoretical analysis indicate that various dynamical situations can occur if we look at tumor growth as a simple competition between proliferating malignant cells and immunocompetent killer cells. Dormancy and immunostimulation of tumor growth are compatible with our descriptions. Their establishment would simply result from the nature of the stable solutions of the kinetic equation describing the competitive mechanism. A similar approach has been used by other investigators to look at tumor growth.[28–31]

Dormancy is particularly interesting, as it is possible that many human tumors go through a stage of dormancy before resuming growth. We have discussed previously the fact that trivial explanations cannot account for dormancy in the case of the P815 mastocytoma.[16] Briefly, it is not due to a different rate of growth of

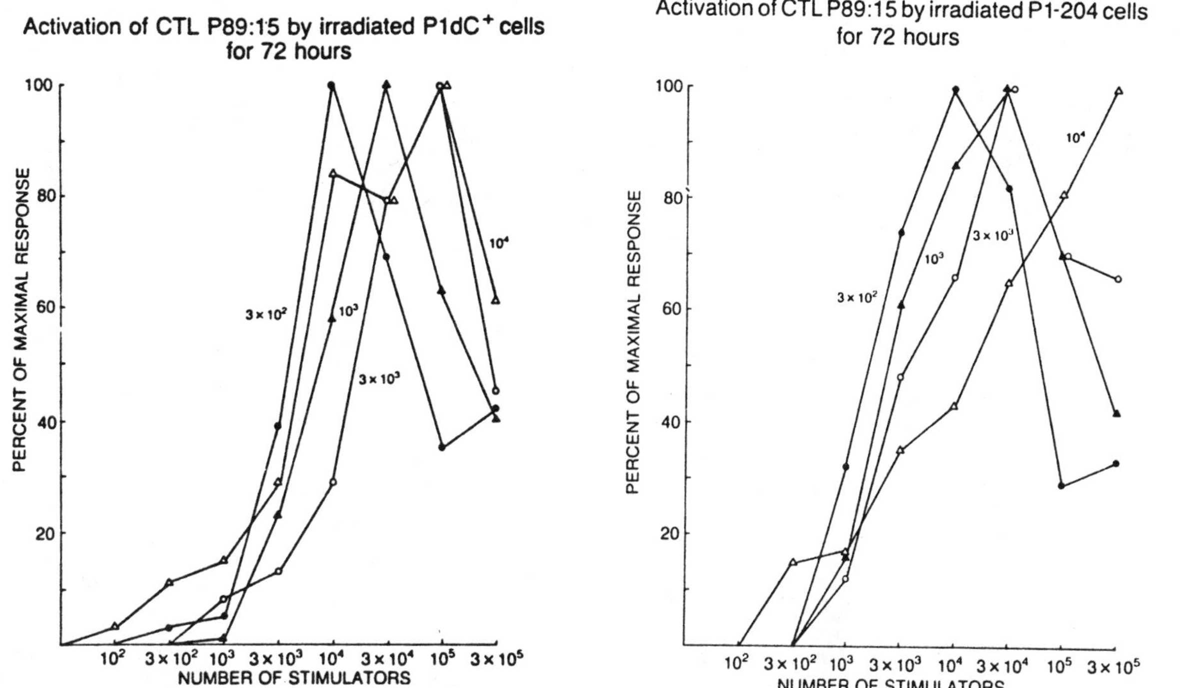

FIGURE 8 Proliferative response of varying numbers of the CTL clone CTL-P89:15 to various numbers of stimulating P1-204 (a) and P1dC$^+$ (b) cells. Proliferation was evaluated by pulsing the cultures with 1 μCi/well of [^{3}H]thymidine.

dormant cells *in vivo*. Dormant and nondormant tumoral cells present the same susceptibility of being lysed by anti-P815 CTL clones. The study of Uyttenhove et al.[14] has shown that P815 expresses a set of tumor-associated antigens designated as A, B, C, and D (see also the article by T. Boon[32] for a review). Usually, escaping P815 cells (i.e., the cells proliferating *in vivo* after partial rejection of the tumor) express the following phenotype: $A^-B^-C^+D^+$. P1-204 cells expressing this variant phenotype have been isolated. Dormant cells slowly evolve from the $A^+B^+C^+D^+$ phenotype to the $A^-B^-C^+D^+$ in the course of dormancy (C. Uyttenhove, personal communication). Dormant cells (i.e., cells which lead to the establishment of dormancy when they are injected i.p. under the form of an ascitic tumor) expressing the $A^-B^-C^+D^+$ phenotype (P1dC$^+$ cells) have been isolated by C. Uyttenhove.[16] So far, no phenotypic differences have been found between P1-204 and P1dC$^+$.cells in terms of the expression of tumor-associated antigens. Now, although both cell types are killed at the same rate by a CTL clone (CTL-P89:15) recognizing the C antigens, it is still possible that the dormant and nondormant cells stimulate the proliferation of anti-C CTL differently. This might be due to a different density of C antigen on the surface of the dormant cells or to a modified configuration of the C molecule of dormant cells. Obviously, we should assume that such modification does not alter the killing step, as P1-204 and P1dC$^+$ cells are killed similarly. Indeed, the kinetics of cytolysis are identical for both cell types. In order to test this hypothesis, we used an assay recently developed by Ashwell et al.[33,34] This assay is based on the fact that, in a T cell proliferation assay, increasing numbers of antigen-specific T cell clones present shifted antigen dose response curves (i.e., optimal stimulation requires a larger dose of antigen when more T cells are present in the assay). If the antigenic structure (which is a complex between a protein and a class II molecule in their case) is modified, they observe a shift of the entire set of dose response curves. This shift can reflect a loss of potency and/or of affinity. These data were obtained with T helper cell clones stimulated by antigen-pulsed antigen-presenting cells. The anti-P815 CTL clones are grown *in vitro* in the presence of irradiated P815 cells (stimulator cells) and feeder cells (spleen cells of DBA/2 origin). The growth medium is supplemented with the supernatant of a secondary allogeneic mixed lymphocyte culture (C57B1/6 anti-DBA/2) as a source of growth factors. Figure 7 presents the result of a typical experiment. As in the case of the stimulation of T_H cell clones,[33,34] we observed a set of dose-response curves. Increasing numbers of stimulators are required in order to obtain optimal stimulation of the proliferation of increasing number of cloned CTL. We use the CTL clone CTL-P89:15 for those experiments. It recognizes the C tumor-associated antigen which is present on P1-204 and PldC$^+$ cells. When those two cell lines are used as stimulators, no striking differences were observed for the distribution of the stimulation curves, as shown in Figures 8a and b. From those experiments, we conclude that there is no critical difference in the density of C antigens on P1-204 and P1dC$^+$ cells. Also, there should be no modification of the three dimensional structure of the C antigen able to alter its recognition by CTL-P89:15.

The data of the proliferation assay and of cytolysis indicate that there should be no difference for the binding and lytic constants characterizing the interaction between the CTL and dormant as well as nondormant targets. This is compatible with the existence of multiple growth regimes for a given set of parameter values. Our model suggests that such a situation can exist if we look at tumor growth as a process resulting from the dynamics of interacting cell populations. Nevertheless, it remains paradoxical that the dormant variant always induces dormancy, as if the choice of growth pattern was already programmed in the P1dC^{+} cell line independently from the status of the mouse and the initial conditions of the system. The solution to this paradox remains an open question.

Generally speaking, our theoretical ideas could be tested experimentally by evaluating the values of the different kinetic parameters during tumor growth *in vivo*. We are currently developing experimental techniques allowing the determination of binding and lytic constant.

ACKNOWLEDGMENTS

We thank Professor Prigogine for providing continuous support and interest in this research. This work is conducted pursuant to a contract with the National Foundation for Cancer research. We thank J. Thierie for drawing Figure 4. We also thank Dr. C. Uyttenhove for providing the various cell lines used in those studies.

REFERENCES

1. Wheelock, E. F. (1981), "The Tumor Dormant State," *Adv. in Cancer Research* **34**, 107–140.
2. Old, L. J., E. A. Boyse, D. A. Clarke, and F. A. Carswell (1962), "Antigenic Properties of Chemically Induced Tumors," *Ann. N.Y. Acad. Sci.* **101**, 80–106.
3. Klein, E. (1972), "Tumor Immunology: Escape Mechanisms," *Annales de L'Institut Pasteur (Paris)* **113**, 593–602.
4. Bonmassar, E., E. Menconi, A. Goldin, and G. Cudkowicz (1974), "Escape of Small Numbers of Allogeneic Lymphoma Cells from Immune Surveillance," *J. Natl. Cancer Inst.* 53, 475–479.
5. Gorcyznski, R. M. (1974), "Evidence for *In Vivo* Protection against Murine Sarcoma Virus Induced Tumors by T Lymphocytes from Immune Animals," *J. Immunol.* **112**, 532–536.
6. Adams, D. O., and R. Snyderman (1979), "Do Macrophages Destroy Nascent Tumors?" *J. Natl. Cancer Inst.* **62**, 1341–1345.
7. Heberman, R. B., and J. R. Ortaldo (1981), "Natural Killer Cells: Their Roles in Defense against Disease," *Science* **214**, 24–30.
8. Adams, D. O., T. Hall, Z. Steplewski, and H. Koprowski (1984), "Tumors Undergoing Rejection Induced by Monoclonal Antibodies of IgG2a Isotype Contain Increased Numbers of Macrophages Activated by a Distinctive Form of Antibody-Dependent Cytolysis," *Proc. Natl. Acad. Sci. USA* **81**, 3506–3510.
9. Moscovitch, M., Z. Grossman, D. Rosen, and G. Berke (1986), "Maturation of Cytolytic T Lymphocytes," *Cell. Immunol.* **102**, 52–67.
10. Broder, S., L. Muul, and T. A. Waldmann (1978), "Suppressor Cells in Neoplastic Disease," *J. Natl. Cancer Inst.* **61**, 5–11.
11. Greene, M. I. (1980), "The Genetic and Cellular Basis of Regulation of the Immune Response to Tumor Antigens," *Contemporary Topics in Immunobiology* **11**, 81–116.
12. North, R. J. (1984), "The Murine Antitumor Immune Response and its Therapeutic Manipulation," *Adv. Immunol.* **35**, 89–155.
13. Bear, H. D. (1986), "Tumor-Specific Suppressor T-Cells which Inhibit the *In Vitro* Generation of Cytolytic T-Cells from Immune and Early Tumor-Bearing Host Spleens," *Cancer Research* **46**, 1805–1812.
14. Uyttenhove, C., J. Maryanski, and T. Boon (1983), "Escape of Mouse Mastocytoma P815 after Nearly Complete Rejection is due to Antigen-Loss Variants rather than Immunosuppression," *J. Exp. Med.* **157**, 1040–1052.
15. Prehn, R. T. (1977), "Immunostimulation of the Lymphodependent Phase of Neoplastic Growth," *J. Natl. Cancer Inst.* **59**, 1043–1049.
16. Hiernaux, J. R., R. Lefever, C. Uyttenhove, and T. Boon (1986), in *Paradoxes in Immunology*, Eds. G. W. Hoffmann, J. Levy, and G. T. Nepom (Boca Raton: CRC Press), 95–106.

17. Tuttle, R. L., V. C. Knick, C. R. Stopford, and G. Wolberg (1983), "*In Vivo* and *In Vitro* Antitumor Activity Expressed by Cells of Concomitantly Immune Mice," *Cancer Research* **43**, 2600–2605.
18. Garay, R., and R. Lefever (1978), "A Kinetic Approach to the Immunology of Cancer: Stationary States Properties of Effector-Target Cell Reactions," *J. Theor. Biol.* **73**, 417–438.
19. Lefever, R., and T. Erneux (1984), in *Nonlinear Electrodynamics in Biological Systems*, Eds. W. R. Adey and A. F. Lawrence (New York: Plenum), 287–305.
20. Berke, G., K. A. Sullivan, and D. B. Amos (1972), "Tumor Immunity *in vitro*: Destruction of a Mouse Ascites Tumor through a Cycling Pathway," *Science* **177**, 433–434.
21. Rothstein, T. L., M. Mage, G. Jones, and L. L. McHugh (1978), "Cytotoxic T Lympocyte Sequential Killing of Immobilized Allogeneic Tumor Target Cells Measured by Time-Lapse Microcinematography," *J. Immunol.* **121**, 1652–1656.
22. Thorn, R. M., and C. S. Henney (1976), "Kinetic Analysis of Target Cell Destruction by Effector T Cells. I. Delineation of Parameters Related to the Frequency and Lytic Efficiency of Killer Cells," *J. Immunol.* **117**, 2213–2219.
23. Callewaert, D. M., D. F. Johnson, and J. Kearney (1978), "Spontaneous Cytotoxicity of Culture Human Cell Lines Mediated by Normal Peripheral Blood Lymphocytes. III. Kinetic Parameters," *J. Immunol.* **121**, 710–717.
24. Merrill, S. J. (1982), "Foundations of the Use of an Enzyme-Kinetic Analogy in Cell-Mediated Cytotoxicity," *Mathematical Biosciences* **62**, 219–235.
25. Zagury, D., J. Bernard, P. Jeannesson, N. Thiernesse, and J. C. Cerottini (1973), "Studies of the Mechanism of the T Cell-Mediated Lysis at the Single Effector Cell Level. I. Kinetic Analysis of Lethal Hits and Target Cell Lysis in Multicellular Conjugates," *J. Immunol.* **123**, 1604–1609.
26. Machen, C. A., and A. S. Perelson (1984), "A Multistage Model for the Action of Cytotoxic T Lymphocytes in Multicellular Conjugates," *J. Immunol.* **132**, 1614–1624.
27. Perelson, A. S., and C. A. Machen (1984), "Kinetics of Cell-Mediated Cytotoxicity: Stochastic and Deterministic Multistage Models," *Mathematical Biosciences* **70**, 161–194.
28. DeLisi, C., and A. Rescigno (1977), "Immune Surveillance and Neoplasia. I. A Minimal Mathematical Model," *Bull. Math. Biology* **39**, 201–221.
29. Albert, A., M. Freedman, and A. S. Perelson (1980), "Tumors and the Immune System: the Effects of a Tumor Growth Modulator," *Math. Biosci.* **50**, 25–58.
30. Grossman, Z., and G. Berke (1980), "Tumor Escape from Immune Elimination," *J. Theor. Biol.* **83**, 267–296.
31. DeBoer, R. J., and P. Hogeweg (1985), "Tumor Escape from Immune Elimination: Simplified Precursor Bound Cytotoxicity Models," *J. Theor. Biol.* **113**, 719–736.

32. Boon, T. (1983), "Antigenic Tumor Cell Variants Obtained with Mutagens," *Adv. in Cancer Research* **39**, 121–151.
33. Ashwell, J. D., and R. H. Schwartz (1986), "T-Cell Recognition of Antigen and the Ia Molecule as a Ternary Complex," *Nature* **320**, 176–179.
34. Ashwell, J. R., B. Fox, and R. H. Schwartz (1986), "Functional Analysis of the Interaction of the Antigen-Specific T Cell Receptor with its Ligand," *J. Immunol.* **136**, 757-768.

SETH MICHELSON
Research Biomathematician, Department of Radiation and Biology Research,
Rhode Island Hospital, 593 Eddy St., Providence, RI 02903

Immune Surveillance: Towards a Tumor-Specific Model

ABSTRACT

Two models of tumor escape from an active anti-tumor response have been developed. The models describe cell populations which exhibit random lifelengths distributed in accordance with a gamma distribution. They have been simulated using Monte Carlo techniques which model the lifelength as a series of maturation chambers each distributed in accordance with an exponential distribution. The malignantly transformed cells are characterized by three binary variables: Escape Route, Target Structure Stability, and End Susceptibility. The simulated results show that the ability of a tumor to escape depends on the activity generated by the immune system in the anti-tumor response. However, tumor escape is also a function of tumor cell type.

INTRODUCTION

Two describing tumor escape within an active anti-tumor environment have been developed. The models describe two different escape routes for the tumors. The first is termed "intragenerational" and the second is termed "intergenerational." In the intragenerational scenario, cells can alter the concentration of their surface target structures without benefit of mitosis. The intergenerational escape route describes a more Darwinian phenomenon. Tumor cells do not alter the concentration of their surface target structures, but at mitosis, we allow for a mutational probability such that either one or both daughters can alter the surface concentration of immunologically recognizable target structures.

By classifying tumor cells into categories which describe the escape route they employ, the concentration to which they escape (what we will call the "non-susceptible state"), and the speed with which they can alter their surface concentrations (what we call target structure stability), we can analyze the immune surveillance theory with respect to the tumor. We describe the immune response in phenomenological terms as two different, but harmonious, response waves. One wave represents immunologic elements which require the complete recognition, recruitment and amplification machinery classically defined for the T cell effector population, and is called the "inductive response wave." The other wave represents the class of immunologic effectors that do not require such machinery (e.g., NK cells). This wave is called the "non-inductive response wave." Each is described as a 6-parameter function model of time. Using Monte Carlo methods, we simulate tumor growth for each of eight types of tumors (described in detail below), across a number of anti-tumor response scenarios. We show that the type of escape mechanism, the adaptability of the tumor, and the susceptibility level to which the tumors escape all fundamentally affect the success rate of the immune surveillance mechanism. From these results we conclude that the debate over the effectiveness of the immune surveillance mechanism should not be limited to the control processes within the immune system, but rather, it should be expanded to include an active target system that adapts to the assaults.

THE MODELS

Boyse et al.[1] have described the phenomenon of "modulation." In their experimental system, the concentration of target structures can decrease within the presence of an active anti-tumor response, and can then be re-established once the anti-tumor response has abated. We term this second phenomenon "demodulation." Wolf et al.[2] showed similar results in the DBA/2J P815-X2 mastocytoma. Further, modulation need not result in the total removal of target structures from the surfaces of the tumor cells. Though still expressed, the concentration of target structures upon the surface is decreased in modulated populations.[3,4] Therefore, in our models we associate intragenerational escape with a decrease in surface target concentration to a level we arbitrarily dub "non-susceptible" meaning that the cells are less likely to be recognized, conjugated with, and programmed for lysis by active elements of the

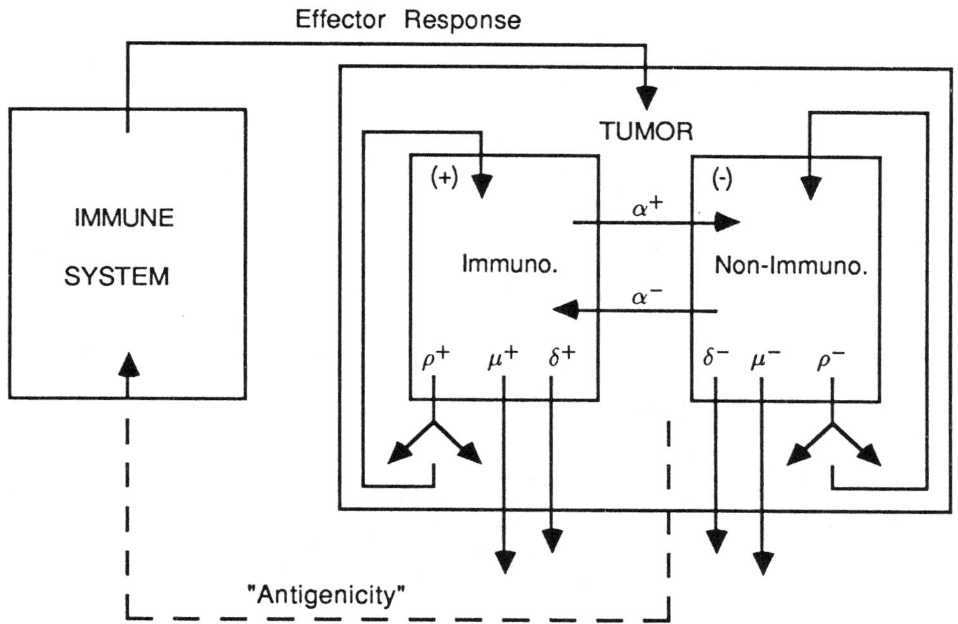

FIGURE 1 Modulation (intragenerational) model.

immune system. These changes in the cell population are not permanent, and the cell populations can revert to there "susceptible" target structure concentrations once the anti-tumor response has abated.

Data implicating intergenerational escape mechanisms have also been observed. MacDougall and colleagues[5] were able to select sublines of NK resistant K562 cells, and this resistance appears to be permanent. Kimber and Moore were also able to select a subline of NK resistant K562 cells.[6] Fidler et al.,[7] were able to select a number of permanent variants of the B16 melanoma system. These variants represented lymphocyte resistance in both poorly and actively metastatic sublines of the original tumor. They also showed that these selected sublines were permanent in their expression of response.

Two models have been developed to describe these two tumor escape phenomena.[8] Briefly, the tumors may escape either intergenerationally or intragenerationally. The models are branching processes embedded into a two-compartment system which characterizes the susceptibility of the cell populations (See Figures 1 and 2). These models represent typical birth-death-migration processes from which Kolmogorov-Chapman differential equations can be derived[8,9] The typical differential equations for the respective models were derived by Michelson:[8]

$$
\begin{aligned}
\frac{dP_{n,m}(t)}{dt} =& (n+1)\big[P_{n+1,m}(t)\mu^{+}(t;t_0) + P_{n+1,m}(t)\delta^{+} \\
&+ P_{n+1,m-1}(t)\alpha^{+}(t;t_0)\big] + (n-1)P_{n-1,m}(t)\rho^{+}(t) \\
&+ (m+1)\big[P_{n,m+1}(t)\mu^{-}(t;t_0) + P_{n,m+1}(t)\delta^{-} \\
&+ P_{n-1,m+1}(t)\alpha^{-}(t;t_0)\big] + (m-1)P_{n,m-1}(t)\rho^{-}(t) \\
&- P_{n,m}(t)\big[n\big(\mu^{+}(t;t_0) + \delta^{+} + \alpha^{+}(t;t_0) + \rho^{+}(t)\big) \\
&+ m\big(\mu^{-}(t;t_0) + \delta^{-} + \alpha^{-}(t;t_0) + \rho^{-}(t)\big)\big]
\end{aligned}
\tag{1}
$$

$$
\begin{aligned}
\frac{dP_{n,m}(t)}{dt} =& (n+1)\big[P_{n+1,m}(t)\mu^{+}(t;t_0) + P_{n+1,m}(t)\delta^{+} \\
&+ P_{n+1,m-2}(t)\rho^{+}(t)(1-\sigma^{+})^{2}\big] \\
&+ 2nP_{n,m-1(t)}\rho^{+}(t)(\sigma^{+}(1-\sigma^{+}) \\
&+ (n-1)P_{n-1,m}(t)\rho^{+}(t)(\sigma^{+})^{2} \\
&+ (m+1)\big[P_{n,m+1}(t)\mu^{-}(t;t_0) + P_{n,m+1}(t)\delta^{-} \\
&+ P_{n-2,m+1}(t)\rho^{-}(t)(1-\sigma^{-})^{2}\big] \\
&+ 2mP_{n-1,m}(t)\rho^{-}(t)(\sigma^{-})(1-\sigma^{-}) \\
&+ (m-1)P_{n,m-1}(t)\rho^{-}(t)(\sigma^{-})^{2} \\
&- P_{n,m}(t)\big[n\big(\mu^{+}(t;t_0) + \delta^{+} + \rho^{+}(t)\big) \\
&+ m\big(\mu^{-}(t;t_0) + \delta^{-} + \rho^{-}(t)\big)\big]
\end{aligned}
\tag{2}
$$

where $P_{n,m}(t)$ represents the probability that n susceptible (i.e., cells with high enough surface concentrations of target structures to be adequately recognized by immune effectors) and m non-susceptible cells (i.e., cells which have decreased their surface concentrations to levels which afford some protection against the anti-tumor response) are present in the tumor at time t. The superscripts "+" and "-" represent the fact that the transitions probabilities act either on the susceptible subpopulation or the (arbitrarily defined) non-susceptible subpopulation. The transition probabilities $\mu^{\cdot}(t;t_0)$ represent cell death due to immunologic causes at time t, given that the anti-tumor response begins at time t_0. (We use the notation "·" to indicate either "+" or "-." Similarly, $\delta^{\cdot}$ represents cell death due to random environmental causes. The transition probabilities $\alpha^{\cdot}(t;t_0)$ in Eq. (1) represent migration between the susceptible and non-susceptible compartments at time t given that the anti-tumor response begins at time t_0. The parameter $(1-\sigma^{\cdot})$ in Eq. (2) represents the mutation rate as expressed at mitosis, i.e., the chance that a cell of one character yields a daughter of the opposite character at mitosis. The parameter $\rho^{\cdot}$(t) represents the reproduction rates of the two subpopulations.

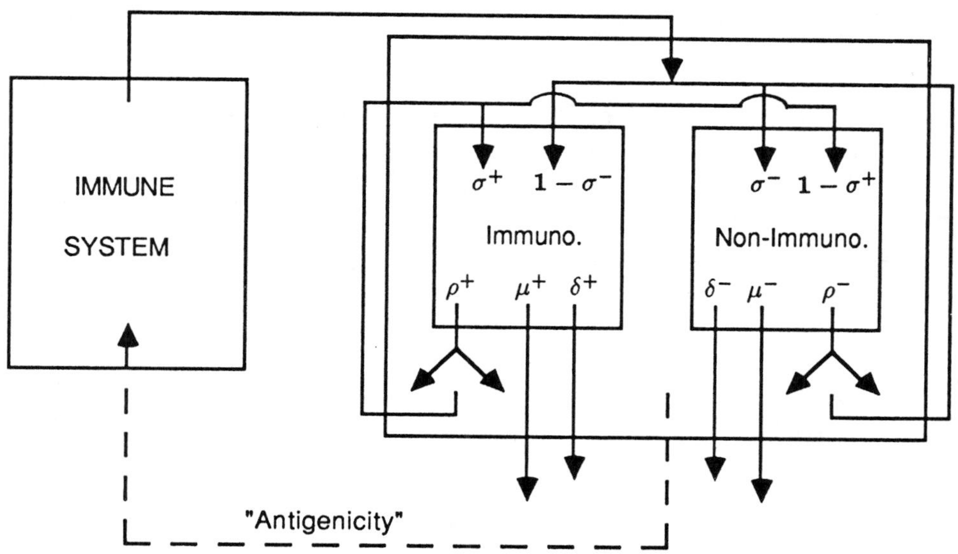

FIGURE 2 Darwinian selection (intergenerational) model.

From these differential equations, one can derive partial differential equations:

$$\begin{aligned}\frac{\partial G(s,z,t)}{\partial t} &= \left[(1-s)\left(\mu^+(t;t_0)+\delta^+-\rho^+(t)s\right)+(z-s)\alpha^+(t;t_0)\right]\partial G/\partial s\\ &\quad+\left[(1-z)\left(\mu^-(t;t_0)+\delta^--\rho^-(t)z\right)+(s-z)\alpha^-(t;t_0)\right]\partial G/\partial z.(3)\\ \frac{\partial G(s,z,t)}{\partial t} &= [(1-s)(\mu+(t;t_0)+\delta^+)\\ &\quad+\rho^+(t)((s(\sigma^+)+z(1-\sigma^+)big)^2-s)]\partial G/\partial s\\ &\quad+[(1-z)(\mu^-(t;t')+\delta^-)\\ &\quad+\rho^-(t)((z(\sigma^-)+s(1-\sigma^-))^2-z)]\partial G/\partial z.\end{aligned} \tag{4}$$

where $G(s,z,t)$ is the joint probability generating function for the stochastic process.

Eqs. (3) and (4) can be differentiated with respect to s and z to yield expressions for the first and second moments of the (random) subpopulation sizes as functions of time. Define at $s=1$ and $z=1$

$$\begin{aligned}M^+(t) &= \partial G/\partial s\\ M^-(t) &= \partial G/\partial z\end{aligned} \tag{5}$$

Then from Eq. (3), one derives

$$\begin{aligned}\frac{dM^+(t)}{dt} =& 1\left(\mu^+(t;t_0) + \delta^+\alpha^+(t;t_0) - \rho^+(t)\right)M^+(t) + \alpha^-(t;t_0)M^-(t)\\ \frac{dM^-(t)}{dt} =& -\left(\mu^-(t;t_0) + \delta^- + \alpha^-(t;t_0) - \rho^-(t)\right)M^-(t) + \alpha^+(t;t_0)M^+(t)\end{aligned} \tag{6}$$

From Eq. (4), one derives

$$\begin{aligned}\frac{dM^+(t)}{dt} =& M^+(t)\left[-\left(\mu^+(t;t_0) + \delta^+\right) + \rho^+(t)\left(2(\sigma^+) - 1\right)\right]\\ &+ M^-(t)2(1-\sigma^-)\rho^-(t),\\ \frac{dM^-(t)}{dt} =& M^-(t)\left[-\left(\mu^-(t;t_0) + \delta^-\right) + \rho^-(t)\left(2(\sigma^-) - 1\right)\right]\\ &+ M^+(t)2(1-\sigma^+)\rho^+(t),\end{aligned} \tag{7}$$

with initial conditions $M^+(0) = 1$, $M^-(0) = 0$.

If we define

$$\begin{aligned}M_2{}^+(t) =& \frac{\partial^2\left[G(s,z,t)\right]}{\partial s^2}\\ M_2{}^-(t) =& \frac{\partial^2\left[G(s,z,t)\right]}{\partial z^2}\\ M_2{}^0(t) =& \frac{\partial^2\left[G(s,z,t)\right]}{\partial s\,\partial z}\end{aligned} \tag{8}$$

then Eq. (3) yields

$$\begin{aligned}\frac{dM_2{}^+(t)}{dt} =& \rho^+(t)M^+(t) - 2\beta^+(t;t_0)M_2{}^+(t) + \alpha^-(t;t_0)M_2{}^0(t)\\ \frac{dM_2{}^-(t)}{dt} =& \rho^-(t)M^-(t) - 2\beta^-(t;t_0)M_2{}^-(t) + \alpha^+(t;t_0)M_2{}^0(t)\\ \frac{dM_2{}^0(t)}{dt} =& -\left[\beta^+(t;t_0) + \beta^-(t;t_0)\right]M_2^0(t) + \alpha^+(t;t_0)M_2^+(t) + \alpha^-(t;t_0)M_2^-(t)\end{aligned} \tag{9}$$

where

$$\beta^+(t;t_0) = \left(\mu^+(t;t_0) + \delta^+(t;t_0) - \rho^+(t)\right)$$

and similarly for $\beta^-(t;t_0)$.

Eq. (4) yields

$$
\begin{aligned}
\frac{dM_2{}^+}{dt} =& \rho^+(t)s(\sigma^+)^2M^+(t) + 2M_2{}^+(t)\big[-\big(\mu^+(t;t_0)+\delta^+\big) \\
&+\rho^+(t)\big(2(\sigma^+)-1\big)\big] + 4M_2{}^0(t)\big(1-(\sigma^-)\big)\rho^-(t) \\
\frac{dM_2{}^-}{dt} =& \rho^-(t)2(\sigma^-)^2M^-(t) + 2M_2{}^-(t)\big[-\big(\mu^-(t;t_0)+\delta^-\big) \\
&+\rho^-(t)\big(2(\sigma^-)-1\big)\big] + 4M_2{}^0(t)\big(1-(\sigma^+)\big)\rho^+(t) \\
\frac{dM_2{}^0}{dt} =& 2\rho^+(t)(\sigma^+)(1-\sigma^+)M^+ + 2\rho^-(t)(\sigma^-)(1-\sigma^-)M^-(t) \\
&+2(1-\sigma^+)\rho^+(t)M_2{}^+(t) + 2(1-\sigma^-)\rho^-(t)M_2{}^-(t)+ \\
&M_2{}^0(t)\big[-\mu^+(t;t_0)-\delta^+ + \rho^+(t)\big(2(\sigma^+)-1\big) \\
&+M_2{}^0(t)\big[-\mu^-(t;t_0)-\delta^- + \rho^-(t)\big(2(\sigma^-)-1\big)
\end{aligned}
\tag{10}
$$

with initial conditions $M_2{}^+(0)=0$, $M_2{}^-(0)=0$, and $M_2{}^0(0)=0$.

MODEL SIMULATION

For the constant coefficients case, one can solve Eqs. (6) and (7) and derive asymptotic stability criteria for the subpopulation sizes.[8] However, Eqs. (3), (4), (6), (7), (9), and (10) are defined for general functions μ, α, and ρ, all of which may be time- and/or response-dependent. Therefore, rather than solve these systems of equations for a series of generally specified functions, we have developed a Monte Carlo simulation which allows one to experimentally manipulate the 6-parameter functions defining the anti-tumor responses of the NK cell compartment (the non-inductive response wave) and the T-cell or IgM compartment (the inductive wave). One may also set the rate at which the tumor alters its susceptibility ("target structure stability"), the means of escape ("escape route"), and the susceptibility level to which escaping cells escape ("susceptibility level"). The cell cycle is modeled as a pipeline of identical maturation chambers, each described by a constant transition rate to the next pipe element. This configuration allows for the specification of the total cell life length as a random variable distributed in accordance with a gamma distribution.

The simulation is described in Figures 3A and 3B. The intragenerational simulation is depicted in Figure 3A. Because demodulation is a time-consuming process, the cells may pass through a series of susceptibility levels. Therefore, a cell that is in the process of demodulating faces a greater and greater risk of being recognized, conjugated with, programmed, and lysed. On the other hand, the intergenerational simulation describes a permanent, genetic mutation, which alters a daughter cell's susceptibility at mitosis (see Figure 3B). Similar mathematical models describing the development of Multi-Drug Resistance (MDR) in evolving tumor populations have been developed.[10–12] These simulations represent the mutation rate as the quantity $(1 - \sigma)$ where σ is the probability that a cell produces a daughter like itself at mitosis. Therefore, only two pipes are needed to simulate intergenerational escape.

The mechanistic description of the simulation is presented in detail elsewhere.[8] Briefly, at each time step (30 minutes) a cell faces a risk of being recognized and lysed (as a function of the level of the immune response), migrating (in the intragenerational model this rate is a function of the immune response level and the cell's susceptibility), or maturing (i.e., progressing through the chambers of the pipe).

For each cell in each maturation chamber of each pipeline, we determine whether a transition should occur as follows: Each type of transition is represented as a transition probability for a thirty minute time step (Δt). Each event can be treated as a Bernoulli trial so that each cell may be processed, in turn, and its fate is determined by the "success probability" from a standard random-number generator. However, rather than calculating a transition probability to determine a "success" for each cell in each chamber, one can treat each chamber (containing, say, k cells) as k

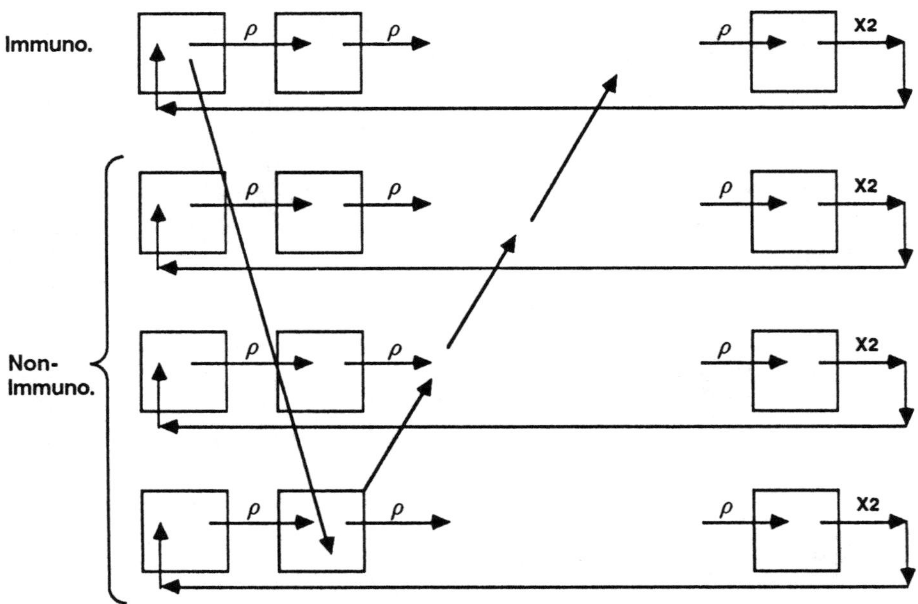

FIGURE 3A Intergenerational pipelines—normal immunosusceptible cells at top. Non-susceptible cells at bottom. Return is via intermediate susceptibility levels (demodulation).

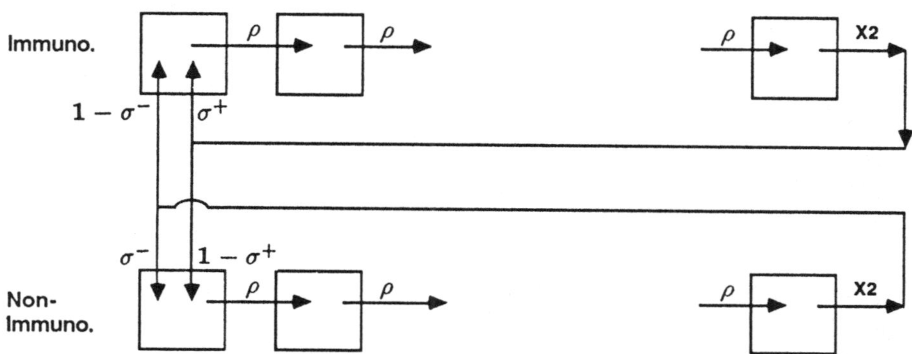

FIGURE 3B Intragenerational pipelines—normal immunosusceptible at top. At mitosis, there is a $(1 - \sigma)$ probability a cell changes its susceptibility status.

Bernoulli trials, and calculate the probability of x successes from k trials (i.e., the standard binomial distribution can be used for event determination). Then a standard random number generator can be used to determine how many of the k cells in the given chamber "succeeded," and how many will move to some new state. For example, if 2 cells reside in a chamber, and if the transition rate for immunologically induced cell death is 0.25, then the probability that no cells will die is 0.5625, the probability that both cells will die is 0.0625, and the probability that one cell will die is 0.3750. Then a random number generated from a uniform random number generator that falls below 0.5625 results in no transition to death (i.e., the cells are not removed from the chamber). A number falling between 0.5625 and 0.9375 results in one cell being removed from the chamber, and a number falling above 0.9375 results in both cells being removed from the chamber.

Each simulated experiment represents at least 50 realizations of the simulation for each tumor type. When deemed necessary, the statistical power to discern small differences was calculated and more realizations were run as needed. Similar to calculations performed for comparison of proportions in large-scale randomized clinical trials, we posited "true" differences between experiments, and derived the number of realizations needed to have an 80% chance of observing that difference in our simulation. Each realization begins with a single immunosusceptible cell in the first maturation stage of the cell cycle (i.e., the first chamber of the pipe). The realization was followed for 20 days, until the population was deemed to have "escaped," or until the population was forced to extinction. Therefore, the system can stop in one of three states: Extinction, End of Simulated Time, or Escaped. Only the realizations that were deemed to have escaped can yield an equivocal result (i.e., could the realization, had it continued, gone on to force the population into extinction?). A simple statistical study has been presented elsewhere,[8] and it was shown that in every case in which a stopping rule was applied, the population had achieved exponential growth, and the null hypothesis that the population was not growing could be rejected and that the ratio of daily population sizes was significantly greater than 1.

PARAMETERIZATION OF THE SIMULATIONS

The simulations require that the cell population be characterized as to its target structure stability, escape route, and susceptibility level to which it escapes. We also require that the mean and standard deviation of the random life length distribution be provided to the simulation. Finally, the user must also provide the simulation with six parameters which describe the phenomenological shape of a two wave (non-inductive and inductive) immune response. The shapes of the two waves are presented in Figure 4. They scale an *in vitro* response level to the time-dependent *in vivo* response. The user provides the start time, the time it takes the

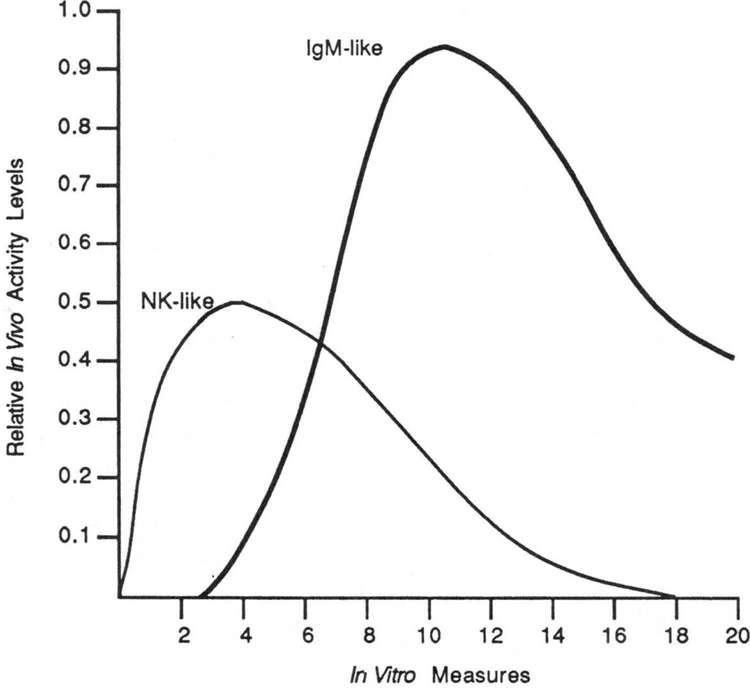

FIGURE 4 "Typical" anti-tumor response profile—non-inductive (NK-like) and inductive (IgM-like) response waves.

immune response to reach a maximum activity level (response length), the maximum activity level scaling value (e.g., 95% of *in vitro* activity), the minimum activity level scaling value, the time it takes for the response to decline to preresponse levels (declining response time), and the time a response can maintain its maximum activity. These twelve parameters, six for each wave, completely describe the phenomenological anti-tumor response.

The death rate of cells in an *in vitro* ^{51}Cr release assay was taken as an exponentially decreasing curve with half-life of 12 hours. From this approximation a constant hazard for recognition, conjugation and programming was calculated per thirty-minute period. This hazard rate was then scaled by the activity level *in vivo* and the concentration of target structures on the surface of the cells as a function of which pipe was being simulated. Because we assume that modulation also requires recognition, conjugation and programming, the hazard rate for modulation in the intragenerational model was calculated in a similar manner. As a measure of target structure stability, half-lives of 12 and 3 hours (as observed in the literature, e.g., Boyse,[1] Liang and Cohen[3]) were used for the simulation runs. Demodulation was calculated as a constant hazard of transition in the absence of effector activity, and was thus scaled by a probability that the target cells were not recognized, conjugated with, and programmed given their target structure concentrations and activity level of anti-tumor response.

The intergenerational model requires a mutation rate rather than a modulation rate. Goldie and Coldman[10] suggest that the rate of viable mutation to MDR is on the order of 1 mutation per million mitoses. In these studies, the mutation rates are taken to be 1 per thousand, 1 per hundred, or 5 per hundred. Michelson[8] shows that at 1 per thousand, even in 100 realizations of any simulation the proportion of realizations going extinct is 100%.

Fenyo et al.[13] observed a ten-fold decrease in surface antigen concentration in the YAC-IR line. Therefore, a candidate for the susceptibility level to which cells will escape is 10% (i.e., cells achieving a 10% surface concentration of their target structures are considered to be non-susceptible). However, Young and Hakomori[14] suggest that in the murine lymphoma L5178Y, the glycolipid gangliotriosylceramide acts as a target structure for IgM antibody, and that the concentration of this antigen upon the surface of escaping cells is one half its value in the normal population. Therefore, a second candidate for the arbitrary boundary which we will call non-susceptible will be 50%.

RESULTS

The simulation studies presented in this paper classify the tumors by three binary factors. The motivation for the use of these factors has been given above. The values of the cut points for stability and susceptibility have been described in the parameterization discussion just concluded. The tumors are thus classified by:

ESCAPE ROUTE: "A" = intragenerational; "R" = intergenerational

TARGET STRUCTURE STABILITY:
Intragenerational Stable = 12 hour half-life (e.g., Boyse[1])
Unstable = 3 hour half-life (e.g., Liang and Cohen[3])
Intergenerational Stable = 1 mutation per 100 mitoses
Unstable = 5 mutations per 100 mitoses

ESCAPE TO SUSCEPTIBILITY: 50% (Young and Hakomori[14]) vs. 10% (Fenyo et al.[13]) of original target structure concentration

The tumors are, thus, classified into 8 categories:

A1 = Intragenerational, stable, 50% target structure
A2 = Intragenerational, stable, 10% target structure
A3 = Intragenerational, unstable, 50% target structure
A4 = Intragenerational, unstable, 10% target structure
R1 = Intergenerational, stable, 50% target structure
R2 = Intergenerational, stable, 10% target structure
R3 = Intergenerational, unstable, 50% target structure
R4 = Intergenerational, unstable, 10% target structure

A variety of anti-tumor responses were tested in detail[8] and a summary of these simulations are presented here. The immune response profiles are summarized in Table 1 for each simulation run, and the results (proportion of realizations that go extinct) for each simulated experiment for each type of tumor are presented in Table 2.

Looking at these two tables in detail one can identify a number of cross-tumor variations within a single response profile. For example, consider the first three sets of responses termed "Typical," "Enhanced NK," and "Very Enhanced NK." These experiments were designed to identify a range of activity levels that NK cells may reach in normal hosts. From Table 1 one notices that the inductive response parameters are held constant for each simulated experiment. However, the non-inductive response parameters all change in the variable λ which represents the maximum *in vivo* activity achieved by NK cells as a percentage of *in vitro* activity. The results in Table 2 for these three simulated experiments show that the proportion of tumor clones going extinct increases for each tumor as the NK activity is heightened. However, within a single experiment, a *tumor specific response* is also evident, and the tumor type most likely to escape in each case are the unstable tumors escaping intergenerationally to the 10% susceptibility level.

Another example are the lines labeled "Interferon," "Extended Interferon" and "Interferon + IgM." These profiles were generated in a series of simulated experiments that were designed to describe the typical activity levels observed by Golub and colleagues[15,16] in the NK component of the immune response in melanoma patients who were treated with various regimens of interferon. In the "Interferon"

TABLE 1 Immune Response Profiles Used in Table 2[1]

Response Type	t_0	Π_0	λ	RT	L	DT
TYPICAL						
Non-Inductive	0.0	0.005	0.5	3.5	1.0	13.0
Inductive	3.0	0.005	0.99	8.0	1.0	13.0
ENHANCED NK						
Non-Inductive	0.0	0.005	0.75	3.5	1.0	13.0
Inductive	3.0	0.005	0.99	8.0	1.0	13.0
VERY ENHANCED NK						
Non-Inductive	0.0	0.005	0.94	3.5	1.0	13.0
Inductive	3.0	0.005	0.99	8.0	1.0	13.0
STUTMAN EXPT. 1						
Non-Inductive	0.0	0.005	0.5	3.5	1.0	13.0
Inductive	30.0	0.000	0.0	30.0	0.0	0.0
STUTMAN EXPT. 2						
Non-Inductive	0.0	0.005	0.75	3.5	1.0	13.0
Inductive	30.0	0.000	0.0	30.0	0.0	0.0
T-CELL						
Non-Inductive	0.0	0.005	0.5	3.5	1.0	13.0
Inductive	3.0	0.005	0.99	5.0	1.0	13.0
WIDE VALLEY						
Non-Inductive	0.0	0.005	0.5	3.5	1.0	8.0
Inductive	3.0	0.005	0.99	8.0	1.0	13.0
INTERFERON						
Non-Inductive	1.0	0.005	0.99	6.0	1.0	13.0
Inductive	30.0	0.000	0.00	30.0	0.0	0.0
EXTENDED INTERFERON						
Non-Inductive	1.0	0.005	0.99	6.0	3.0	13.0
Inductive	30.0	0.000	0.00	30.0	0.0	0.0
INTERFERON + IgM						
Non-Inductive	1.0	0.005	0.99	6.0	1.0	13.0
Inductive	3.0	0.005	0.99	8.0	1.0	13.0

[1] where:
t_0 is the time the response began;
Π_0 is the minimum activity level observed *in vivo*;
λ is the maximum activity level observed *in vivo*;
RT is the time it takes the response to achieve its maximum activity level;
Lis the length of time the response remains at its maximum level
DT is the time it takes the response to return to its original activity level.

TABLE 2 Summary Table of Simulated Extinction Proportions

Response Type	Tumor Type by Immune Response A1	A2	A3	A4	R1	R2	R3	R4
TYPICAL RESPONSE	1.00	1.00	1.00	0.98	1.00	0.96	0.96	0.52
	1.00	1.00	1.00	0.98	1.00	0.98	1.00	0.66
	1.00	1.00	1.00	0.98	1.00	0.97	0.98	0.59
ENHANCED NK	1.00	1.00	1.00	0.98	1.00	0.98	0.98	0.70
VERY ENHANCED NK	1.00	1.00	1.00	1.00	1.00	1.00	1.00	0.78
STUTMAN EXPT. 1	0.43	0.31	0.39	0.26	1.00	—	—	0.29
STUTMAN EXPT. 2	0.90	0.72	0.81	0.71	1.00	—	—	0.58
T-CELL	1.00	0.96	1.00	0.94	1.00	1.00	1.00	0.74
WIDE VALLEY	1.00	—	1.00	—	1.00	—	0.92	—
INTERFERON	0.98	0.96	0.98	0.66	1.00	0.96	0.78	0.46
EXTENDED INTERFERON	1.00	0.96	0.96	0.82	1.00	0.98	0.82	0.50
INTERFERON + IgM	1.00	1.00	1.00	1.00	1.00	0.94	1.00	0.54

[1] where Tumor Type is defined by:
A1 = Intragenerational, stable, 50% target structure;
A2 = Intragenerational, stable, 10% target structure;
A3 = Intragenerational, unstable, 50% target structure;
A4 = Intragenerational, unstable, 10% target structure;
R1 = Intragenerational, stable, 50% target structure;
R2 = Intergenerational, unstable, 10% target structure;
R3 = Intergenerational, stable, 50% target structure; and
R4 = Intergenerational, unstable, 10% target structure.

and "Extended Interferon" simulations the inductive response is obliterated because, at present it is not yet clear whether interferon suppresses the activity level within the humoral compartment. Therefore, in two of the simulations we remove the response, while in the third it is restored. The proportion of tumor clones going extinct increases with increasing NK activity, but again, the variety of responses across tumor type is startling. The most interesting result in this series is given in the line "Interferon + IgM" in which the inductive response is restored. Here every tumor type but two exhibited 100% extinction proportions. The two exceptions were both intergenerationally escaping tumors that escaped to the 10% susceptibility level. These results are similar to those observed in the mathematical models of DeBoer.[17–19]

A second aspect of heterogeneity that could affect a tumor's ability to escape the immune surveillance mechanism is the mean lifelength of the cells that are emerging in the tumor clone. Suppose that all cells are escaping intragenerationally, and that the cells can be classified as "slow" or "fast" (24-hour vs. 18-hour mean lifelengths), "quick demodulators" or "slow demodulators" (DEMOD: 12-hour half-life vs. 18-hour half-life) and "stable" or "nonstable" (MOD: 12-hour half-life vs. 3-hour half-life to modulation). In Table 3 we present the results of two replicate simulations for each type of tumor under the "Typical Response" profile (as defined in Table 1). We analyze these extinction results in the 2^3 Factorial Analysis of Variance (ANOVA) presented in Table 4. Results in Table 4 report typical ANOVA style results with each individual factor (MOD, LL, DEMOD) and their respective interactions (MOD X LL, MOD X DEMOD, DEMOD X LL, MOD X DEMOD X LL) testing the null hypothesis that none of the observed effects in mean survival cannot be accounted for by chance alone.

TABLE 3 2^3 Factorial Design—Heterogeneity Test

MOD	DEMOD	2 RUNS (N=50) MEAN LIFELENGTH (LL)	EXTINCT
3	12	24	49, 50
3	12	18	30, 32
3	18	24	50, 49
3	18	18	33, 38
12	12	24	50, 50
12	12	18	47, 42
12	18	24	50, 50
12	18	18	42, 44

TABLE 4 2^3 Factorial Analysis of Variance

SOURCE[1]	SUM SQ'S	D.F.	MEAN SQ.	F	TAIL
Mean	31152.25	1	31152.25	6922.7	0.0000
Mod	121.0	1	121.0	26.90	0.0008
Demod	2.25	1	2.25	0.50	0.4996
Lifelength	506.0	1	506.0	112.50	0.0000
Mod X Demod	9.0	1	9.0	2.00	0.1950
Mod X LL	100.0	1	100.0	22.22	0.0015
Demod X LL	2.25	1	2.25	0.55	0.4996
Mod X LL X Demod	9.0	1	9.0	2.00	0.1950
Error	18.0	1	18.0		

[1] where, MOD represents the rate of modulation as a categorical variable (12 hours vs. 3 hours), DEMOD represents the rate of demodulation as a categorical variable (12 hours vs. 18 hours) and LL represents the mean lifelength of the cell cycle distribution (18 hours vs. 24 hours). Their statistical interactions are represented by "X" in the sense of MOD X DEMOD.

The results in Table 4 show that the rate of modulation, MOD, the mean lifelength, LL, of the cell population, and their interaction, MOD X LL, are all significant classifiers in determining tumor escape. Therefore, heterogeneity as expressed in rates of growth along with the stability of target structures on modulating cells can fundamentally effect the immune system's ability to clear a tumor in the early stages of its evolution.

DISCUSSION

Tumor heterogeneity, by its very nature, imparts a spectrum of response to hostile factors across the entire tumor cell population. Biological heterogeneity within a number of tumor systems has been reviewed by Dexter and Leith.[20] One system in particular has been studied in detail. Termed DLD-1, this system is a human adenocarcinoma of the colon first extracted from a patient and characterized as consisting of two karyotypically and morphologically distinct subpopulations.[21–24] Other tumor systems have also been observed in mice.[25–27]

Tumor heterogeneity with respect to immunosusceptibility may also be expressed across tumor populations. Whether this immunoheterogeneity is expressed as a decrease in a tumor's ability to escape intragenerationally (e.g. modulation) or in a permanent decrease in tumor-specific target structures upon the cell surfaces which facilitate intergenerational escape (e.g., genetically determined at mitosis), a spectrum of response still remains.

Some controversy has raged throughout the immunology literature as to whether an immune surveillance mechanism exists, and if it does, what level of effectiveness it achieves in protecting the normal host (for example, see [28–30]). For the purpose of this discussion then, we define immune surveillance as the process by which a host's immune system continuously and consistently clears malignantly transformed cells at a subclinical stage. We assume that the cells are frequently transformed to their malignant states due to onslaught by carcinogenic agents.

To better understand this controversy we derived two mathematical models which describe tumor escape as a function of tumor cell type. The type of tumor cell has been arbitrarily classified into eight categories, but a wide spectrum of realistic parameter values could be used to characterize tumor subpopulations within the same model structure. In this paper we have extended the two mathematical models to their respective Monte Carlo simulations. The simulation results suggest that for a single cell type, escape from an active immune surveillance mechanism depends upon the activity levels, timing, etc. achieved during the anti-tumor response. However, within a single response profile, tumor escape depends upon the type of tumor cells that are evolving in the host.

DeBoer and Hogeweg[18] and DeBoer et al.[17,19] model the macrophage/T-cell interactions within a normal host to describe the control mechanisms exhibited in an anti-tumor effector response. They show that the antigenicity levels (cell susceptibility in these models) to which the tumor cells escape are the primary factors in determining tumor escape, and that these levels exert their own feedback control over the anti-tumor response. They show that modulation, in and of itself, is a poor mechanism to facilitate tumor escape. If all cells are modulating rapidly (i.e., if the target structures are unstable in this model), tumor escape can be achieved. However, heterogeneous mixtures of modulating and non-modulated cells can be routinely rejected due to concomitant immunization of the host by the fully antigenic subpopulation. Their simulations are deterministic and describe tumor populations achieving sizes of 10^7 cells. Our simulations are stochastic, and usually stop when tumor populations are extinguished early in their evolution (sizes achieved are usually less that 20 cells) or "escape" the anti-tumor response (at sizes usually about 500 cells). Clearly, the extension of each to the other is required. However, the basic conclusion remains, i.e., that tumor escape is as much a function of the tumor population and its characteristics as it is of the anti-tumor response.

REFERENCES

1. Boyse, E. A., E. Stockert, and L. Old (1967), "Modification of the Structure of the Cell Membrane by Thymus Leukemia (TL) Antibody," *Proc. Natl. Acad. Sci. USA* **58**, 954–957.
2. Wolf, J. E., R. B. Fanes, and Y. S. Choi (1977), "Antigenic Changes of DBA/2J Mastocytoma Cells when Grown in BALB/C Mouse," *JNCI* **58**, 1407–1412.
3. Liang, W., and E. P. Cohen (1977), "Detection of Thymus Leukemia Antigen on the Surface Membranes of Murine Leukemia Cells Resistant to Thymus Leukemia Antibodies and Guinea Pig Serum," *JNCI* **58**, 1079–1085.
4. Esmon, N. I., and J. R. Little (1976), "Different Mechanisms for the Modulation of TL Antigens on Murine Lymphoid Cells," *J. Immunol.* **117**, 919–926.
5. MacDougall, S. L., C. Shustik, and A. K. Sullivan (1983), "Target Cell Specificity of Human Natural Killer (NK) Cells. I. Development of an NK-Resistant Subline of K562," *Cellular Immunol.* **76**, 39–48.
6. Kimber, I., and M. Moore (1984), "Clonal Variation of NK Sensitivity Among K562 Lines," *Natural Killer Activity and Its Regulation* (Int'l. Symposium on Natural Killer Activity and Its Regulation), Eds. T. Hashiro, H. S. Koren, and A. Uchida, Excerpta Medica, 308–312.
7. Fidler, I., D. M. Gersten, and M. D. Budmen (1976), "Characterization *In Vivo* and *In Vitro* of Tumor Cells Selected for Resistance to Syngeneic Cytotoxicity," *Cancer Res.* **36**, 3160–3165.
8. Michelson, S. (1987), *Two-Compartment Stochastic Models for Tumor Escape* (Ph.D. Dissertation, University of California, Los Angeles).
9. Goel, N., and N. Richter-Dyn (1974), *Stochastic Models in Biology* (London: Academic Press).
10. Goldie, J. H., and Coldman, A. J. (1979), "A Mathematical Model for Relating the Drug Sensitivity of Tumors to Their Spontaneous Mutation Rate," *Cancer Treat. Rep.* **63**, 1727–1733.
11. Goldie, J. H., A. J. Coldman, and G. A. Gudauskas (1982), "Rationale for the use of Alternating Non-Cross-Resistant Chemotherapy," *Cancer Treat. Rep.* **66**, 439–449.
12. Coldman, A. J., J. H. Goldie, and V. Ng (1985), "The Effect of Cellular Differentiation on the Development of Permanent Drug Resistance," *Math. Biosci.* **74**, 177–198.
13. Fenyo, E. M., E. Klein, G. Klein, and K. Sweich (1968), "Selection of an Immunoresistant Maloney Subline with Decreased Concentration of Tumor Specific Surface Antigens," *JNCI* **40**, 69–89.
14. Young, w. W., and S. Hakomori (1980), "Therapy of Mouse Lymphoma with Monoclonal Antibodies to Glycolipids: Selection of Low Antigenic Variants *In Vivo*," *Science* **212**, 487–489.

15. Golub, S. H., F. Dory, D. Hara, D. L. Morton, and M. W. Burk (1982), "Systemic Administration of Human Leukocyte Interferon to Melanoma Patients. I. Effects on Natural Killer Function and Cell Populations," *JNCI* **68**, 703–710.
16. Golub, S. H., P. D'Amore, and M. Rainey (1982), "Systemic Administration of Human Leukocyte Interferon to Melanoma Patients. II. Cellular Events Associated with Changes in Natural Killer Cell Cytotoxicity," *JNCI* **68**, 711–717.
17. DeBoer, R. J., P. Hogeweg, H. F. J. Dullens, A. DeWeger, and W. Den Otter (1985), "Macrophage T Lymphocyte Interactions in the Antitumor Immune Response: A Mathematical Model," *J. Immunol.* **134**, 2748.
18. DeBoer, R. J., and P. Hogeweg (1986), "Interactions between Macrophages and T-Lymphocytes: Tumor Sneaking through Intrinsic to Helper T Cell Dynamics," *J. Theor. Biol.* **120**, 331–351.
19. DeBoer, R. J., S. Michelson, and P. Hogeweg (1986), "Concomitant Immunization by the Fully Antigenic Counterparts Prevents Modulated Tumor Cells from Escaping Cellular Immune Elimination," *J. Immunol.* **136**, 4319–4327.
20. Dexter, D. L., and J. T. Leith (1986), "Tumor Heterogeneity and Drug Resistance," *J. Clin. Oncol.* **4(2)**, 244–257.
21. Dexter, D.L., E. N. Spremulli, Z. Fligiel, J. A. Barbosa, R. Vogel, A. VanVoorhees, and P. Calabresi (1981), "Heterogeneity of Cancer Cells from a Single Human Carcinoma," *Am. J. Med.* **71**, 949.
22. Dexter, D. L., and P. Calabresi (1982), "Intraneoplastic Diversity," *Biochim. Biophys. Acta.* **695**, 97.
23. Leith, J. T., L. E. Faulkner, S. F. Bliven, E. S. Lee, A. S. Glicksman, and D. L. Dexter (1985), "Disaggregation Studies of Xenograft Solid Tumors Grown from Pure or Admixed Clonal Subpopulations from a Heterogeneous Colon Carcinoma," *Invasion and Metastasis* **5**, 317–335.
24. Leith, J. T. , S. Michelson, L. E. Faulkner, and S. Bliven (1987), "Growth Properties of Artificial Heterogeneous Human Colon Tumors," *Cancer Res.* **47**, 1045–1051.
25. Heppner, G. H., B. E. Miller, and F. R. Miller (1983), "Tumor Subpopulation Interactions in Neoplasms," *Biochim. et. Biophysica. Acta.* **695**, 215–226.
26. Heppner, G. H. (1984), "Tumor Heterogeneity," *Cancer Res..* **44**, 2259–2265.
27. Heppner, G. H., S. E. Loveless, F. R. Miller, K. H. Mahoney, and A. M. Fulton (1984), "Mammary Tumor Heterogeneity," *Cancer Invasion and Metastases: Biologic and Therapeutic Aspects*, Eds. G. L. Nicolson and L. Milas (New York: Raven Press).
28. Allison, A. C. (1977), "Immunological Surveillance of Tumors," *Cancer Immunol. Immunotherapy* **2**, 151–155.
29. Moller, G., and E. Moller (1975), "Considerations of Some Current Concepts in Cancer Research," *JNCI* **55**, 755–759.
30. Kripke, M. L., and T. Borsos (1974), "Immune Surveillance Revisited," *JNCI* **52**, 1393–1394.

Mathematical Models of HIV Infection

STEPHEN J. MERRILL
Marquette University, Department of Mathematics, Statistics and Computer Science, Milwaukee, WI 53233, U.S.A.

AIDS: Background and the Dynamics of the Decline of Immunocompetence

ABSTRACT

A mixed stochastic-deterministic model of the infection of $T4$ (helper-inducer) lymphocytes by the HIV virus is proposed. It is used to explore the loss of immunocompetence in AIDS as it is affected by 1) rate of external antigenic stimulation of $T4$ cells, 2) the effectiveness of defenses and 3) initial size of infection.

INTRODUCTION

In 1981, clusters of several rare opportunistic infections and cancers associated with immunodeficiency were reported.[1–4] These reports triggered the formation of a task force by CDC (Centers for Disease Control) charged with investigating these unusual occurrences. As a result of their investigation, a new disease was identified, characterized by severe immunodeficiency, and was designated AIDS (Acquired Immune Deficiency Syndrome). Apparently, the first cases in the United States were in the late 1970's and since that time, the number of cases has increased geometrically. In certain populations (especially homosexual men and intravenous drug users), the disease is prevalent and, in other populations, the fear of AIDS has prompted the call for drastic changes in lifestyle. The reaction to the disease has been made more frantic with the realization that no cure is currently available, and the disease is always fatal within a fairly short period once diagnosed.

By 1983,[5] a virus had been associated with AIDS patients and after continuous culture,[6] the link between AIDS and the virus, now called HIV (Human Immunodeficiency Virus), was quickly confirmed. HIV is a retrovirus with affinity for the $T4$ (helper-inducer) lymphocyte. The eventual death of the $T4$ lymphocytes resulting from the viral infection may well be one way in which immunocompetence is reduced.

Three central problems in AIDS are natural questions for modeling approaches. The first is to describe the progressive nature of the disease and the resulting loss of immunocompetence. This model would be valuable in designing treatment to minimize the effect of the process.

The second is the determination of the probability of developing AIDS given that the individual is infected. As the estimate of the number of exposed individuals in the United States alone is now over a million, it is important for society and the individuals involved to know the true nature of the risk they face. An associated virus, feline leukemia virus, will produce the typical chronic disease in about 30% of challenged cats.[7] A similar number is unknown for HIV.

The third, related strongly to the second, is to describe the epidemiology of the disease. Given asymptomatic carriers and the possibly long latent period for the disease, this problem may be the most difficult. This one, unfortunately, is also the most pressing as it is necessary to describe the expected number of cases in order to insure that appropriate medical care will be available when required and so that the true scope of the disease can be recognized.

In this paper, we will only examine the first question. In addition, we will try to identify important parameters and the variation in the time course of the disease as functions of those parameters. The form of the model will be a mixed stochastic-deterministic model.

1. MECHANISM OF HIV-LYMPHOCYTE INTERACTION

HIV consists of a core of a single strand of RNA covered with a glycoprotein envelope. This envelope binds to the $T4(CD4)$ marker on the helper-inducer T-lymphocyte and certain other cells. This marker is involved in the recognition of MHC class II molecules regulating cellular communication during an immune response. The HIV envelope can also bind to macrophages and other cells of monocytic origin, and B-cells. After binding, penetration of the core material is accomplished, this penetration being more efficient in $T4$ lymphocytes than in the other cells mentioned.

Complexed to the *HIV RNA* is the enzyme reverse transcriptase (*RT*) which transcribes the *RNA* into its circular *DNA* form which migrates to the nucleus and somewhat randomly integrates itself into the host cell's *DNA*. It then waits in a latent state.

2. EXPRESSION OF THE VIRAL GENES AND DEATH OF THE HOST

HIV genes are not expressed until the infected *T*-lymphocyte is stimulated or activated.[8] During the usual beginning of the transcription process to create new *RNA* in an activated cell, viral genes are also transcribed. These genes code for envelope proteins, *RT*, core antigens and *tat-III*, which is responsible for the "trans"-activation of the viral *DNA*. This enhances gene expression. This also explains why *HIV* replication is effective even with the random integration of

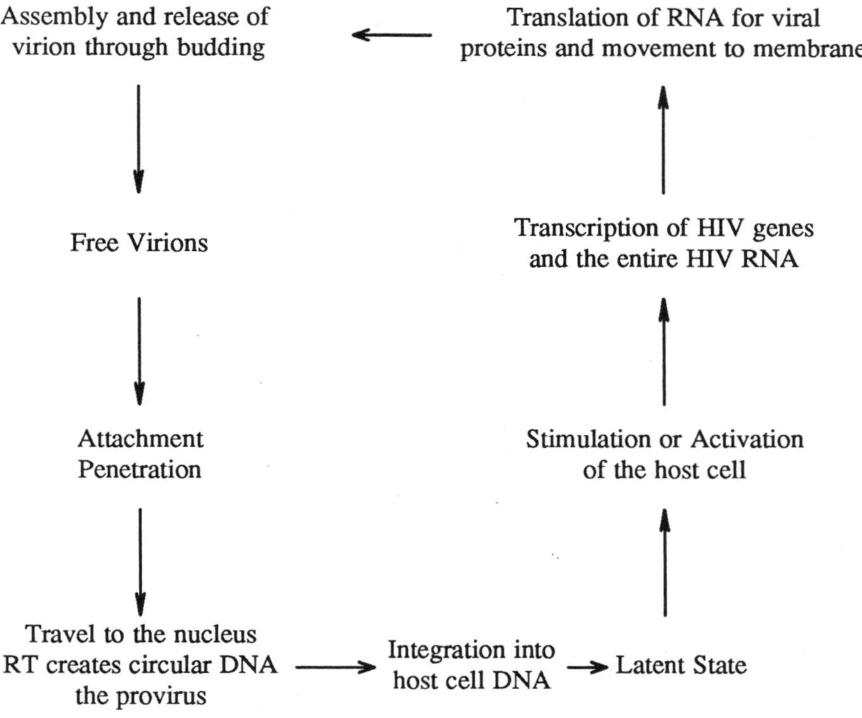

FIGURE 1 The life cycle of the *HIV* virus.

the *HIV-DNA* into the host *DNA*. Also transcribed are copies of the complete viral *RNA* by the host *RNA* polymerase. The *RNA* migrates to the cytoplasm and eventually to the membrane, where assembly into the complete viron is completed. Release of the viral particles are accomplished by budding (see Figure 1).

After making many *HIV* copies, the $T4$ cell dies. The actual mode of death is not known at this time, although *tat-III* is required. *In vitro* and in the brain, the cells fuse into giant multinucleated cells. In other cell types, especially macrophages, *HIV* replication does not generally result in the death of the cell. Infected macrophages, then, form a reservoir for the *HIV* virus.

Several aspects of this process are important to emphasize in order that a model can be constructed. First, multiple infection of a $T4$ with the virus is possible.[9] Secondly, during the latent phase, no viral antigens appear on the host cell's surface (as the genes have not yet been transcribed). Thirdly, antibody made against the *HIV* envelope has been found to contain a low titer of neutralizing antibody.[10,11] Thus, in most cases, the humoral response plays an ineffective role in trying to prevent *HIV* infection of $T4$ cells. Fourthly, the rate of development of the disease depends on the rate of $T4$ stimulation (the rate of infection by other agents)[12] and the release of *HIV* by the reservoir population.

The literature for AIDS is very large and growing rapidly. One reference which describes the molecular biology, epidemiology and clinical features of AIDS is [13]. Some current speculation and great illustrations can be found in [14]. The assumptions in this paper are similar to those presented in [15] (without the reservoir population). A model based on different assumptions was presented by Cooper.[16]

3. THE MODEL OF T4 LYMPHOCYTE INFECTION

Let $V(t)$ be the concentration of infectious *HIV* viral particles at time t and $I(t)$ the population of infected $T4$ lymphocytes. Let $L(t)$ be the population of live $T4$ cells.

We first consider the reservoir population. Let $Mac(t)$ be the total population of circulating and resident macrophages at time t. This population will be considered to be divided into two subpopulations, $M(t)$ and $M_i(t)$. $M(t)$ is the portion uninfected by *HIV* while $M_i(t)$ is the infected population.

Assume that the rate of macrophage infection by *HIV* is proportional to the number of circulating infective virus particles and to the number of uninfected cells. Then

$$M_i'(t) = \gamma_1 V(t)\big(Mac(t) - M_i(t)\big) - \gamma_2 M_i(t)$$

where γ_2 is the natural removal/death rate of infected macrophages. Then

$$\begin{aligned} Mac'(t) =& \Big\{S_1 - \gamma_1 V(t)\big(Mac(t) - M_i(t)\big) - \gamma_2' M(t)\Big\} \\ &+ \Big\{\gamma_1 V(t)\big(Mac(t) - M_i(t)\big) - \gamma_2 M_i(t)\Big\} \\ =& M'(t) + M_i'(t). \end{aligned}$$

Here, S_1 is the source rate of macrophages from the bone marrow and γ_2' is the removal rate of uninfected macrophages. As the HIV virus does not have the same cytopathic effect on macrophages as it does on $T4$ lymphocytes, we will assume $\gamma_2' = \gamma_2$, then

$$Mac'(t) = S_1 - \gamma_2 Mac(t).$$

The main result of this last assumption is that macrophage numbers are not disrupted by the viral infection. For that reason, it is reasonable to assume that $Mac(t)$ is maintained at or near its equilibrium value, S_1/γ_2.

As the rate of new particle production depends on the number of infected cells and the rate at which they are stimulated or activated, we assume that this rate is proportional to the size of the infected population. The viral particles are cleared by nonspecific mechanisms and by neutralization. We assume that the nonspecific mechanisms are dominant, at least in the relatively early stages. Thus,

$$\frac{dV}{dt} = N\alpha I(t) + N\delta M_i(t) - N\beta V(t) \tag{1}$$

where N is the expected number of viral particles released per cell. α and β are positive constants. $1/(\alpha I(t))$ is the expected time to the next stimulation of an infected T4 cell by outside antigens while $\ln 2/\beta N$ is the half-life of an HIV viral particle in the body. By increasing β, the effect of neutralizing antibody can be simulated. $N\delta$ is the relative rate at which infected macrophages produce viral particles.

Dividing (1) by N, one has this differential equation in the form

$$\frac{1}{N}\frac{dV}{dt} = \alpha I(t) + \delta M_i(t) - \beta V(t).$$

As N is large (50–1000), this suggests that $V(t)$ changes on a much faster time scale than other quantities in the model (especially $I(t)$). From the theory of singularly perturbed systems, for almost all time

$$\frac{dV}{dt} \approx 0.$$

This leads to our approximation,

$$V(t) = \frac{\alpha}{\beta} I(t) + \frac{\delta}{\beta} M_i(t). \tag{2}$$

$I(t)$ is a stochastic quantity satisfying the following:

$$\begin{aligned} &Pr\{I(t+\Delta t) - I(t) = 1 | I(t) = j\} = k_1 V(t)\big(L(t) - j\big)\Delta t + o(\Delta t) \\ &Pr\{I(t+\Delta t) - I(t) = -1 | I(t) = j\} = \alpha j \Delta t + o(\Delta t) \quad \text{and} \\ &Pr\{I(t+\Delta t) - I(t) = 0 | I(t) = j\} = 1 - [k_1 V(t)\big(L(t) - j\big) + \alpha j]\Delta t + o(\Delta t). \end{aligned} \tag{3}$$

As such, $I(t)$ can be seen to be a non-homogeneous birth-and-death process.[17] Using (2) and the forward Kolmogorov equations, if

$$p_i(t) = Pr\{I(t) = i \,|\, I(0) = I_0\} \qquad \text{for all } i = 0, 1, 2, \ldots, L$$

then

$$p_i' = \begin{cases} -\big[\alpha i + k_1\big(\frac{\alpha}{\beta} i + \frac{\delta}{\beta} M_i(t)\big)\big(L(t) - i\big)\big] p_i \; + [\alpha(i+1)] p_{i+1} \\ \qquad + \big[k_1\big(\frac{\alpha}{\beta}(i-1) + \frac{\delta}{\beta} M_i(t)\big)\big(L(t) - (i-1)\big)\big] p_{i-1}, \\ \qquad \text{for } 1 \le i \le L(t) \le L \\ 0, \qquad \text{for } L(t) < i \le L \end{cases}$$

Also $p_0' = \alpha p_1$ and $p_L' = -\alpha L p_L + k_1\big(\frac{\alpha}{\beta}(L-1) + \frac{\delta}{\beta} M_i(t)\big) p_{L-1}$ where L is the upper bound of $L(t)$.

The expected value of $I(t)$, $E(t)$, is

$$E(t) = \exp\big(I(t)\big) = \sum_{i=0}^{L} i p_i(t).$$

Differentiating,

$$\begin{aligned} E'(t) &= \sum_{i=0}^{L} i p_i'(t) \\ &= \sum_{i=0}^{L} -i\big[\alpha i + k_1\big(\frac{\alpha}{\beta} i + \frac{\delta}{\beta} M_i(t)\big)\big(L(t) - i\big)\big] p_i \\ &\quad + \sum_{i=1}^{L} (i-1)(\alpha i) p_i + \sum_{i=0}^{L} (i+1) k_1\big(\frac{\alpha}{\beta} i + \frac{\delta}{\beta} M_i(t)\big)\big(L(t) - i\big) p_i \end{aligned}$$

$$= \sum_{i=0}^{L} \{ [-\alpha + k_1 \frac{\alpha}{\beta}(L(t) - i)] i + k_1 \frac{\delta}{\beta} M_i(t) \ (L(t) - i) \} p_i$$

$$= \sum_{i=0}^{L} \{ (-\alpha + k_1 \frac{\alpha}{\beta} L(t) - k_1 \frac{\delta}{\beta} M_i(t)) i - k_1 \frac{\alpha}{\beta} i^2 \} p_i$$

$$+ \sum_{i=0}^{L} [k_1 \frac{\delta}{\beta} M_i(t) L(t)] p_i$$

As the variance of the process $Var(t) = \sum_{i=0}^{L} i^2 p_i - \left(\sum_{i=0}^{L} i p_i \right)^2$,

$$\begin{aligned} E'(t) = & (-\alpha + k_1 \frac{\alpha}{\beta} L(t) - k_1 \frac{\delta}{\beta} M_i(t)) E(t) \\ & - k_1 \frac{\alpha}{\beta} (Var(t) + (E(t))^2) + k_1 \frac{\delta}{\beta} M_i(t) L(t). \end{aligned} \tag{4}$$

If the variance is known, (4) could be solved for $E(t)$. As that is unlikely, a second quantitative argument linking the two quantities is necessary. One the most important quantities in the progression of AIDS is the number of functioning $T4$ lymphocytes (and the $T4/T8$ ratio), as the decline of this number mirrors the loss of immunocompetence and eventual disease symptoms.

Assume that the cytopathic effect adds an extra death term to the homeostatic equation governing $L(t)$, the number of functioning $T4$ lymphocytes. Here, we will examine $\hat{L}(t)$ the expected number

$$\hat{L}'(t) = S_2 - \left(\varepsilon \hat{L} + \alpha E(t) \right) \tag{5}$$

where S_2 is the source, ε is the natural removal rate and excess death depends on stimulation of the infected population.

Consider $\mathcal{R}(t)$ viral particles distributed among $L(t)$ cells, each particle chosen independently of the others. Then, under the conditions present (at least early in the infection), the distribution of $p(k)$, the probability of a T-lymphocyte being infected with k particles, follows a Poisson distribution with parameter $\lambda = \lambda(t)$.

Thus, $p(k) = e^{-\lambda} \lambda^k / k!$ for $k = 0, 1, 2, \ldots$ In particular, $p(0) = e^{-\lambda}$. This is the probability of picking a noninfected cell. By a binomial argument,

$$E(t) = L(t) \left(1 - e^{-\lambda(t)} \right). \tag{6}$$

Similarly,

$$Var(t) = L(t) \left(1 - e^{-\lambda(t)} \right) e^{-\lambda(t)}.$$

Replacing $L(t)$ by $\hat{L}(t)$, and differentiating (6),

$$E'(t) = \hat{L}'(1 - e^{-\lambda(t)}) + \hat{L}e^{-\lambda(t)}\lambda'(t). \tag{7}$$

Using (4), (5) and (7),

$$\begin{aligned} E'(t) =& \hat{L}'(1 - e^{-\lambda(t)}) + \hat{L}e^{-\lambda(t)}\lambda'(t) \\ =& (-\alpha + k_1\frac{\alpha}{\beta}\hat{L}(t) - k_1\frac{\delta}{\beta}M_i(t))E(t) \\ & - k_1\frac{\alpha}{\beta}(Var(t) + (E(t))^2) + k_1\frac{\delta}{\beta}M_i(t)\hat{L}(t) \end{aligned} \tag{8}$$

After simplification and solving for λ', (8) becomes

$$\begin{aligned} \lambda'(t) =& (1 - e^{-\lambda(t)})\Big(-S_2 + \hat{L}(t)\Big\{\varepsilon - (\alpha e^{-\lambda(t)} + k_1\frac{\delta}{\beta}M_i + k_1\frac{\alpha}{\beta}e^{-\lambda(t)})\Big\} \\ & + (\hat{L}(t))^2\Big\{k_1\frac{\alpha}{\beta} + (1 - e^{-\lambda(t)})\Big\}\Big) + k_1\frac{\delta}{\beta}M_i(t)\hat{L}(t). \end{aligned} \tag{9}$$

The model developed here consists of (5), (9) and an equation for $M_i(t)$. They are listed below

$$\begin{aligned} M_i' &= \gamma_1(\frac{\alpha}{\beta}\hat{L}(1 - e^{-\lambda(t)}) + \frac{\delta}{\beta}M_i(t))(\frac{S_1}{\gamma_2} - M_i) - \gamma_2 M_i \\ \hat{L}' &= S_2 - (\varepsilon\hat{L} + \alpha\hat{L}(1 - e^{-\lambda(t)})) \\ \lambda' &= (1 - e^{-\lambda(t)})\Big(-S_2 + \hat{L}(t)\Big\{\varepsilon - \Big(\alpha e^{-\lambda(t)} + k_1\frac{\delta}{\beta}M_i + k_1\frac{\alpha}{\beta}e^{-\lambda(t)}\Big)\Big\} \\ & \quad + (\hat{L}(t))^2\Big\{k_1\frac{\alpha}{\beta} + (1 - e^{-\lambda(t)})\Big\}\Big) + k_1\frac{\delta}{\beta}M_i(t)\hat{L}(t) \\ M_i(0) &= M_0 \geq 0, \hat{L}(0) = L_0 \geq 0, \lambda(0) \geq 0. \end{aligned} \tag{10}$$

This system will examined more fully at a later time; here we will examine an associated system, actually the disease process proposed in [15]. Here, we also only consider the excess deaths of $T4$ cells caused by the infection. The system to be studied (no macrophage reservoir) is

$$\begin{aligned} \lambda' &= (1 - e^{-\lambda(t)})\Big\{\frac{k_1\alpha}{\beta}(\hat{L} - 1) - \alpha\Big\} \\ \hat{L} &= -\alpha\hat{L}(1 - e^{-\lambda(t)}) \\ \lambda(0) &\geq 0, \hat{L}(0) = L_0 \geq 0. \end{aligned} \tag{11}$$

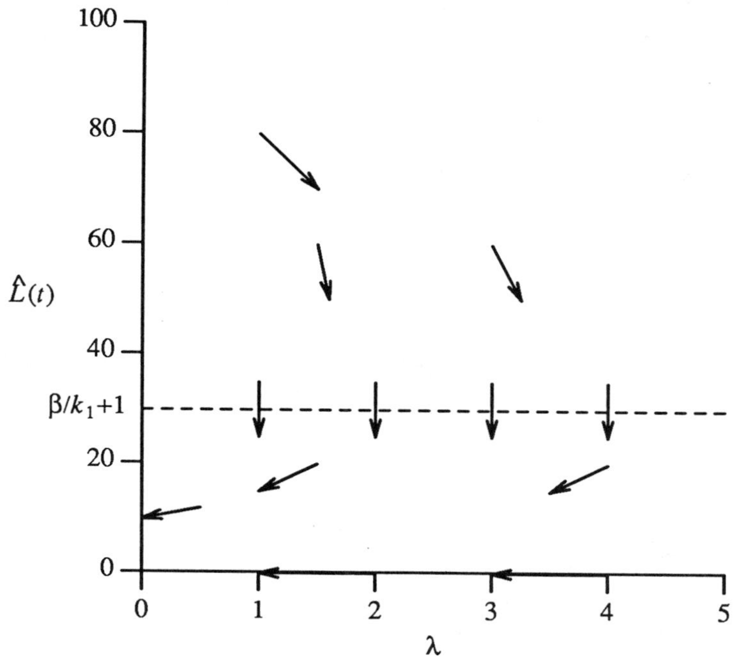

FIGURE 2 The phase plane for Eq. (11).

In examining (11), we find that whenever $\lambda = 0$ and $\hat{L} \geq 0$, there is an equilibrium. The phase plane appears as in Figure 2.

That all trajectories approach the line segment $(0, (\beta/k_1) + 1)$ in the L-axis is shown in

THEOREM 1. With $\lambda(0) = \varepsilon > 0$ and $\hat{L}(0) > 0$, $\hat{L}(t)$ and $\lambda(t)$ associated solutions of (11)

i. $\lim_{t\to\infty} \hat{L}(t)$ exists, is different from zero and lies in $\big(0, (\beta/k_1) + 1\big)$, and

ii. $\lim_{t\to\infty} \lambda(t) = 0$.

PROOF As $\hat{L}(t)$ is monotone decreasing and bounded below by zero, $\lim_{t\to\infty} \hat{L}(t)$ exists. Moreover from the phase-plane, $\lim_{t\to\infty} \lambda(t) = 0$. Thus, the w-limit must lie in $[0, L_0]$. From (5),

$$\hat{L}(t) = L(0)e^{-\int_0^t \alpha\left(1-e^{-\lambda(s)}\right) ds}.$$

If $\hat{L} \to 0$ as $t \to \infty$, $\int_0^t \left(1 - e^{-\lambda(s)}\right) ds \to +\infty$ as $t \to \infty$. From (9),

$$\lambda(t) - \lambda(0) = \int_0^t (1 - e^{-\lambda(s)}) \frac{k_1 \alpha}{\beta} \hat{L}(s) ds - \int_0^t (1 - e^{-\lambda(s)}) (\frac{k_1 \alpha}{\beta} + \alpha) ds. \quad (12)$$

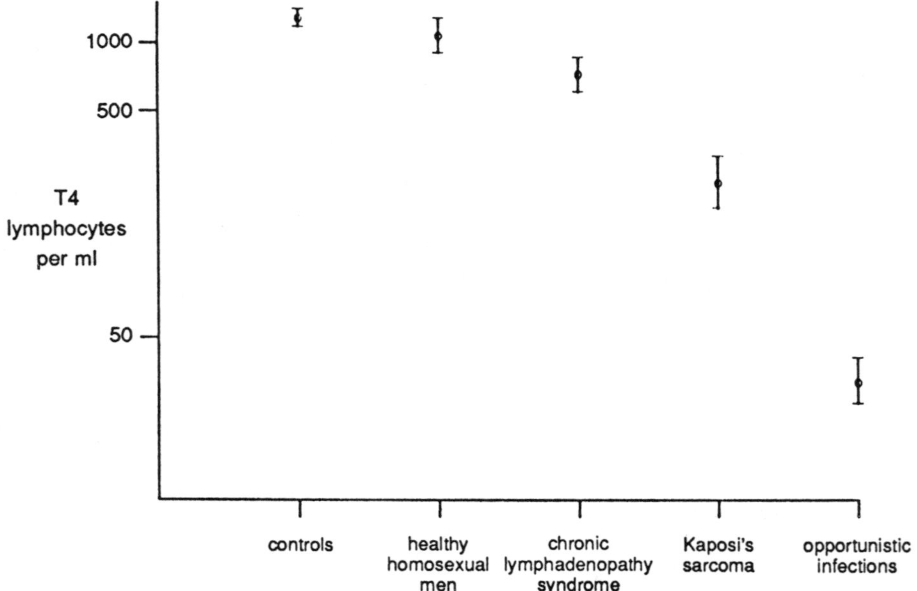

FIGURE 3 The relationship between $T4$ numbers and the stage of AIDS as seen associated clinical disease. Redrawn from H. C. Lane and A. S. Fauci (1987), "Infectious Complications of AIDS," *AIDS: Modern Concepts and Therapeutic Challenges*, Ed. S. Broder (New York: Marcel Dekker), 185–203.

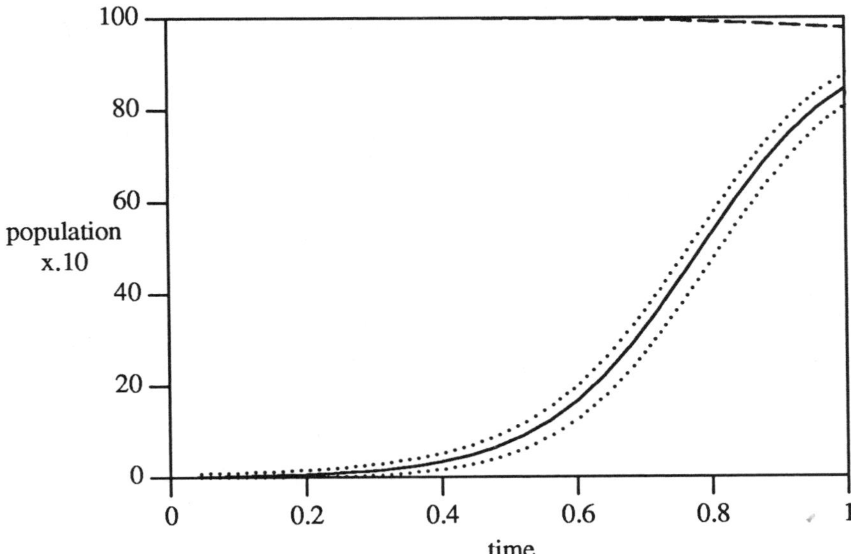

FIGURE 4 $\alpha = .1$ and $(k_1\alpha)/\beta = .9$. Solid line is the infected $T4$ population and dashed line is the total $T4$ population. Time in years.

As $\lambda(t) - \lambda(0)$ is bounded, if $\int_0^t \left(1 - e^{-\lambda(s)}\right) ds \to +\infty$, then both the first and second integrals in (12) above are infinite as $t \to \infty$. But, also from (5), $-(1/\alpha)\left(\hat{L}(t) - \hat{L}(0)\right) = \int_0^t \left(1 - e^{-\lambda(s)}\right)\hat{L}(s)ds$ has a finite limit, thus $\lim_{t\to\infty} \hat{L}(t) > 0$.

That this limit is in $\left(0, (\beta/k_1) + 1\right)$, follows from examination of the phase-plane.

4. PRELIMINARY RESULTS

The model Eqs. (11) developed in the previous section, while simplistic, hold a large amount of information. First, note that essentially only two parameters are present, α and $k_1\alpha/\beta$. A further reduction to a one-parameter problem is possible by a change in time scale, but here we wish to explore the result both of the rate of antigen exposure (represented by α) and the effect of neutralizing antibody and nonspecific defenses (the parameter β).

For this short series of simulations, we will assume L, the initial total $T4$ count (per ml) is 1000. Then if $\alpha = 1$, this corresponds roughly to 2.7 stimulations by

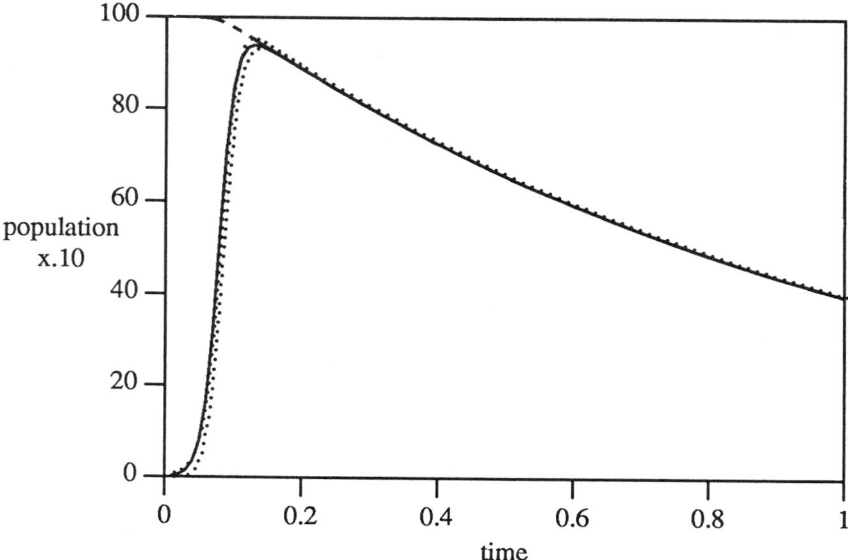

FIGURE 5 $\alpha = 1$ and $(k_1\alpha)/\beta = .9$. Solid line is the infected $T4$ population and dashed line is the total $T4$ population. Time in years.

external antigens per day. Other multiples of $\alpha = 1$ give rise to the same factor times 2.7 (if $\alpha = .1$, we have .27 stimulations per day or one every 4 days).

For the second parameter, $k_1\alpha/\beta$, from the phase plane and Theorem 1, it has been noted that the eventual resulting $L(t)$ value will be below $(\beta/k_1) + 1$. If this value is small, then the patient is ripe to develop certain of the characteristic symptoms, malignancies and infections found in AIDS, see Figure 3. Assuming k_1 (the affinity of HIV for $T4$ cells) is a constant, we vary β to display the effect of that parameter.

The last parameter to vary in the problem is the initial member of infected cells or in our case, $\lambda(0)$, the mean number of viral particles per cell. Here, we begin with $\lambda(0) = .001$ or $1000(1 - e^{-.001}) = 1$ infected cell. The effect of varying $\lambda(0)$ will also be demonstrated.

All illustrations use the time scale of years on the horizontal and population/10 on the vertical. Figures involving infected $T4$ population display the expected value of all sample paths plus or minus one standard deviation.

THE EFFECT OF α

In Figure 4, $\alpha = .1$ and $k_1\alpha/\beta = .9$. Here the infected fraction of $T4$ (+ and − one S.D.) is shown. As $(\beta/k_1) + 1 = 1.01$, eventually the $T4$ population will be nearly completely depleted. However, after one year, little change in the $T4$ population is noted. Certainly no symptoms will result even though positive antibody to HIV would have been noticed midyear.

In Figure 5, $\alpha = 1$ and $k_1\alpha/\beta = .9$. With external antigens arriving at 10 times the previous rate, rapid infection of $T4$ cells is noted and a corresponding rapid decline. After one year there would probably be an inverted $T4/T8$ ratio, and $T4$ numbers would have been noticeably depleted. Here, $(\beta/k_1) + 1 = 2.1$.

THE EFFECT OF β

If $\alpha = 1$ and $k_1\alpha/\beta = .01$, no infection was noted after 1 year $((\beta/k_1) + 1 = 101)$. In Figure 6, $\alpha = 4$ and $k_1\alpha/\beta = .025$. Here, even though $(\beta/k_1) + 1 > 100$, some decline has begun, even though only a small percentage of $T4$ cells are infected. This is shown even more clearly in Figure 7, where $\alpha = 4$ and $k_1\alpha/\beta = .05$. Here the decline is very rapid.

This model clearly shows the marked influence of the rate of external infections as a cofactor in the rate of disease development.

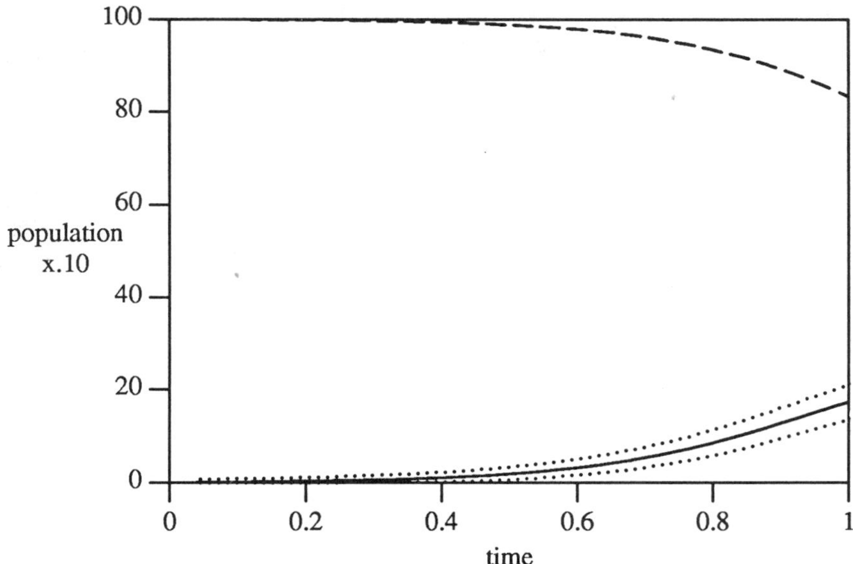

FIGURE 6 $\alpha = 4$ and $(k_1\alpha)/\beta = .025$. Solid line is the infected $T4$ population and dashed line is the total $T4$ population. Time in years.

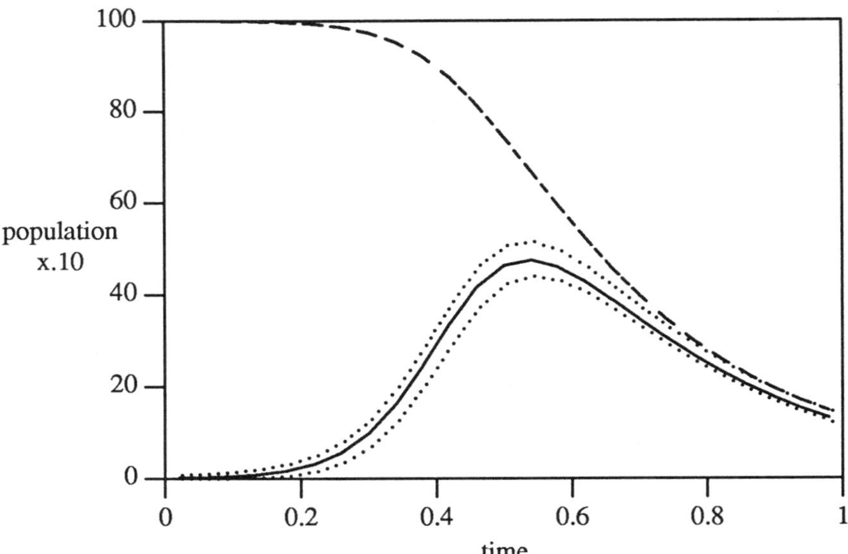

FIGURE 7 $\alpha = 4$ and $k_1\alpha/\beta = .05$. Solid line is the infected $T4$ population and dashed line is the total $T4$ population. Time in years.

THE EFFECT OF $\lambda(0)$

This is easiest to show by reference to the phase plane. As the solution curves cannot cross, if $\lambda_1(0) = \varepsilon_1$ and $\lambda_2(0) = \varepsilon_2$ with $_0 < \varepsilon_1 < \varepsilon_2$, and λ_1, L_1 and λ_2, L_2 solutions of (11) then

$$\lim_{t\to\infty} L_1(t) > \lim_{t\to\infty} L_2(t).$$

DISCUSSION

The assumptions in this model become reasonable after the acute phase of HIV infection. This generally erupts a few week after exposure and lasts about 2 weeks. The patient has a positive antibody test after this event. The cells infected after this acute stage represents the initial condition for this model. This simple model of an HIV infection displays many of the features clinically recognized in AIDS. Although the results are still preliminary, it is hoped that further work will more closely parallel longitudinal clinical studies already undertaken (such as [18]). This will require a more careful description of the humoral response and the role of $T8$ suppressor cells as well as the role of the reservoir population.

In numeric tests, the model (11) compared to the associated deterministic model (Eq. (4) without the variance term) shows little difference except as $\lambda(t)$ gets larger. Then the deterministic model becomes (and stays) markedly larger. The deterministic model would then overestimate the disease progression.

REFERENCES

1. Gottleib, M. S., R. Schroff, H. M. Schanker, et al. (1981), "*Pneumocystis Carinii* Pneumonia and Mucosal Candidiasis in Previously Healthy Homosexual Men," *N. Engl. J. Med.* **305**, 1425–1431.
2. Masur, H., M. A. Michelis, J. B. Greene, et al. (1981), "An Outbreak of Community-Acquired *Pneumocystis Carinii* Pneumonia: Initial Manifestation of Cellular Immune Dysfunction," *N. Engl. J. Med.* **305**, 1431–1438.
3. Siegal, F. P., C. Lopez, G. S. Hammer, et al. (1981), "Severe Acquired Immunodeficiency in Male Homosexuals Manifested by Chronic Perianal Ulcerative Herpes Simplex Lesions," *N. Engl. J. Med.* **305**, 1439–1444.
4. Gottlieb, G. J., A. Rayaz, J. V. Vogel, et al. (1981), "A Preliminary Communication on Extensive Disseminated Kaposi's Sarcoma in Young Homosexual Men," *Am. J. Dermatopathol.* **3**, 111–114.
5. Barré-Sinoussi, F., J. Chermann, F. Rey, et al. (1983), " Isolation of a T-Lymphocyte Retrovirus from a Patient at Risk of Acquired Immune Deficiency Syndrome (AIDS)," *Science* **220**, 868–870.
6. Popovic, M., M. G. Sarngadharan, E. Read, and R. C. Gallo (1984), "Detection, Isolation and Continuous Production of Cytopathic Retroviruses (HTLV-III) from Patients with AIDS and Pre-AIDS," *Science* **224**, 497–500.
7. Hardy, Jr. W. J. (1980), *Feline Leukemia Virus 33* (North Holland: Elsevier).
8. Montegnier, L., J. C. Chermann, F. Barré-Sinoussi, et al. (1984), "A New Human T-Lymphotropic Retrovirus: Characterization and Possible Role in Lymphadenopathy and Acquired Immune Deficiency Syndromes," *Human T-Cell Leukemia Lymphoma Virus*, Eds. R. C. Gallo, M. Essex, and L. Gross (Cold Spring Harbor Laboratory), 363–370.
9. Hahn, B. H., G. M. Shaw, S. K. Arya, et al. (1984), "Molecular Cloning and Characterization of HTLV-III Virus Associated with AIDS," *Nature* **312**, 166–169.
10. Weiss, R. A., P. R. Clapham, R. Cheingsong-Popov, et al. (1986), "Neutralization of Human T-Lymphotropic Virus Type III by Sera of AIDS and AIDS-Risk Patients," *Nature (London)* **316**, 69–72.
11. Robert-Guroff, M., M. Brown, and R. C. Gallo (1986), "HTLV-III Neutralizing Antibodies in Patients with AIDS and AIDS-Related Complex," *Nature (London)* **316**, 72–74.
12. Quinn, T. C., P. Piot, J. B. McCormick, et al. (1987), "Serologic and Immunologic Studies in Patients with AIDS in North America and Africa: The Potential Role of Infectious Agents as Cofactors in Human Immunodeficiency Virus Infection," *Jour. Am. Med. Assoc.* **257**, 2617–2621.
13. Broder, S. (ed.) (1987), *AIDS: Modern Concepts and Therapeutic Challenges* (New York: Marcel Dekker).
14. Klatzmann, D., and J. C. Gluckman (1986), "*HIV* Infection: Facts and Hypotheses," *Immunol. Today* **7**, 291–296.

15. Zagury, D., J. Bernard, R. Leonard, et al. (1986), "Long-Term Cultures of HTLV-III-Infected T-Cells: A Model of Cytopathology of T-Cell Depletion in AIDS," *Science* **231**, 850–853.
16. Cooper, L. N. (1986), "Theory of an Immune System Retrovirus," *Proc. Natl. Acad. Sci., U.S.A.* **83**, 9159–9163.
17. Karlin, S., and H. M. Taylor (1975), *A First Course in Stochastic Processes* (New York: Academic Press), 2nd Edition.
18. Melbye, M., R. J. Biggar, P. Ebbesen, et al. (1986), "Long-Term Seropositivity for Human T-Lymphotropic Virus Type III in Homosexual Men without the Acquired Immunodeficiency Syndrome: Development of Immunologic and Clinical Abnormalities," *Ann. Int. Med.* **104**, 496–500.

ANGELA MCLEAN
Parasite Epidemiology Research Group, Department of Pure and Applied Biology, Imperial College, London University, London SW7 2BB, England

HIV Infection from an Ecological Viewpoint

ABSTRACT

A model is presented which describes the dynamics of a population of T_H cells in response to infection with HIV. Preliminary results are presented concerning threshold criteria for the establishment of HIV. A second model is considered which describes the impact of a second replicating pathogen upon an established HIV infection, and, vice-versa, the effect of a new HIV infection upon a pre-existing, persistent infection.

INTRODUCTION

The human immunodeficiency virus (HIV) presents an interesting ecological problem. The population of T_H cells of an infected individual act as host to the virus,[1] and in their role as orchestrators of the cell-mediated immune response[2] can also be viewed as predators upon the virus. Thus, a situation exists where two populations are both stimulated to expansion by each other, yet both act to decrease the size of the other. The aim of the project described here is to study the dynamics of the interactions between populations of T_H cells and HIV. As always in mathematical biology, there is a tension between the desire to construct simple, analytically tractable models, and the wish to include a large number of realistic complications. The approach used here is to opt for simplicity and tractability. The model that is presented is an antigen-driven model of the dynamics of two populations of T_H cells, (healthy and infected), and a population of circulating virus. Having established an understanding of the simple model described here, it will be possible to progress to the consideration of a number of more complex models.

GENERAL MODEL

The investigations are based upon a general model which is expressed as a set of three ordinary differential equations containing three auxiliary functions: Γ, Ω, and Δ. The first part of each equation represents basic assumptions concerning the dynamics of each population, whilst the auxiliary functions are there to allow great flexibility in the assumptions built into the model. The equations describing the dynamics of the healthy T_H cell (X), infected T_H cell (Y) and virus (V) populations are as follows:

$$\frac{dX}{dt} = \Lambda - \mu X - \beta XV + \Gamma(X,Y,V) \tag{1}$$

$$\frac{dY}{dt} = \beta XV - (\mu + \alpha)Y + \Omega(X,Y,V) \tag{2}$$

$$\frac{dV}{dt} = \lambda\alpha Y - dV - \beta XV + \Delta(X,Y,V) \tag{3}$$

Healthy T_H cells migrate into the circulation at a constant rate Λ and have a life expectancy $1/\mu$. Thus, in the absence of HIV, there is a T_H cell population of fixed size Λ/μ, experiencing constant turnover. In the presence of virus, healthy cells are infected at per capita rate βV. The stimulating effect of HIV on the healthy T_H cell population is included in the term $\Gamma(X,Y,V)$. The dynamics of the infected T_H cell population are described in the second equation. Influx to the class is from the mass action term βXV representing infection of healthy cells by circulating virus. It is assumed that infected cells have a decreased life expectancy; hence, they are removed at per capita rate $(\mu+\alpha)$. Once again stimulation by the presence of virus is included in the auxiliary function $\Omega(X,Y,V)$. The third equation describes the dynamics of the population of circulating virus. The term $\lambda\alpha Y$ represents the assumption that it is in a burst of viral reproduction that the host T_H cell is killed. Removal of circulating virus by non-specific immune mechanisms is represented by the term $-dV$, and entry of virus into healthy T_H cells by the term $-\beta XV$. The removal of virus by the action of the specific immune response is included in the

TABLE 1

$\Gamma(X,Y,V)$	
Virus stimulates proliferation of susceptible cells	$+\gamma XV$
or	$\frac{+\gamma XV}{D+V}$
Syncitia formation[3]	$-\nu XY$
Cell-cell infection[4]	$-\kappa XY$
Stimulation of susceptible cells by infected cells	$+\rho XY$
or	$\frac{+\rho XY}{G+Y}$
$\Omega(X,Y,V)$	
Clonal expansion of infected cells stimulated by virus	$+\phi YV$
or	$\frac{+\phi YV}{F+V}$
Killing of infected cells mediated by susceptible cells	$-\omega XY$
Cell-cell infection[4]	$+\chi XY$
$\Delta(X,Y,V)$	
Removal of virus regulated only by healthy T_H cells	$-\sigma XV$
Removal of virus regulated by both subsets of T_H cells	$-(\sigma XV + \pi YV)$

Table 1 describes some of the possible components of the auxiliary functions Γ, Ω and Δ. Clearly very many combinations would be possible. However, in some cases, two different biological assumptions would be expressed in the same algebraic form (e.g., syncitia formation and cell-cell infection) and this acts to reduce the number of different possibilities requiring investigation.

function $\Delta(X,Y,V)$. Table 1 lists some of the interactions that could be included in the auxiliary functions Γ, Ω and Δ.

MODEL 1

The first model that is considered sets $\Gamma(X, Y, V) = +\gamma XV$, $\Omega(X, Y, V) = 0$ and $\Delta(X, Y, V) = -\sigma XV$. This model has an equilibrium point at the point $(\Lambda/\mu, 0, 0)$ whose local stability is determined by the composite parameter

$$R_o = \frac{\Lambda\beta\lambda\alpha}{(\mu+\alpha)[\mu d + \Lambda(\beta+\sigma)]} \tag{4}$$

If $R_o < 1$ $(\Lambda/\mu, 0, 0)$ is locally stable and if $R_o > 1$ $(\Lambda/\mu, 0, 0)$ is unstable and there is a second, locally stable equilibrium at:

$$X^* = \frac{(\mu+\alpha)d}{\beta\lambda\alpha - (\beta+\sigma)(\mu+\alpha)} \tag{5}$$

$$Y^* = \frac{\beta}{(\beta-\gamma)(\mu+\alpha)}\left[\frac{\beta\lambda\alpha\Lambda - \Lambda(\beta+\sigma)(\mu+\alpha) - d\mu(\mu+\alpha)}{\beta\lambda\alpha - (\beta+\sigma)(\mu+\alpha)}\right] \tag{6}$$

$$V^* = \frac{\beta\lambda\alpha\Lambda - \Lambda(\beta+\sigma)(\mu+\alpha) - d\mu(\mu+\alpha)}{\beta\lambda\alpha - (\beta+\sigma)(\mu+\alpha)} \tag{7}$$

the parameter R_o can be interpreted (in a very similar way to an epidemiological basic reproductive rate[5,6]) as the average number of new virions produced by one virion introduced into a population of T_H cells all of which are healthy. The second equilibrium which exists and is locally stable when $R_o > 1$ represents an established HIV infection with a depressed number of healthy T_H cells and an increased number of infected T_H cells.

MODEL 2

The second model investigated here considers the impact on this sytem of the introduction of a second pathogen.

$$\frac{dX}{dt} = \Lambda - \mu X - \beta XV + \gamma XV + \zeta XA \tag{8}$$

$$\frac{dY}{dt} = \beta XV - (\mu+\alpha)Y + \theta YA \tag{9}$$

$$\frac{dV}{dt} = \lambda\alpha Y - dV - \beta XV - \sigma XV \tag{10}$$

$$\frac{dA}{dt} = RA - bXA - cYA \tag{11}$$

The size of the population of the second pathogen at time t is represented by $A(t)$. In the absence of HIV, the equations describing the interaction between healthy T_H

cells (X) and pathogen 2 (A) are simply the Lotka-Volterra predator-prey equations with constant immigration of healthy T_H cells.[7,8] Thus, the sort of second pathogen that can be described is one that is either incapable of establishment, or once it has established, is maintained within the body at a stable level. As before there is an equilibrium at $(\Lambda/\mu, 0, 0, 0)$ and this is stable to invasion by HIV if $R_o < 1$, and stable to invasion by pathogen 2 if $A_o < 1$, where

$$R_o = \frac{\Lambda\beta\lambda\alpha}{(\mu+\alpha)[\mu d + \Lambda(\beta+\sigma)]} \tag{12}$$

and

$$A_o = \frac{R\mu}{b\Lambda} \tag{13}$$

However, the equilibrium $(X^*, Y^*, V^*, 0)$, representing a persistent HIV infection in the absence of pathogen 2, (thus, X^*, Y^* and V^* are as given in equations 5, 6 and 7) is unstable to invasion by pathogen 2 if

$$X^* + cY^*/b < \Lambda A_o/\mu \tag{14}$$

Equation 14 gives the condition under which HIV depresses the immune response to such an extent that a pathogen that would normally be excluded is able to invade. Such a situation is illustrated in Figure 1. The figure shows solutions

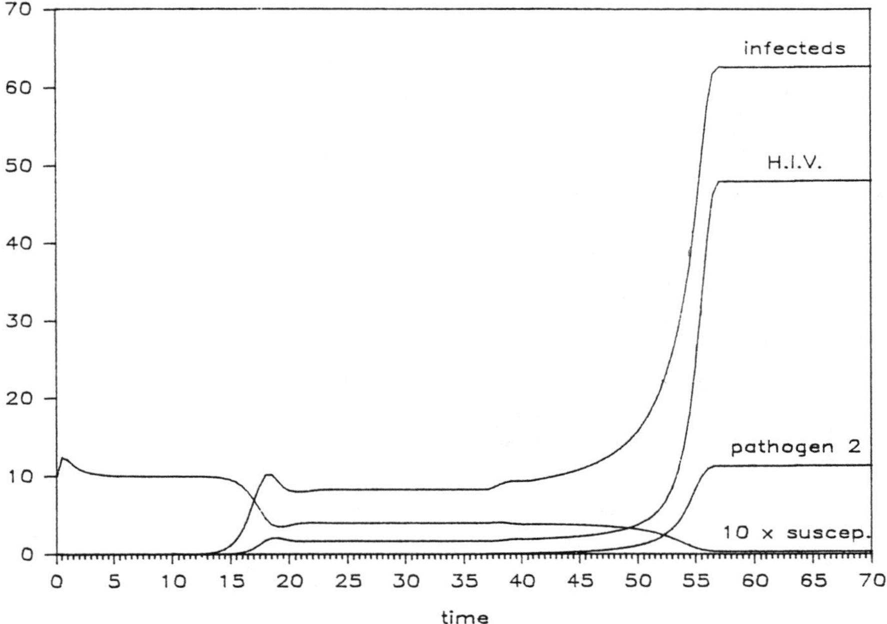

FIGURE 1 Introduction of a second pathogen to a system with an existing HIV infection.

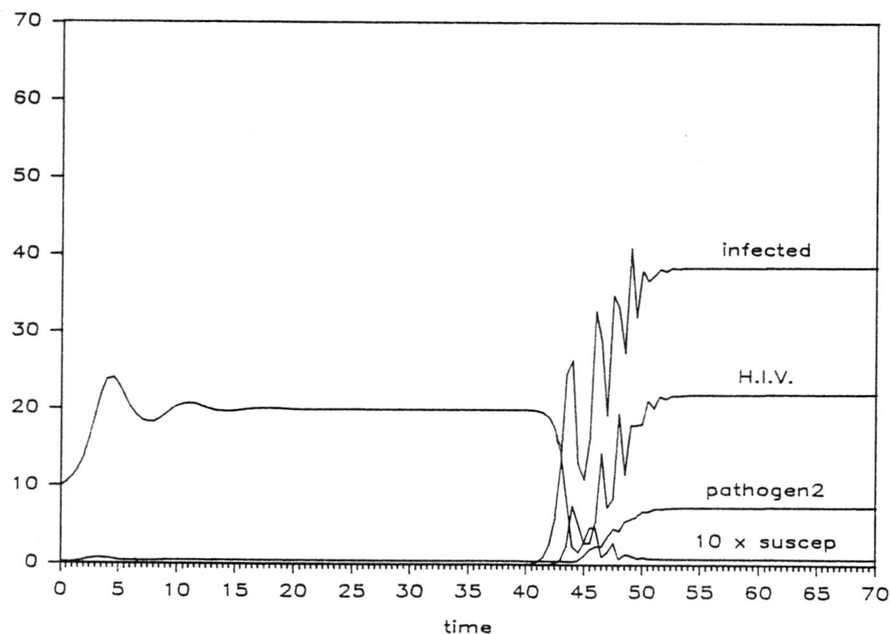

FIGURE 2 Introduction of HIV to a system with an established second pathogen.

for scaled variables that should be interpreted in the following way. The number of healthy T_H cells is drawn scaled by $\Lambda/10\mu$. So the "virgin" level of healthy T_H cells is represented in the graph by $X = 10$. The number of infected T_H cells is drawn scaled by $\Lambda/100\mu$ so, when $Y = 100$ on the graph, there as many infected T_H cells as there were healthy T_H cells before any pathogens were introduced. The unit of time in the figure is $1/\mu$, the expectation of life of a healthy T_H cell. Figure 1 illustrates the following sequence of events. At time zero pathogen 2 is introduced to the system in its virgin state, $(\Lambda/\mu, 0, 0, 0)$. $A_o < 1$ so the pathogen is quickly cleared and the system returns to the original equilibrium. At time 37 units, HIV is introduced and, as $R_o > 1$, the system eventually settles to an endemic HIV infection $(X^*, Y^*, V^*, 0)$, (where X^*, Y^* and V^* are as given in equations 5, 6 and 7). On reintroduction of pathogen 2, because of the immunosuppression caused by the HIV infection, establishment is possible and a new equilibrium level is established with all 4 populations co-existing.

Another situation that is of interest is the effect of a new HIV infection upon an established infection with a second pathogen. Figure 2 illustrates a situation where an existing infection is stimulated by the introduction of HIV and a new equilibrium is established with all four populations co-existing, the second pathogen at greatly increased prevalence to that prior to the introduction of HIV. Variables are scaled in the same way as in Figure 1.

CONCLUSIONS

Using a very simple model, two threshold conditions have been derived. The first determines whether or not an HIV infection can establish, and the second gives the degree of immunosuppression that is necessary for an established HIV infection to allow the invasion of a second pathogen that would normally be excluded. This simple model has been able to mimic the initial viraemia and settling to a state with slightly depressed T_H cell counts of which only a small proportion are infected. The other observed events that this simple model can mimic are the invasion and establishment of an opportunistic infection which would normally be excluded and the reactivation of an established infection. This model is, however, unable to mimic the situation where total T_H cell counts tend to zero, but only very few cells are infected,[9] nor does this model mimic the slow drift down towards such a situation.

Having come to an understanding of the behaviour of this simple model, the impact of further complications can now be investigated with a view to seeing which, if any, of the interactions described in Table 1 could explain these last two observed properties.

REFERENCES

1. Klatzmann, D., F. Barre-Sinoussi, M. T. Nugeyre, C. Dauget, E. Vilmer, C. Griscelli, F. Brun-Vezinet, C. Rouzioux, J. C. Gluckmaan, J. C. Chermaann, and L. Montagnier (1984), "Selective Tropism of Lymphadenopathy Associated Virus for Helper-inducer Lymphocytes," *Science* **225**, 59 - 63.
2. Roitt, I.M. (1971), *Essential Immunology* (Oxford: Blackwell Scientific Publications).
3. Sodroski, J., W. C. Goh, C. Rosen, K. Campbell, and W. A. Haseltine (1986), "Role of the HTLVIII/LAV Envelope in Syncytium Formation and Cytopathicity," *Nature* (London) **322**, 470 - 474.
4. Mitssuuya, H. and S. Broder (1987), "Strategies for Anti-viral Therapy in AIDS," *Nature* (London) **325**, 773 - 778.
5. Macdonald, G. (1952), "The Analysis of Equlibrium in Malaria," *Trop. Dis. Bull.* **49**, 813 - 829.
6. Anderson, R.M. and R. M. May (1982), "Directly Transmitted Infectious Diseases: Control by Vaccination," *Science* **215**, 1053 - 1060.
7. Lotka, A.J. (1925), *Elements of Physical Biology* (Baltimore: Williams and Wilkins).
8. Volterra, V. (1926), "Variazioni e fluttuazioni deel numero d'individui in specie animali conviventi," *Mem. Acad. Lincei* **22**, 31 - 113.
9. Harper, M.E., L. M. Marselle, R. C. Gallo, and F. Wong-Staal (1986), "Detection of Lymphocytes Expressing Human T-Lymphotropic Virus Type III in Lymph Nodes and Peripheral Blood from Infected Individuals by *In Situ* Hybridization," *Proc. Natl. Acad. Sci. U.S.A.* **83**, 772 - 776.

NATHAN INTRATOR, GREGORY P. DEOCAMPO, and LEON N. COOPER
Center for Neural Science, Department of Physics, and Division of Applied Mathematics
Brown University, Providence, RI 02912

Analysis of Immune System Retrovirus Equations

INTRODUCTION

An attempt has been made to mathematically model the patterns of viral growth and the associated immune system response characteristically associated with infection by the human immunodeficiency virus (HIV). By employing a very simple model of viral growth and the humoral immune system response, the interaction between a "normal" virus and the immune system response is compared with that of a viral entity called an immune system retrovirus (ISRV) (Cooper, 1986). A model treating the cellular response of the immune system (Reibnegger et al., 1987) appears to lead to similar results.

NORMAL VIRUS GROWTH

The case of a "normal," non-immune system specific virus is considered first.

A B cell specific to antigen X (B_X) encounters a normal virus presenting antigen X (X). The B cell–virus complex is formed, which shall be called ($B_X X$). For convenience, subscripts shall be suppressed where no confusion will result. The B cell–virus complex, BX, then participates in an antigenically specific interaction with a T cell specific to antigen X, forming the B cell–virus–T cell complex, BXT. It is assumed that it is this complex which triggers the T cell clonal expansion. See Figure 1.

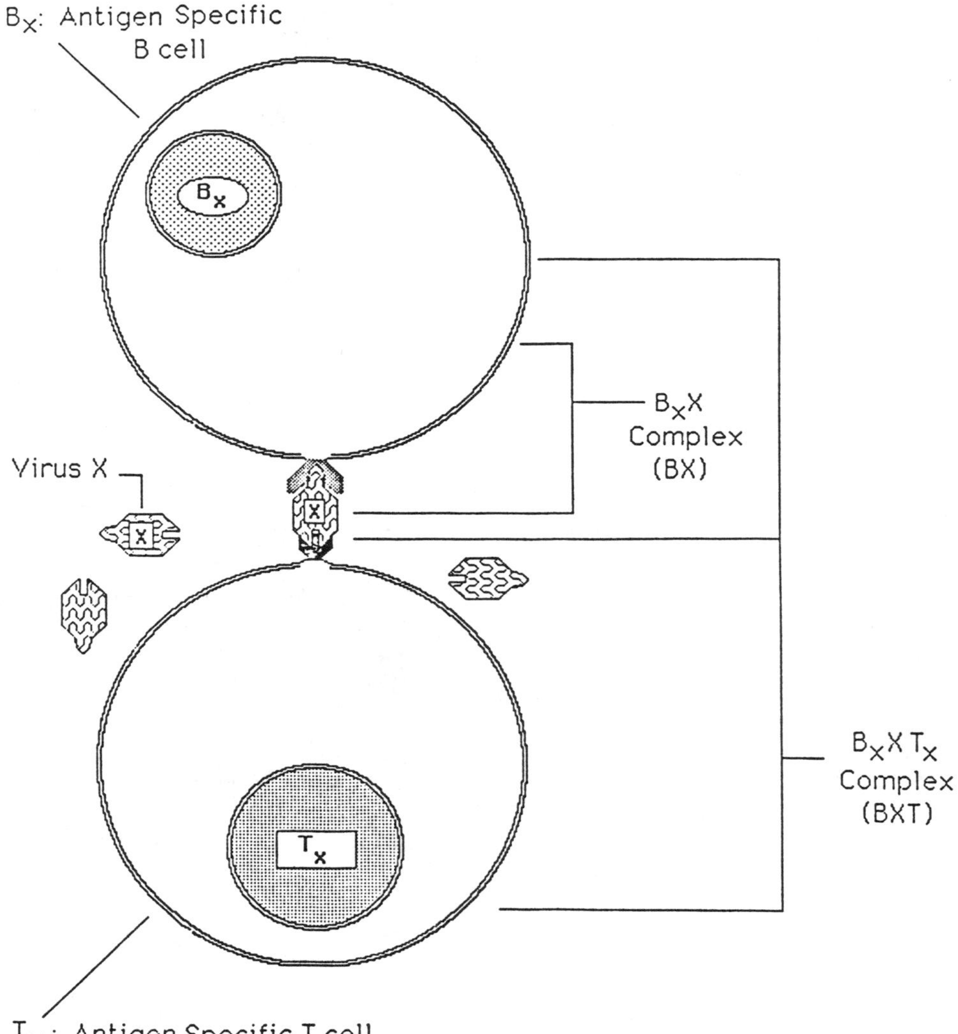

FIGURE 1 Normal Virus: Biological Entities Followed by the Model.

The non-linear integral differential equations of the model are presented in Eqs. (1a-f) (Cooper, 1986).

$$\frac{dBX}{dt} = \gamma_1 BX - \gamma_2 (BX)T - \lambda_1 (BX) \tag{1a}$$

$$\frac{dBXT}{dt} = \gamma_2 (BX)T - \lambda_2 (BXT) \tag{1b}$$

$$\frac{dB}{dt} = \gamma_3 (BXT) + \varepsilon_B - \gamma_1 BX - \lambda_3 B \tag{1c}$$

$$\frac{dT}{dt} = \gamma_4 (BXT) + \varepsilon_T - \gamma_2 (BX)T - \lambda_4 T \tag{1d}$$

$$\frac{dJ}{dt} = \gamma_5 JX - \lambda_5 J \tag{1e}$$

$$\frac{dX}{dt} = \gamma_6 J' - \lambda_6 X - I(X,t) \tag{1f}$$

B and T represent the concentration of antigen-X specific B and T cells, and X represents the concentration of normal virus X. (BX) represents the concentration of the B cell–virus X complex, while (BXT) represents the concentration of the B cell–virus X–T cell complex. The various γ's and λ's represent rates of growth and loss, and the ε_B and ε_T terms represent natural B and T cell production. J represents the concentration of virus X target cells, and J' denotes infected target cells.

To simplify analysis of these equations, it was assumed that the concentrations of the reacting B cells and T cells were approximately equal. Thus, in what follows, only the B cell response is explicitly discussed.

It is in this context that an equation is derived describing the rate of change of a population of normal virus X. The equation includes a term describing the immune system response, taken to reflect only the specific immune response. Having obtained an equation for viral growth, what is of interest is determining whether or not there exists a potential turning point in the growth of the virus population, that is, a point at which the immune system response is able to either control the infection, or eliminate it entirely.

A functional form is assigned to the immune antibody response term, $I(X,t)$ (see Appendix, Cooper, 1986). Qualitatively, it is assumed that some proportion of proliferating B cells are plasma cells, and in turn each plasma cell produces some number n of antigen-specific immunoglobulin. The rate at which individual Ig molecules attach themselves to the virus is denoted by b. On the average, s number of Ig molecules are said to be required for phagocytotic labelling or virus immobilization. The functional form of the immune response term is considered under the limiting condition where the immune system is able to control the infection; if this condition exists at all, it will exist for $s = 1$. With these assumptions, it was shown (Cooper, 1986) that the production term in the equation for normal virus growth will be bounded.

Given a monotonically increasing immune system response, a point can be found such that $(d/dt)X = 0$. This represents a turning point in viral growth and the maximum concentration of normal virus, $X_{\max}$ (Cooper, 1986).

The salient points in deriving an expression for $X_{\max}$ are that the region of rapid virus growth, and a slow immune response, is dominated by the singularity around the turning point, and the rate of change of virus X behaves like X. Using this singularity, an expression for the maximum concentration of normal virus X is obtained:

$$X_{\max} \approx \frac{\gamma_x 3}{\gamma_B B_0} \tag{1g}$$

γ_x is the rate of growth of the normal virus, and γ_B is the rate of growth of the specific immune response. B_0 is the initial specific immune response.

With these results, assumptions, and methods in mind, the case of an immune system retrovirus (ISRV) is considered.

ISRV GROWTH

In this case, the biological scenario considered is the following. An antigen-specific T cell encounters an ISRV. Through an interactive process involving the outer protein coat of the ISRV and the T_4 receptor protein of the T cell membrane, the ISRV is able to infect the cell, presumably via receptor-mediated endocytosis. The cell then is presented with B cell-processed antigen and growth factors, and mitosis in the T cell is induced. Viral RNA is then "reverse transcriptased" to form the DNA provirus, and becomes transcriptionally active. This stimulates the ISRV into rapid replication. The viral RNA and associated proteins are then rapidly produced. The viral products self-assemble into complete viruses, and in a way not yet fully understood (possibly related to the concentration of T_4 protein and a process of membrane "blebbing"), escape the cell to further infect the system and continue replicating. The host cell often dies in this process, resulting in a severely stunted clonal expansion.

The model proceeds then from the following assumptions (Cooper, 1986).

1. The causative agent of infection is a retrovirus.
2. The retrovirus has an infectious affinity for T_4 positive T lymphocytes.
3. The retrovirus neither integrates nor replicates unless the T_4 cell is mitotically active.
4. The mitotic activation of the T_4 cell, and the subsequent clonal expansion, requires a binary event resulting from stimulation by a B cell processed antigen.

As the ISRV attacks the immune system itself, the natural approach would seem to be to combine the equations for viral growth with those describing the

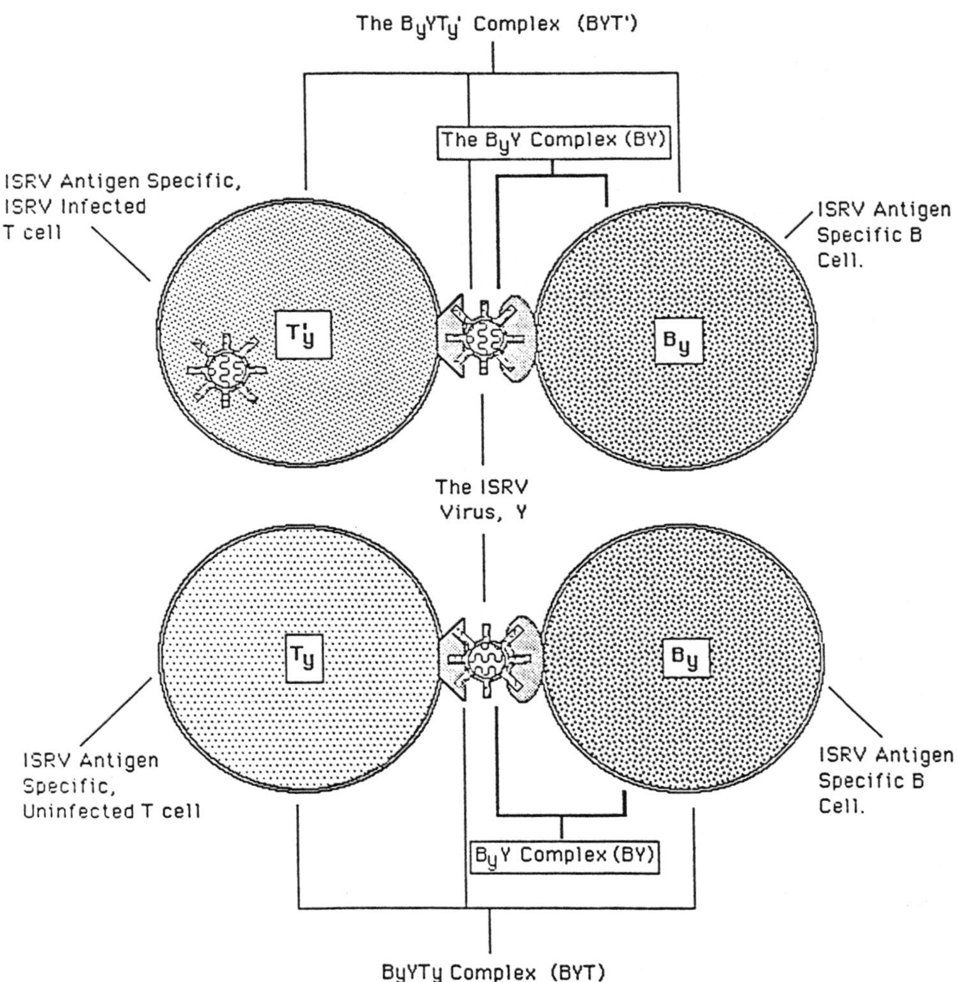

FIGURE 2 Immune System Retrovirus: Biological Entities Followed by the Model.

immune system response. There are two distinct cases of infection: (i) infection of the organism due purely to the presence of ISRV, and (ii) a mixed infection by ISRV and some other, normal virus.

Case (i), a pure ISRV infection without any other stimulation of the immune system, shall be considered first.

Reproduction of ISRV requires infection of T_Y, an ISRV-antigen specific T cell, by Y, the ISRV, resulting in T_Y' (Figure 2). The formation of T_Y' is followed by the binary recognition event in which $(B_Y Y T_Y')$ is formed, where B_Y is the ISRV antigen-specific B cell. Thus, $(B_Y Y T_Y')$ denotes the B cell–ISRV–T cell complex, where the B and T cells are specific to ISRV antigen. Notational conventions and model construction is analogous to the case of a normal virus. B and T are the concentrations of cells reactive to ISRV antigen, and Y is the concentration of ISRV. The "prime" notation denotes ISRV infection of a host cell. The γ, ε, and λ are as above. Again, a set of Eqs. (2a-g) can be written describing these interactions.

$$\frac{dBY}{dt} = \gamma_1 BY - (BY)(\gamma_2 T + \gamma' 2T') - \lambda_1 (BY) \tag{2a}$$

$$\frac{dBYT}{dt} = \gamma_2 (BY)T - \lambda_2 (BYT) \tag{2b}$$

$$\frac{dBYT'}{dt} = \gamma_2' (BY)T' - \lambda_2' (BYT') \tag{2c}$$

$$\frac{dB}{dt} = \gamma_3 (BYT) + \varepsilon_B - \gamma_1 BY - \lambda_3 B \tag{2d}$$

$$\frac{dT}{dt} = \gamma_4 (BYT) + \varepsilon_T - \gamma_2 (BY)T - \gamma_4' TY - \lambda_4 Ty \tag{2e}$$

$$\frac{dT'}{dt} = \gamma_4' TY - \gamma_2' (BY)T' - \lambda_4' T' \tag{2f}$$

$$\frac{dY}{dt} = \gamma_6' (BYT') - \lambda_6 Y - I(Y,t) \tag{2g}$$

Case (ii) is probably more realistic: the immune system is simultaneously excited by the presence of both ISRV and another antigen system–a growing normal virus X, for example, or some other situation in which the immune system is subject to turnover. Schematically, this situation is presented in Figure 3. To the previous equations describing ISRV infection are added expressions describing the situation of an antigen X-specific cell hosting an ISRV (Eqs. 2h-i). Having described this mixed interaction, it is now possible to obtain a fundamental equation for ISRV growth (Eq. 2j). Here, the "double prime" notation indicates a mixed-type infection; T_x'' denotes an ISRV-infected T cell specific to antigen X.

$$\frac{dT_x''}{dt} = \gamma_4'' T_x Y - \gamma_2'' (B_x X) T_x'' - \lambda_4'' T_x'' \tag{2h}$$

$$\frac{dB_x X T_x''}{dt} = \gamma_2'' (B_x X) T_x'' - \lambda_2'' (B_x X T_x'') \tag{2i}$$

$$\frac{dY}{dt} = \gamma_6' (B_y Y T_y') + \gamma_6'' (B_x X T_x'') - \lambda_6 Y - I(Y,t) \tag{2j}$$

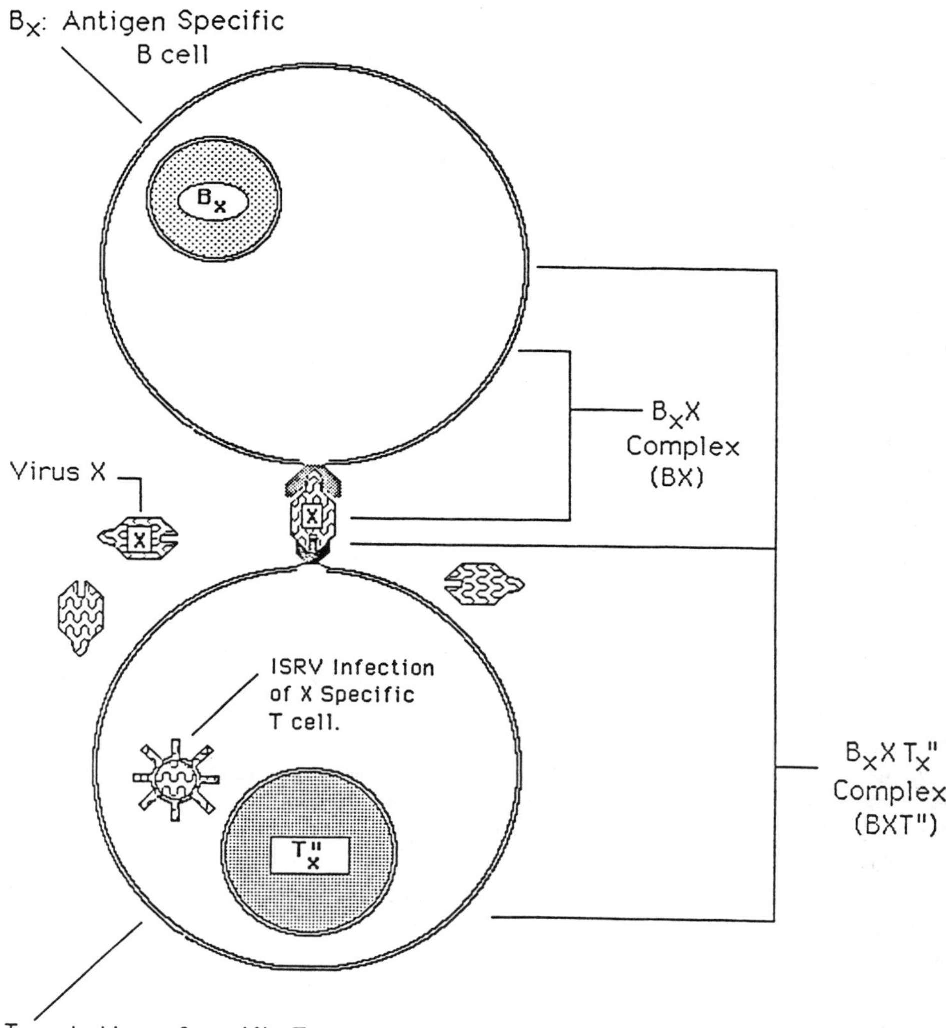

FIGURE 3 Mixed Interactions. Infected Normal Virus: Biological Entities Followed by the Model.

RESULTS

Examining the ISRV growth equation, it is found that the first production term is non-linear in Y, the concentration of ISRV. The second production term is linear in Y, and is due to the fact that ISRV growth can occur because of ISRV infection of T cells specific for antigen X in addition to cells specific to ISRV antigen.

A mixed interaction, when the immune system is simultaneously excited by the presence of both ISRV and normal virus X, is considered. There are two possible scenarios; the first is when X is a rapidly growing normal virus, the second when X represents a chronically controlled infection, a roughly constant allergic reaction, or simply represents the natural turnover of the immune system.

When the immune system is excited by the presence of another antigenic stimulant in addition to the ISRV, or is otherwise subject to a constant turnover, and the concentrations of ISRV, B, and T cells reactive to it are all low, then the dominant ISRV production term is linear in Y. This could be the case of infection resulting from a low dose of the virus, and with a not-overly-active immune system in an otherwise healthy individual.

ISRV growth behaves, under such circumstances, like the growth of a normal virus X discussed previously. Y_{max} is strongly dependent on the rate of growth of the virus and is inversely proportional to the immune system response. This situation is demonstrated in Figures 4 and 5. The results are similar to what would be expected for a normal virus: Figure 4 shows the inverse relationship between Y_{max} and the level of immunization, B_0. The greater the level of initial B cells responsive to Y, the lower Y_{max}. Figure 5 demonstrates the relatively flat response of Y_{max} to increases in initial infecting dose, Y_0. This is not a surprising result if Y_0 represents an infecting dose capable of at least initially evading the specific immune response. This situation would appear to be controllable, at least potentially.

These results contrast sharply with the situation resulting when X represents not a chronically controlled infection or slowly growing virus, but instead is a rapidly growing normal virus. Since in this case the effective 'γ_x' of Eq. (1g) is itself growing exponentially.

Consider now the case of a pure ISRV infection where growth is dominated by the non-linear term. This is analogous to the case for a mixed infection for which Y and/or B_Y have reached sufficiently high concentrations. The derivation of the Y_{max} expression is similar to that of the turning point and maximum concentration of a normal virus (Appendix, Cooper, 1986).

Examining the effect of the level of initial infecting dose (Y_0) upon Y_{max} in this region dominated by non-linear ISRV production, an interesting dependency is observed (Figure 6). Increasing the level of the initial infecting dose results in an increasing maximum concentration of ISRV. The model itself suggests that in this region of non-linear growth and a large concentration of either virus or B cells, virus growth will be no larger than BY, with the immune response term, $I(Y,t)$ also growing as BY (Cooper, 1986). This is suggestive of a slight degree of infectious

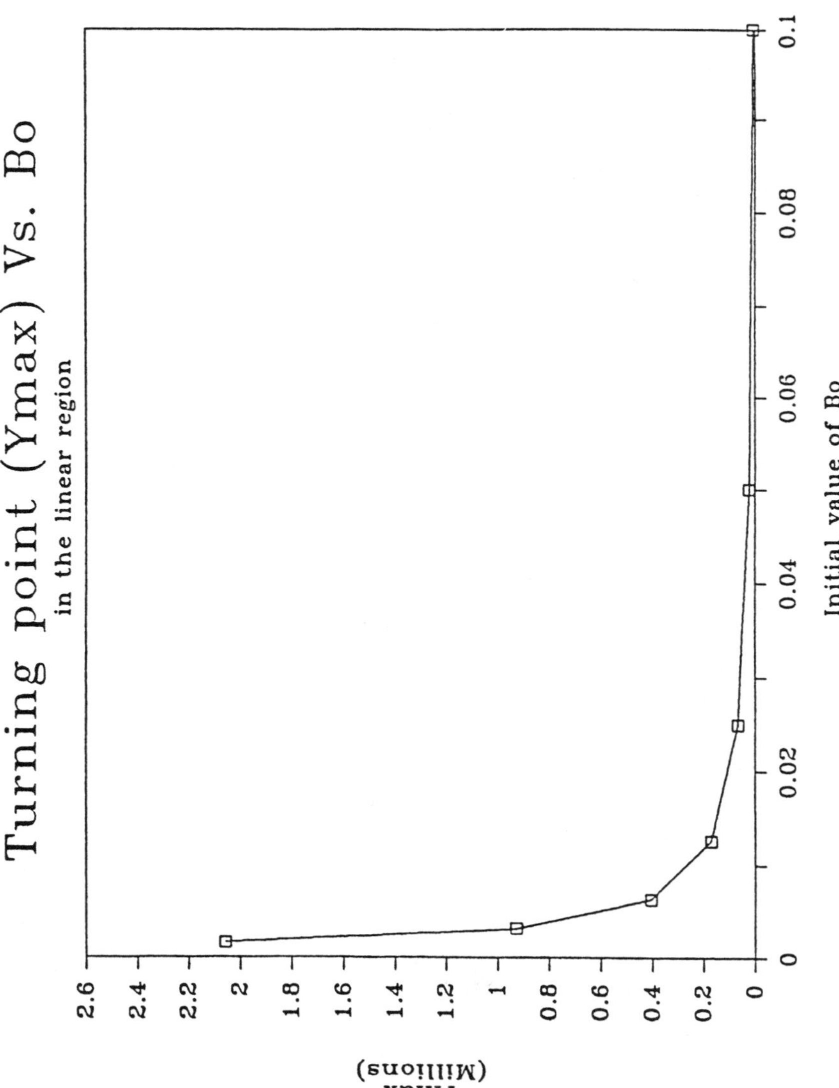

FIGURE 4 Dependence of turning point in ISRV growth on initial antibody concentration, B_{0Y}, in the linear region.

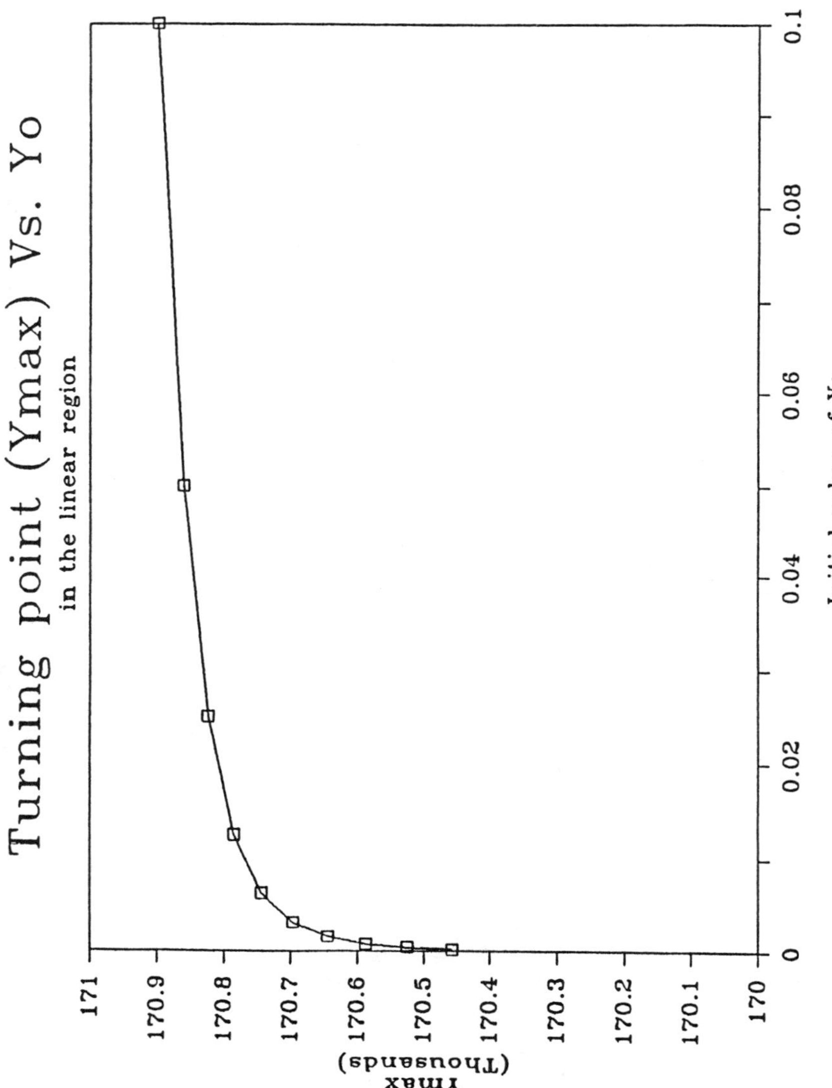

FIGURE 5 Dependence of turning point in ISRV growth on initial virus concentration, Y_0, in the linear region.

control by the immune system, critically dependent upon the various rate factors and the depleting effect of T-cell destruction by ISRV.

Figure 7 is a plot of Y_{max} as a function of B_0, again produced from the non-linear region. Y_{max} is strikingly dependent upon B_0, with an increasing level of immunization dramatically raising the maximum concentration of ISRV. This dependency of ISRV growth upon the concentration of B implies that once T' reaches some critical value, increasing B induces an even more rapid rate of viral growth, making clinical matters worse. This is in sharp contrast to a normally growing virus X, where bolstering the immune system serves to limit the severity of the infection.

Examining Figures 6 and 7 for low concentration of Y and/or B, however, suggests that until some value of B or some value of the product YB causes the non-linearity to dominate, non-linear ISRV production can be considered to be very low. This result is largely due to the fact that the T_4 cells which become infected as a result of ISRV growth are not themselves stimulated to grow and reproduce the virus. The immune response term, $I(Y,t)$ is also linear in Y. Further, for low enough concentrations of Y and/or B, virus production will increase no more rapidly than B^2Y^2 (Cooper, 1986). Thus, for low values of Y and/or B and a pure ISRV infection, there is not likely to be any significant virus growth. The details depend upon how rapidly BY, TY, and BYT' decay. It appears that a pure ISRV infection will grow slowly until the specific immune system response is engaged. This immune system action limits continued ISRV growth, until some value of YB causes the non-linearity to dominate.

To summarize, for low ISRV concentration, either from a new or chronic ISRV infection, ISRV growth is dominated by the linear term. This represents a potentially controllable situation. When faced with a rapidly growing normal virus, however, ISRV growth behaves like a normal virus with an increasing rate of growth. The non-linear term eventually becomes important, dominating virus production. This, in turn, leads to explosive ISRV growth, with concomitant destruction of the T_4 T helper cells and an immobilization of the immune system.

CONCLUSIONS AND IMPLICATIONS

There exist low-growth linear and non-linear regions which result in very slow ISRV production. It is reasonable to expect situations of infection with (a) no antibody response and (b) with detectable antibody response with no obvious symptoms of infection.

ISRV exists in a genetic continuum of related strains (Gallo et al., 1986). If there exists a common region of antigenic identification, then an initial infection controlled by the immune system can result in the build-up of appropriate B cells. Weak infections by other genetic strains could then be contained (until the situation gets out of control, perhaps due to a simultaneous normal infection). What

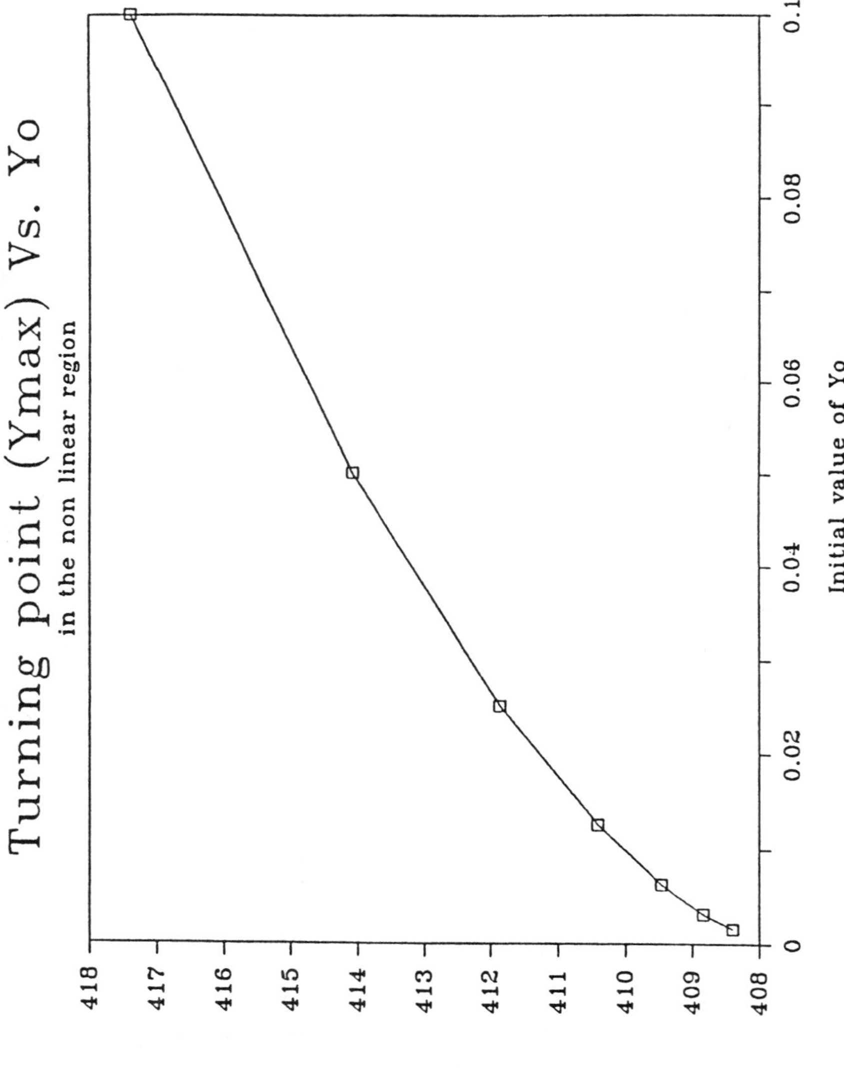

FIGURE 6 Dependence of turning point in ISRV growth on initial ISRV concentration, Y_0, in the non-linear region.

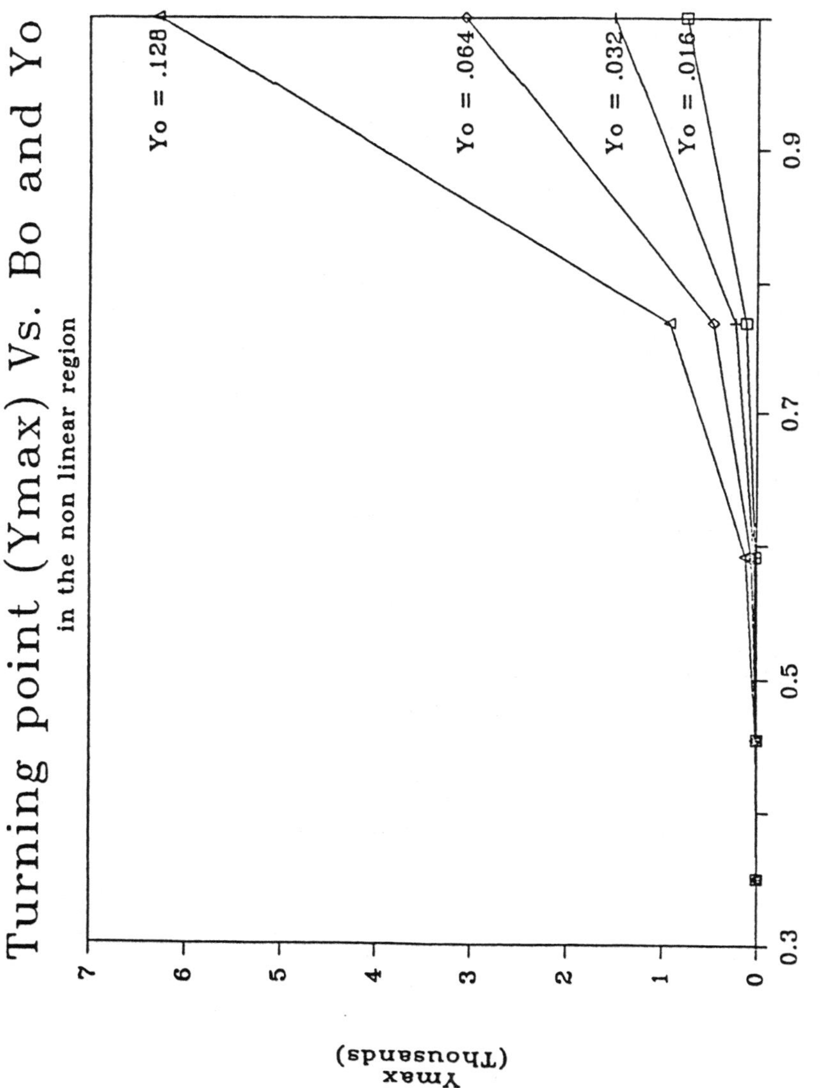

FIGURE 7 Dependence of turning point in ISRV growth on initial antibody concentration and initial ISRV concentration in the non-linear region.

experimental evidence exists (Kulstead, 1986; Gallo, 1986) seems to indicate the possibility of some immunization to infection.

ISRV infection simultaneous with infection by a rapidly growing normal virus, or other infection, leads to rapid ISRV production resulting in almost certain destruction of the immune system response.

In contrast to a normal virus, no level of initial B and T memory cells can provide complete protection. This vulnerability is due to the risk of simultaneous infection by ISRV and normal virus, which almost inevitably leads to rapid ISRV production and T_4 cell destruction.

After the population of ISRV-infected T cells grows large compared to the population of uninfected T cells, for an ISRV infection, increasing the concentration of B cells reactive to ISRV antigen can become severely counter productive, as the ISRV growth rate is dependent on the concentration of ISRV-antigen specific B cells.

Since ISRV antibodies help control infection, but the presence of B and T cells specific to ISRV, the natural source of ISRV antibodies, actually boosts the severity of infection, the exogenous supply of ISRV antibodies while suppressing the immune response may be an aid in controlling infection.

No assumptions have been made with respect to the existence of biological co-factors, latency periods, or incubation periods for the virus. However, significant delays between infection and symptoms are suggested by the model, without assuming any intrinsic, prolonged period of latency. In such circumstances, any immunostimulant will act as a co-factor.

This latency is a result of the harboring of ISRV by non-antigenically stimulated T cells. A crucial question, then, is: what is the lifetime of an infected but non-activated T_4 cell?

As a sample application of these ideas, consider the well-noted correlation between HIV antibody presence and the increased virulence of Kaposi's sarcoma (Cooper, 1986). It is believed that Kaposi's sarcoma is virus derived; spread of this virus seems to be mediated as much by the virus as by the migration of malignant cells. In a patient, not infected with ISRV (HIV), the spread of Kaposi's sarcoma is inhibited by the normal functioning of the immune system. In the AIDS patient, however, the virus that produces the sarcoma acts as a normal virus described in the model. The stimulation this virus provides to the immune system increases the rate of growth for the ISRV. This then destroys the capacity of the immune system to respond, thus allowing for the virulent spread of this normally controllable virus. This scenario is applicable to other infections, and correlates well with the observed increase in severity of the opportunistic infections to which the AIDS patient is subject.

This model also suggest negative consequences for vaccination procedures for individuals unknowingly harboring an ISRV infection. Vaccinations, especially live vaccines, act as powerful immunostimulants. In an ISRV-infected individual, such vaccinations could activate infected T_4 cells. Before any immunization, it would seem prudent to check first for HIV antibodies.

Several critical questions are raised: what is the lifespan of an ISRV-infected T cell? This could help determine how long an individual is at risk for developing the syndrome, given the presence of antibodies. Are there co-factors, beyond those required for T_4 stimulation? Do other T_4 protein positive target cells contribute to virus production?

The model distinguishes between the different classes of human retroviruses (HTLV-I, II, III, HIV, etc.) based on their different rates of production. Y_{max} is very sensitive to the rates of viral production.

REFERENCES

1. Cooper, Leon N. (1986), "Theory of an Immune System Retrovirus," *Proc. Natl. Acad. Sci. USA* **83**, 9159–9163.
2. Gallo, Robert C. (1986), "The First Human Retrovirus," *Scientific American*, December 1986.
3. Hahn, B. H., G. M. Shaw, M. E. Taylor, R. R. Redfield, P. D. Markham, S. Z. Salahuddin, F. Wong-Staal, R. C. Gallo, E. S. Parks, and W. P. Parks (1986), "'Genetic Variation in HTLV-III/LAV over Time in Patients with AIDS or at Risk for AIDS," *Science* **232**, 1548–1553.
4. Kulstead, Ruth, Ed. (1986), *AIDS, Papers from Science 1982-1985* (AAAS, Washington, D.C.).
5. Reibnegger, G., D. Fuchs, A. O. Hausen, E. R. Werner, M. P. Dierich, and H. Wachter (1987), "Theoretical Implications of Cellular Immune Reactions against Helper Lymphocytes Infected by an Immune System Retrovirus," to appear in *Proc. Natl. Acad. Sci. USA*.

Complexities of Antigen-Antibody Systems

BOYD A. WAITE and EDDIE L. CHANG
Department of Chemistry, United States Naval Academy, Annapolis, Maryland 21402 and Bio/molecular Engineering Branch, Code 6190, Naval Research Laboratory, Washington, D. C. 20375, respectively

Models of Immunolysis Assays: A Vesicle-Based Approach for Direct Binding

I. INTRODUCTION

Many biological regulatory phenomena, especially those related to the immune system, involve interactions between soluble ligands and receptor molecules embedded within membranes of a structure which might generally be referred to as a "receptor-carrier." For example, prior to release of histamine by basophils, the ligand (i.e., the allergen) interacts with the IgE receptor sites which are anchored to the receptor-carrier superstructure, in this case, a basophil.[1] Another system (which is the primary context for this study) consists of antibody ligands (IgG's or IgM's) interacting with antigens (i.e., receptors) covalently attached to the membrane molecules of liposomes (i.e., receptor-carriers). Such a system plays a key role in liposome-based immunolysis assays which have been developed in recent years.[2–4]

Many detailed theoretical studies of ligand/receptor binding phenomena have been presented recently.[5,6] These have included kinetic as well as equilibrium studies of monovalent, bivalent, and multivalent ligands interacting with monovalent or bivalent receptors. Crosslinking effects have been modeled, as well as branching structures leading to patch formation and gel-phase transition phenomena. Nearly all of these studies have focused directly on ligand/receptor interactions, assuming the membranes in which the receptors are bound to be of infinite extent, thus precluding the possibility of addressing questions, such as the number of bound receptors or crosslinks *per cell* or *per vesicle* (other than average numbers). In other words, most theoretical models have taken a *ligand* or *receptor* counting approach, rather than focusing directly on *receptor-carrier* counting approaches.

The purpose of this study is to propose a receptor carrier-based (i.e., vesicle-based) approach for studying the effects of ligand/binding interactions for a variety of simple systems, all of which involve monovalent receptor sites embedded within the vesicle membranes. Such an approach ties closely with actual experimental probes, many of which involve nothing more than vesicle-counting methods. If the binding phenomenon of interest manifests itself differently for vesicles bound in distinct ways (e.g., with different numbers of bound ligands or crosslinks per vesicle), then such a model should prove useful for interpreting and verifying certain binding mechanisms as well as effector mechanisms.

Some results applying a receptor-carrier based approach have been presented previously. For example, Brendel and Perelson[7] have recently presented a kinetic analysis of the viral attachment problem wherein the attaching particles (antibodies) are effectively monovalent and the viral particles are assumed to possess monovalent receptor sites. Another example of a carrier-based study involves multiple-reactive site polymers which are catalogued based on distribution functions describing the probability of polymers having a certain number of sites unreacted.[8]

The explicit context involved in these studies is that of immunolysis assays using liposomes and complement-mediated lysis.[2–4] These liposome-based immunoassays have shown a great deal of promise as a highly sensitive and selective tool for determining the presence of low levels of antigen or antibody.[9] The basic experimental approach involved is to covalently bind a certain number of antigens to the surface of phospholipid vesicles (liposomes) and then immunologically release the encapsulated contents of the sensitized vesicles by complement-mediated lysis. Several types of encapsulated contents have been used in past studies, including spin-labelled compounds,[10] enzymes[11] or enzyme substrates,[12] and fluorescent compounds (e.g., carboxyfluorscein[13]).

The important idea is that the release of the encapsulated contents produce a measurable change in the system, thus resulting in a *counting* of the vesicle. If vesicles bound in different ways (e.g., different numbers of bound ligand) respond differently to the detection mechanism involved (e.g., complement-mediated lysis), then these techniques should provide a way, given a valid model of the system, to determine answers to several important questions. For example, how many vesicles have antibodies bound to them; of these how many have exactly one, two, ..., i antibodies bound; and how do these numbers change with vesicle or antigen concentration, or with antibody affinity? Certain details of the complement-mediated lysis mechanism should also be revealed using such a scheme. For example, it should be possible to devise experiments which can determine whether one, two, a few, or many antibodies are needed to lyse a vesicle.

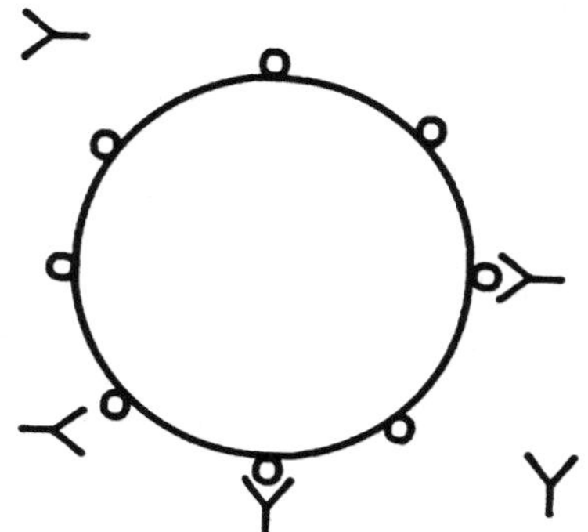

FIGURE 1 Receptor-Carrier (Vesicle) Involved in Immunolysis Assay. Each vesicle is coated with n monovalent receptor sites (antigens). Monovalent ligand (e.g., Fab fragments) specific to the receptor sites binds giving rise to a distribution of receptor-carrier concentrations, V_i $(i = 1, \ldots, n)$. Bound ligands subsequently effect the lysis of the vesicles, releasing their contents which are then detected.

Of critical importance in many ligand/receptor binding studies is the effect of crosslinking of receptor sites by multivalent ligands.[1] At least for the case of bivalent ligands interacting with monovalent receptor sites, this question can be addressed within the context of this receptor-carrier framework. More complicated systems (e.g., multivalent receptor sites) will not be treated here and represent enormous difficulties due to the large number of different types of receptor/ligand structures.

The increased complexity of the multivalent ligand systems manifests itself in the form of several new questions which a model of such a system should be able to address. For example, does the effector mechanism (e.g., complement lysis) depend on whether the antibodies are bound via an unrestricted number of sites (i.e., one or two or more) or a restricted number of sites (i.e., only two or more)? Are there ranges of parameter space which allow the multivalent interaction to be modeled in terms of the monovalent framework, simply replacing the association constant representing the antibody/receptor interaction with a functional affinity?[14] Does the mechanism of vesicle lysis require proximate binding of antibody molecules, and if so, can this mechanism be distinguished from a less restricted mechanism?

Section II includes an alternative derivation of the results obtained by us previously[15] for the receptor-carrier-based approach for the monovalent ligand/ monovalent receptor system. Section III presents a preliminary discussion of some of the principle features expected from a multivalent ligand/monovalent receptor model, the detailed quantitative derivations of which will be presented elsewhere.[16] Section IV includes a summary as well as possible future studies involving this receptor-carrier-based approach.

II. THE MONOVALENT DIRECT BINDING MODEL

Consider the system depicted in Figure 1, consisting of liposomes covered with an average number, n, of monovalent antigens per vesicle, suspended in a buffered medium. Monovalent antibodies (i.e., Fab fragments) specific to the antigen are introduced to the suspension of vesicles. For the purpose of constructing a framework for this simple case (i.e., monovalent antibody interacting with monovalent liposome-based antigen), it is assumed that addition of a certain lysis-mediating serum after a short period of incubation results in the lysis of a certain percentage of the vesicles. For the system involving multivalent antibodies (e.g., IgG's or IgM's) described in section III, the lysis-mediating serum is specifically assumed to be a complement-based serum.

The detailed questions which need to be addressed for such a system include the following: What fraction of vesicles have i antibodies bound, where $i = 1, \ldots, n$? What is the average number of antibodies bound per vesicle? How many antibodies must bind to a vesicle in order to initiate the lysis mechanism?

Let Ab^0, V^0, n, and K represent the initial (total) antibody concentration, total vesicle concentration, average number of antigens per vesicle, and intrinsic associative binding constant for the antigen/antibody interaction, respectively. The most detailed receptor-carrier-based solution involves the set of functions, $f_i(i = 1, \ldots, n)$, representing the fraction of vesicles having i (monovalent) antibodies bound. By definition,

$$f_i = V_i/V^0 \tag{1}$$

where V_i is the concentration of vesicles with i antibodies bound.

The kinetic equations governing this system of vesicles are given by

$$\begin{aligned} dV_0/dt &= -\,k_1 a n V_0 + k_{-1}V_1 \\ dV_i/dt &= -\,k_1 a(n-i)V_i + k_{-1}(i+1)V_{i+1} \\ &\quad + k_1 a(n-i+1)V_{i-1} - k_{-1}iV_i \qquad i = 1, \ldots, n-1 \\ dV_n/dt &= k_1 a V_{n-1} - k_{-1} n V_n \end{aligned} \tag{2}$$

where a is the unbound antibody concentration and k_1 and k_{-1} are the forward binding and dissociative rate constants, respectively, which are related to K by

$$K = k_1/k_{-1} \tag{3}$$

Conservation relationships for antibodies and vesicles must also be satisfied, i.e.,

$$Ab^0 = a + \sum_{i=0}^{n} iV_i \tag{4}$$

$$V^0 = \sum_{i=0}^{n} V_i \tag{5}$$

Implicitly involved in these equations is an equivalent binding site assumption (i.e., that the binding constant K is independent of the extent of binding of vesicles).

The complete equilibrium solution is obtained by setting each of the above kinetic equations to zero. Then, by using the conservation relationships, it is possible to express the functions f_i in terms of the experimentally measurable parameters, Ab^0, V^0, n, and K. This approach has been recently described in detail by Brendel and Perelson.[7]

An alternative approach for finding the equilibrium solution involves simple probability arguments (which can subsequently be applied to the more complicated multivalent models). The Law of Mass Action governing the ith antibody binding can be represented by

$$V_{i-1'} + a \overset{K}{\rightleftharpoons} V_{i'} \tag{6}$$

where the prime ($'$) indicates a specific arrangement of the bound antibodies. This is equivalent to the following mathematical relationship:

$$V_{i'} = (aK)V_{i-1'} \tag{7}$$

The total concentration V_i (which includes *all* possible arrangements of the bound antibodies) is simply related to the specific $V_{i'}$ concentration by

$$V_{i'} = V_i/d_i \tag{8}$$

where d_i is the number of arrangements possible for placing i antibodies among n available sites. For this simple monovalent antibody system, the form for d_i is particularly simple, i.e.,

$$d_i = \binom{n}{i} \tag{9}$$

Substituting Eq. 8 into Eq. 7 and repeating the process i times results in the general relationship

$$V_i = \binom{n}{i}(aK)^i V_0 \tag{10}$$

Verification that this is the correct equilibrium relationship among the V_i is seen by substituting back into the kinetic equations (Eqs. 2).

Finally, in order to solve completely in terms of total (initial) vesicle and antibody concentrations, the conservation equations must be solved. Using Eq. 10, the vesicle conservation relationship (Eq. 5) takes the form

$$V^0/V_0 = (1 + aK)^n \tag{11}$$

Combining Eqs. 10 and 11 yields the following (exact) expression for the f_i:

$$f_i = \binom{n}{i}(aK)^i(1 + aK)^{-n} \tag{12}$$

Combining Eq. 10 with the antibody conservation relationship (Eq. 4) produces the following exact functional dependence of Ab^0 on a:

$$Ab^0 = a + \frac{nV^0aK}{1 + aK} \tag{13}$$

This equation can then be exactly solved (given Ab^0) for a, which can then be substituted into Eq. 12 for the f_i. Alternatively, for the important applications where the average number of bound antibodies per vesicle $\nu << n$ (e.g., for the low antibody concentration ranges involved in immunoassays), the antibody conservation relationship can be solved approximately to yield

$$aK \simeq \frac{Ab^0}{K^{-1} + nV^0} \tag{14}$$

which results in the final (approximate) expression for f_i:

$$f_i = \binom{n}{i}\left(\frac{Ab^0}{K^{-1} + nV^0}\right)^i\left(1 + \frac{Ab^0}{K^{-1} + nV^0}\right)^{-n} \qquad \nu << n \tag{15}$$

PRACTICAL CONSIDERATIONS

Figure 2 shows how the distribution functions f_i vary with antibody concentration for several values of i. Partial summations over i of the f_i's can be performed to obtain the fractions of vesicles with one or more antibodies bound, two or more antibodies bound, etc. These are given by

$$\begin{aligned} f(1) &= 1 - f_0 = 1 - (1 + x)^{-n} \\ f(2) &= 1 - f_0 - f_1 = 1 - (1 + nx)(1 + x)^{-n} \end{aligned} \tag{16}$$

where $x = aK$. Applying the exact relation[15] $x = \nu/(n - \nu)$ leads to the approximate form

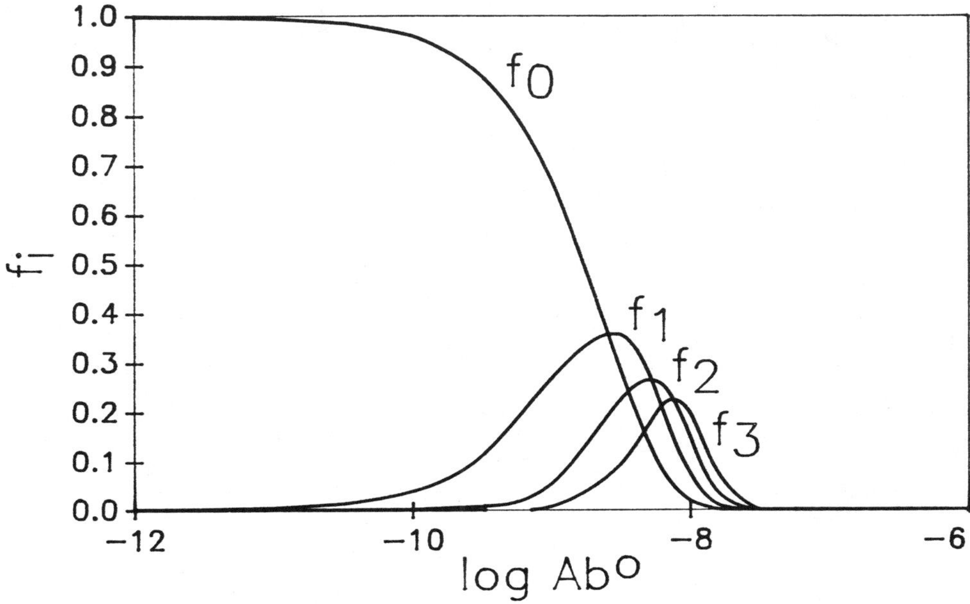

FIGURE 2 The fraction of vesicles with i antibodies bound, f_i, for various values of i, plotted as a function of total antibody concentration Ab^0. Parameters used for this calculation are: $K = 10^5 M^{-1}$, $V^0 = 4.2 \times 10^{-13} M$, $n = 4000$.

$$(1+x)^{-n} = \left(1 - \frac{\nu}{n}\right)^n \simeq e^{-\nu} \qquad \nu << n \tag{17}$$

which is an excellent approximation for values of $nx \leq 10$. Substitution of Eq. 17 into Eqs. 16 produces the approximate forms for $f(1)$, $f(2), \ldots$

$$\begin{aligned} f(1) &= 1 - e^{-nx} \\ f(2) &= 1 - (1 + nx)e^{-nx} \end{aligned} \tag{18}$$

A very important question which can now be addressed concerns the number of bound antibodies (per vesicle) required for initiation of the lysis mechanism. Of course, for this simple system involving monovalent antibody (i.e., Fab fragments), the lysis mechanism must be non-complement based. Nevertheless, as will be seen, it is important to develop this point so as to provide a framework on which to base the more realistic systems involving multivalent antibody in which the complement pathway *is* effective in lysing vesicles.

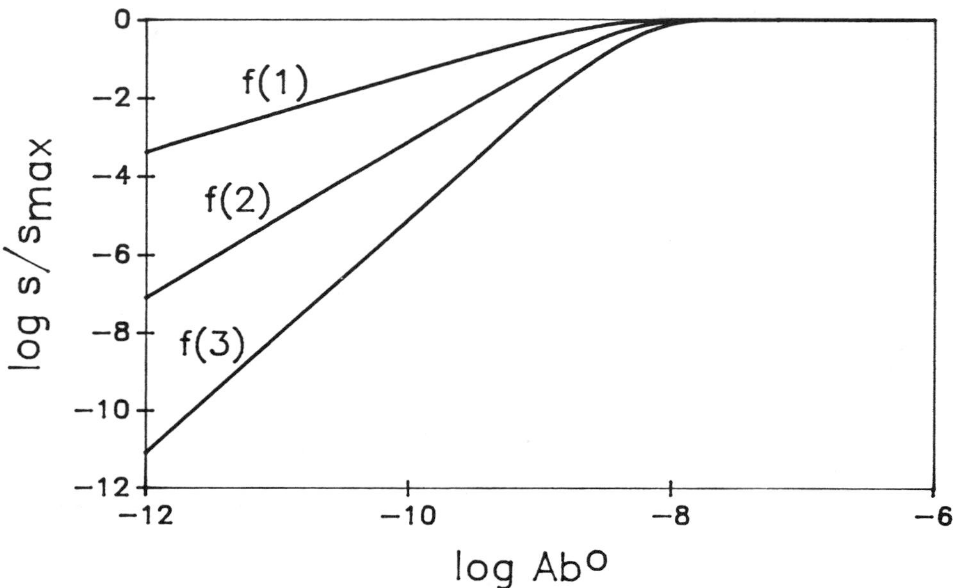

FIGURE 3 The log of the detection signal s (relative to a maximum signal s_{max} corresponding to 100% of the vesicles lysed) plotted as a function of the total antibody concentration Ab^0 for various mechanistic assumptions, i.e., $f(i)$ corresponds to a lytic mechanism where i or more bound antibodies are required for initiation. The mechanisms are experimentally distinguishable due to the distinct slopes associated with each.

The number of bound antibodies required for vesicle lysis can be determined via an assay system in which the logarithm of the final signal (e.g., fluorescence detection for the case of carboxyfluorescein release) is plotted against the logarithm of the antibody concentration (see Figure 3). The slope of the experimental curve can be compared with the slopes for the $f(i)$'s which according to the model are 1, 2, ..., for $f(1)$, $f(2)$, ..., for ranges of the experimental parameters of interest.

III. THE MULTIVALENT ANTIBODY MODEL

A more realistic system for modeling the complement-mediated immunolysis assay and for illustrating the receptor-carrier-based approach to binding studies involves incorporation of multivalent antibodies, e.g., IgG or IgM. For this case, the complexity of the analysis increases, as does the possibility for diverse types of mechanisms of lysis mediation by the bound antibodies. Examples of the increased complexity of the analysis include the fact that multivalent antibodies can, after the initial linkage via one of their binding sites, crosslink to other receptors (antigens) via secondary binding sites. For the case of IgG, the extent of crosslinking is strictly limited to a receptor/antibody complex consisting of two receptors and one antibody. For the case of IgM, it is possible for more receptors to be involved in the crosslinked complex due to the higher valency of IgM. Because of this crosslinking phenomenon, it seems clear that lateral antigen density should have a profound effect on the binding of antibody to vesicles.

In looking towards a quantitative multivalent model analogous to the monovalent treatment presented in Section II, it is worthwhile to catalogue some of the important experimental observations which will have to be accommodated. The purpose of this section is simply to discuss qualitatively certain of these anticipated features. The quantitative development of the model will be presented elsewhere.[16]

Figure 4 shows the relevant structures involved in such a model. A receptor-carrier labeling scheme similar to that incorporated in the monovalent model is utilized, i.e., $V_{i,j}$. In this scheme, the first index represents the number of antibodies which are bound by only one site and the second index represents the number of antibodies which are bound by two sites. For the case of IgG, this indexing system represents the complete set of possible binding arrangements, whereas for the case of IgM, it would be necessary to extend the indexing to include antibodies bound by three sites, four sites, etc.

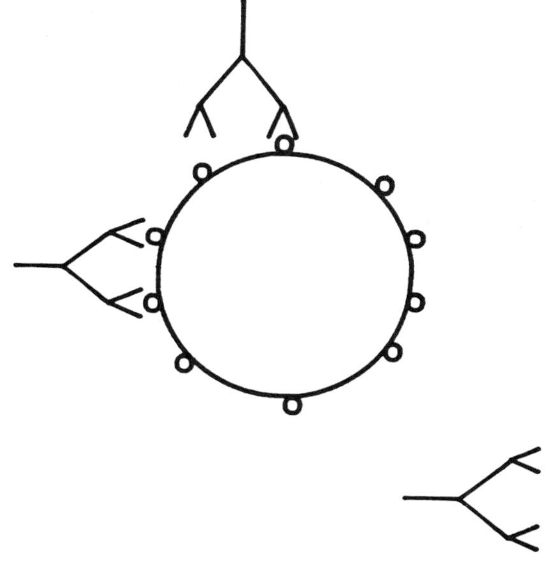

FIGURE 4 The possible structures involved in a bivalent ligand/monovalent receptor immunolysis model. Vesicles are labeled $V_{i,j}$, where i represents the number of antibodies which are bound via a single site and j represents the number of antibodies which are bound via two sites.

As for the monovalent treatment, the equivalent binding site assumption is made. In addition, it will be assumed that the receptor sites (antigens) freely diffuse within the membrane, thus creating the situation where each unbound antigen is available for crosslinking with any unbound antibody site attached to an already anchored antibody molecule.

One of the critical experimental observations for such a multivalent interaction is that the "effective" affinity of the antibody for membrane-bound receptors increases significantly with valency of the antibody.[14] This observation implies several things. First, it implies that the binding process can be described in terms of the monovalent model of section II, replacing the intrinsic association constant K by a functional affinity K_{eff} which incorporates the various effects giving rise to a stronger binding.

It is clear that for the case of low antibody concentration, where the average number of bound antibodies per vesicle is small, there will be little competition for adjacent sites. This would be the limiting situation most conducive to antibodies crosslinking to other receptor sites, effectively resulting in bound species which are all multivalently linked to the vesicle membrane. A quantitative model for multivalent binding ligand should accommodate such a limiting situation.

In addition, the observation of higher apparent binding affinity for IgM over IgG implies that the functional affinity K_{eff} should be proportional to (for the case of low antibody concentration) or more generally nonlinearly dependent on the valency λ of the impinging antibody ($\lambda = 2$ for IgG, $\lambda = 10$ for IgM).

Next, the experimental observations[17] imply that the functional affinity K_{eff} increases with lateral receptor density n. While the exact functional dependence of K_{eff} on n is not immediately obvious from the experimental observations, nevertheless it should be possible to show this dependence from a quantitative model.

Finally, the efficiency of crosslinking should depend on geometric factors which affect the local concentration of adjacent receptor sites. Ignoring the effects of local concentration enhancement on the initial binding step,[18] this geometric factor involved in the enhancement of the crosslinking step can be derived as shown below. A similar factor has been used in previous investigations.[19]

Consider the crosslinking process

$$V_{i+1,j-1} \rightarrow V_{i,j} \tag{19}$$

which represents one monovalently bound antibody crosslinking to a free antigen receptor. For a single vesicle, the reaction volume for such a process is just $4\pi r^2 d$ where r is the radius of the vesicle and d is the thickness of the spherical shell surrounding the vesicle surface in which the antibody binding sites and receptor binding sites migrate and collide (see Figure 5).

The concentration of unbound antibody sites within this region is given by $\beta(i+1)/(N_0 4\pi r^2 d)$ and the concentration of unbound antigen (receptor) sites within the membrane is given by $(n-i-2j+1)/(N_0 4\pi r^2 d)$, where N_0 is Avogadro's number and β is a factor denoting the number of available sites remaining on the singly bound antibody ($\beta = 1$ for IgG). The rate of reaction of these reactants corresponds to the rate of appearance of new vesicles, $V_{i,j}$. Within the entire reaction vessel of volume L^3, there are $V_{i+1,j-1}N_0L^3$ vesicles available to undergo such a transformation. Thus, if $n_{i,j}$ is the number of moles of vesicles available,

$$\frac{d(n_{i,j}/4\pi r^2 d)}{dt} = \frac{k\beta(i+1)(n-i-2j+1)V_{i+1,j-1}N_0L^3}{(N_0 4\pi r^2 d)^2} \tag{20}$$

where k is the bimolecular reaction rate constant. But $V_{i,j} = n_{i,j}/L^3$, and so

$$\begin{aligned}\frac{dV_{i,j}}{dt} &= \frac{k\beta(i+1)(n-i-2j+1)V_{i+1,j-1}}{N_0 4\pi r^2 d} \\ &= (k/\alpha)(i+1)(n-i-2j+1)V_{i+1,j-1}\end{aligned} \tag{21}$$

Thus, the geometric factor which accounts for the enhancement of the crosslinking binding step is given by

$$\alpha = \frac{N_0 4\pi r^2 d}{\beta} \tag{22}$$

or, in other words, it is the molar volume of the spherical shells surrounding the vesicles in which the crosslinking events occur.

The observation of increased apparent binding affinity due to multivalency of the binding ligand is, therefore, summarized as

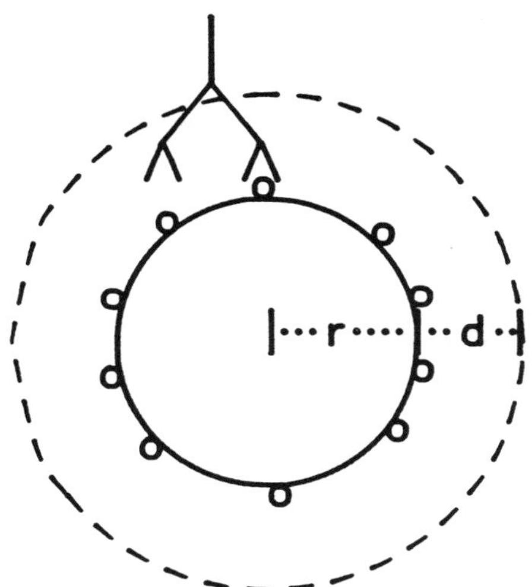

FIGURE 5 The geometric structural considerations giving rise to the factor α which describes the enhanced local concentration of receptor sites as seen by potentially crosslinked antibody. d represents the effective thickness of the layer surrounding the vesicle in which the crosslinking "collisions" occur. r is the radius of the vesicle.

$$K_{\text{eff}} \propto \frac{\lambda n^{y} K_{\text{int}}{}^{z}}{\alpha} \tag{23}$$

where K_{int} is the intrinsic binding affinity for antibody site/receptor site interactions, y and z are arbitrary parameters (perhaps λ-dependent) to be obtained, and λ and α are as defined above.

A quantitative receptor-carrier-based multivalent model should also be able to address certain mechanistic questions. For example, suppose that the lytic mechanism occurs either in an unrestricted fashion (i.e., where the effector function of the bound antibody is independent of whether it is bound by one or more sites) or in a restricted fashion (i.e., where the effector function of the bound antibody requires that it be attached by at least two sites). First, it is intuitive that in the low antibody concentration regime, since nearly all bound antibodies will be uninhibited in their crosslinking tendency, these two mechanisms would be indistinguishable. However, for higher antibody concentration ranges, it should be possible to force each receptor site to be occupied by singly attached antibody. For the unrestricted mechanism, lysis would still occur, whereas for the restricted mechanism, lysis would be effectively inhibited. A quantitative model should be able to predict such effects.

Another significant question which would require a quantitative receptor-carrier based model involves the need for proximity of adjacently bound antibodies in order to effect lysis. For example, Perelson and Wiegel[20] have calculated the probability that receptor-carriers having i bound species would have two or more of these within a distance ϵ of each other, given by

$$P_i = 1 - e^{-i(i-1)/2M} \tag{24}$$

where $M = A/\pi\epsilon^2$ and A is the surface area of the vesicle. With this formula, it should be possible, given a correct multivalent model, to calculate the fraction of lysed vesicles (whether restricted or unrestricted) by the formula

$$f_{\text{lysed}} = \sum_{i=0}^{n} P_i f_i \tag{25}$$

where f_i is the fraction of vesicles having i appropriately effective antibodies bound.

IV. DISCUSSION

Many experimental assay methods involve either cell counting procedures or vesicle counting procedures. In order to model such systems, it becomes important to be able to construct ways of labelling vesicles (or cells) so as to include all of the possible structures and distributions of such structures on the surface. Obviously, this is only a simple task for the case in which the receptor sites on the vesicle surface are monovalent. Thus, although this does apply in many contexts, it does not address many important questions raised by the possibility of multivalent receptor sites (e.g., Ig's on B-cell surfaces, IgE's on basophils, etc.). Such structures include long chains of clustered receptors, rings of clustered receptors, and branched clusters of receptors. Analysis of such systems has been prevalent in the theoretical literature,[2–4] although it is clear that such studies make no reference to receptor-carrier counting. Interfacing such studies with experiments requires a different sort of immunoassay for counting (or monitoring) events which result from such binding.

The lysis techniques are simple and experimentally straightforward. Thus, the modeling presented here (while simple in that it involves only monovalent receptors) has a very good chance of assisting experimentalists as they endeavor to understand mechanisms and structures involved in immunoassay techniques.

Future work will extend these studies to the case of indirect binding models, including the sandwich assay.[21] Again, so long as the vesicle- based binding site (Fab fragment in this case) is monovalent, it should be possible to solve for the distribution of vesicles having various numbers of bridging antigens and lysis mediating antibodies bound.

ACKNOWLEDGMENT

Boyd A. Waite acknowledges partial support of this work from the NRL/USNA Cooperative Program for Scientific Interchange.

REFERENCES

1. Sterk, A. R., and T. Ishizaka (1982), "Binding Properties of IgE Receptors on Normal Mouse Mast Cells," *J. Immunol.* **128**, 838-843.
2. Kinsky, S. C., D. A. Haxby, D. A. Zopf, C. R. Alving, and C. B. Kinsky (1969), "Complement-Dependent Damage to Liposomes Prepared from Pure Lipids and Forssman Hapten," *Biochemistry* **8**, 4149.
3. Six, H. R., K. Uemura, and S. C. Kinsky (1973), "Effect of Immunoglobulin Class and Affinity on the Initiation of Complement-Dependent Damage to Liposomal Model Membranes Sensitized with Dinitrophenylated Phospholipids," *Biochemistry* **12**, 4003.
4. Leserman, L., J. Barbet, F. Kourilsky, and J. Weinstein (1980), "Targeting to Cells of Fluorescent Liposomes Covalently Coupled with Monoclonal Antibody or Protein A," *Nature (London)* **288**, 602.
5. Dower, S. K., J. A. Titus, and D. M. Segal (1984), "The Binding of Multivalent Ligands to Cell Surface Receptors," *Cell Surface Dynamics*, Eds. A. S. Perelson, C. DeLisi, and F. W. Wiegel (New York: Dekker), 277–328.
6. Perelson, A. S. (1984), "Some Mathematical Models of Receptor Clustering by Multivalent Ligands," *Cell Surface Dynamics*, Eds. A. S. Perelson, C. DeLisi, and F. W. Wiegel (New York: Dekker), 223–276.
7. Brendel, V., and A. S. Perelson (1987), "Kinetic Analysis of Adsorption Processes," *SIAM J. Appl. Math.*, in press.
8. Tanford, C. (1961), *Physical Chemistry of Macromolecules* (New York: Wiley), 526.
9. Ishimori, Y., T. Yasuda, T. Tsumita, M. Notsuki, M. Koyama, and T. Tadakuma(1984), "Liposome Immune Lysis Assay (LILA): A Simple Method to Measure Anti-Protein Antibody using Protein Antigen-Bearing Liposomes," *J. Immunol. Methods* **75**, 351.
10. Wei, R., C. R. Alving, R. L. Richards, and E. S. Copeland (1975), "Liposome Spin Immunoassay: A New Sensitive Method for Detecting Lipid Substances in Aqueous Media," *J. Immunol. Methods* **9**, 165.
11. Kataoka, T., J. R. Williamson, and S. C. Kinsky (1973), "Release of Macromolecular Markers (enzymes) from Liposomes Treated with Antibody and Complement," *Biochim. Biophys. Acta* **298**, 158.
12. Kinsky, S. C. (1972), "Antibody-Complement Interaction with Lipid Model Membranes," *Biochim. Biophys. Acta* **265**, 1.
13. Weinstein, J. N., E. Ralston, L. D. Leserman, R. D. Klausner, P. Dragsten, P. Henkart, and R. Blumenthal (1984), in *Liposome Technology III*, Ed. G. Gregoriades (Boca Raton: CRC Press), 183.
14. Hornick, C. L., and F. Karush (1972), "Antibody Affinity. III. The Role of Multivalency," *Immunochemistry* **9**, 325–340.
15. Chang, E. L., and B. A. Waite (1987), "Analysis of Vesicle Immunolysis Assays: The Direct Binding Model," *J. Immunol. Methods* **102**, 33.
16. Waite, B. A., and E. L. Chang, unpublished.

17. Gopalakrishnan, P. V., and F. Karush (1974), "Antibody Affinity. VII. Multivalent Interaction of Anti-Lactoside Antibody," *J. Immunol.* **113**, 769–778.
18. DeLisi, C., and H. Metzger (1976), "Physical Chemical Aspects of Receptor-Ligand Interactions," *Immun. Commun.* **5**, 417–436.
19. Reynolds, J. A. (1979), "Interaction of Divalent Antibody with Cell Surface Antigens," *Biochemistry* **18**, 264–269.
20. Perelson, A. S., and F. W. Wiegel (1979), "A Calculation of the Number of IgG Molecules Required per Cell to Fix Complement," *J. Theoret. Biol.* **79**, 317–332.
21. Maiolini, R., B. Ferrua, J. F. Quaranta, A. Pineoteau, L. Euller, G. Ziegler, and R. Masseyeff (1978), "A Sandwich Method of Enzyme Immunoassay. II. Quantification of Rheumatoid Factor," *J. Immunol. Methods* **20**, 25.

ALBERTO GANDOLFI† and ROBERTO STROM‡
†Istituto di Analisi dei Sistemi ed Informatica del CNR, Viale Manzoni 30, 00185 Roma, Italy and ‡ Dipartimento di Biopatologia Umana, sez. di Biochimica Clinica, Università di Roma "La Sapienza," Viale Regina Elena 324, 00161 Roma, Italy

Pseudo-Cooperativity Resulting from Ring Closure in Divalent Antibody—Divalent Antigen Interactions

INTRODUCTION

Following Valentine and Green's[1] experimental demonstration of the presence of cyclic complexes among the products of the reaction between a di-DNP hapten and divalent anti-DNP antibody, several authors[2–9] have taken into account, in the mathematical description of antigen-antibody interactions, the formation of intramolecular bonds leading to ring closure.

Formation of cyclic complexes is likely to account for the increase in binding affinity, often indicated as a "premium effect," which has indeed been frequently observed in the binding of antibody to multivalent antigens.[10–14] Intramolecular bonds should also be taken into consideration in quantitative descriptions of the network theory of the humoral immune response, since both idiotypic and anti-idiotypic specificities are borne by antibodies which are at least divalent. Moreover, as extensively discussed in [6,7,15], cyclization of immune complexes is also an important factor in the binding of multivalent effectors to cell membrane Ig-receptors and subsequent signals generation in lymphocyte activation and basophil degranulation.

In a previous work,[16] we have considered the reaction in solution between a divalent ligand and a divalent receptor, giving a mathematical description of the equilibrium state in terms of concentrations of free sites, and we have analyzed in detail the resulting binding curves in the Scatchard representation. In the present paper we re-consider those results, showing that the occurrence of cyclic complexes can lead to considerable alterations in the epitope binding curves, the extent of these alterations and also their qualitative behavior being strongly dependent on the possible existence of "monogamous" rings (i.e., of complexes in which both combining sites of a same antibody are bound to the epitopes of a same antigen molecule). When formation of "monogamous" rings is hindered, the binding curves can achieve shapes which mimic, but are not strictly identical to, those originating from a true positive cooperativity.

Our approach can also be extended to the study of a ternary system in which two different divalent antigens interact with hybrid antibodies. We will show how the binding of one antigen can be enhanced by the presence of the other one, since such a presence allows the formation of circular complexes, thus outlining the rationale for the design of an enzyme immunoassay.

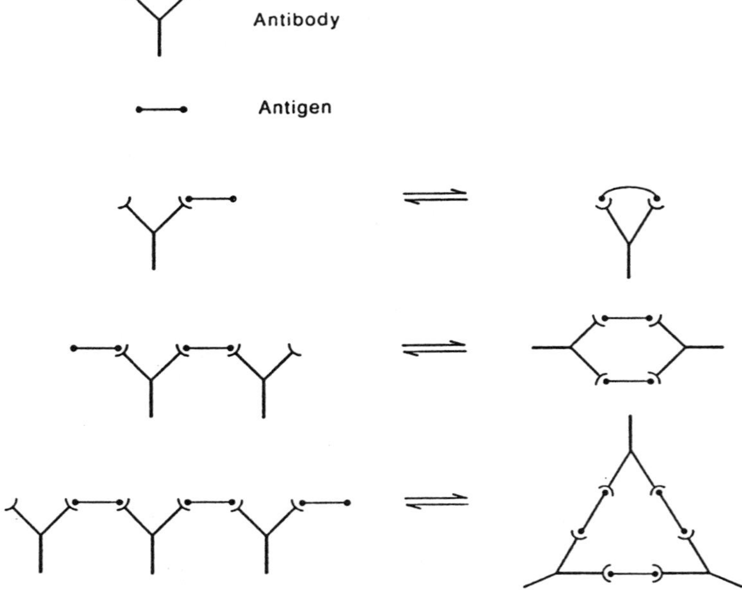

FIGURE 1 Closure of rings from linear chains in the interaction of divalent antigen with divalent antibody.

MATHEMATICAL DESCRIPTION OF THE EQUILIBRIUM STATE

The divalency of antibody and antigen molecules can easily be expected to lead to formation of cyclic complexes, so that linear chains of various sizes are in equilibrium with the corresponding rings (Figure 1). Ring closure occurs by a monomolecular reaction within a single antigen-antibody complex, i.e., by a process which is quite distinct from the bimolecular bond formation.

The model we have used[16] to describe the equilibrium state of the system is based on the following assumptions: i) the epitopes on the antigen molecules (two *per* antigen molecule) are all identical, and the same is true for the antibody combining sites; ii) all bimolecular reactions between antigen epitopes and antibody sites are independent and equivalent, irrespective of the size of the complexes involved; iii) the equilibrium constant of the monomolecular reactions leading to ring closure depends on the number of antibody molecules in the complex.

Let L be the concentration of free antigen molecules, H be the concentration of free epitopes, let $c(n)$ be the concentration of linear chains containing n antibody molecules and terminating on one end with a free epitope and on the other end with a free antibody site. Moreover, let R be the concentration of free antibody molecules, S be the concentration of free antibody sites, K the intrinsic equilibrium constant for bond formation in a bimolecular reaction and let $K'(n)$ be the ratio between the n-ring closure rate constant and the dissociation rate constant of any single bond in the n-ring (the equilibrium constant between a linear chain and the corresponding ring being then equal to $K'(n)/2n$).

Since the concentration of bound antibody sites in linear chains is given at equilibrium by KSH, and since the concentration of rings containing n antibody molecules is equal to $[K'(n)/2n]c(n)$, the concentration S_t of total antibody sites is given by

$$S_t = S + KSH + \sum_{n=1}^{\infty} K'(n)c(n). \tag{1}$$

The law of microscopic reversibility implies, on the other hand, that at equilibrium

$$c(n) = 4^n K^{2n-1} L^n R^n, \qquad n = 1, 2, \dots\ . \tag{2}$$

In terms of concentrations of free sites, the concentrations of free antigen and free antibody molecules can be derived, as shown in [16], by considering the equilibrium concentrations of the possible binding states of the antigen molecules, as proposed by Perelson and DeLisi,[7] together with the equilibrium concentrations of the binding states of the antibody molecules. Alternatively, the same result can be achieved by the following probabilistic argument.

Let us consider a mixed population of free antigens, of free antibody molecules and their complexes, and let N be the total number of antigen molecules in this population. The probability that two epitopes, selected by two independent random trials, belong to a same free antigen molecule is the probability $P(\alpha, \beta)$ of the simultaneous occurrence of event α (both epitopes being free) and of event β (both epitopes occurring on a same antigen molecule). If the population size is large, we have then

$$N_f = P(\alpha, \beta)\binom{2N}{2} \tag{3}$$

where N_f is the number of free antigen molecules, and $\binom{2N}{2}$ expresses the number of distinct pairs of epitopes. If the considered population contains all the free molecules and all the complexes *except* rings at equilibrium, then, according to assumption (ii), the events α and β are independent, so that $P(\alpha, \beta) = P(\alpha)P(\beta)$. Since

$$P(\alpha) = \left(\frac{H}{H + KSH}\right)^2,$$

and

$$P(\beta) = N/\binom{2N}{2},$$

by taking into account that, in the unit volume, $N = (H + KSH)/2$, from equation (3) we have

$$L = \frac{H}{2(1 + KS)}. \tag{4}$$

A similar argument leads to

$$R = \frac{S}{2(1 + KH)}. \tag{5}$$

Inserting equations (2), (4) and (5) into equation (1), we get

$$S_t = S + KHS + \frac{1}{K}\sum_{n=1}^{\infty} K'(n)\left(\frac{KH}{1 + KH}\,\frac{KS}{1 + KS}\right)^n \tag{6}$$

and, since the concentration $S_t - S$ is equal to the concentration of bound epitopes,

$$H_t = H + S_t - S, \tag{7}$$

where H_t is the concentration of total epitopes. Equations (6) and (7) describe the equilibrium state of the system in terms of free antigen epitopes and of free antibody sites concentrations.

THE BINDING CURVE IN THE SCATCHARD PLANE

The effect of ring formation can be most clearly evidenced in the so-called Scatchard binding plot,[17] which offers some advantages for the evaluation of binding parameters and for the detection of positive or negative cooperativity[18] without requiring (at variance from the Hill plot) critical assumptions concerning the total number of binding sites.

Let us consider the normalized Scatchard variables $X = H_b/S_t$ and $Y = H_b/(HS_t)$, where H_b is the concentration of bound epitopes. According to the above description of the equilibrium state, it is possible to study some characteristics of the Y *vs.* X binding curve, obtained with a constant concentration of antibody and varying concentration of antigen. Taking into account in equation (6) that

$$S = S_t(1 - X),$$

and

$$H = \frac{X}{Y},$$

we obtain, for $0 < X < 1$,

$$\frac{K(1-X)}{Y} + \frac{1}{KS_tX}\sum_{n=1}^{\infty} K'(n)\left[\frac{KX}{Y+KX}\,\frac{KS_t(1-X)}{1+KS_t(1-X)}\right]^n = 1 \tag{8}$$

which is the implicit equation of the binding curve in the Scatchard plane. In the absence of ring formation ($K'(n) = 0$ for each n), equation (8) reduces to $Y = K(1 - X)$ which is the well known expression for the case of interaction between equivalent and independent sites. If, instead, ring formation is allowed, we always have $Y > K(1 - X)$, and a higher level of binding is thus obtained.

Due to the implicit form of (8), a detailed study of the Y *vs.* X curve is not straightforward. The asymptotic characteristics can however be derived,[16] allowing some considerations on the importance of $K'(1)$ and of $K'(2)$, i.e., of the ring closure constants for the smallest ring.

As $X \to 0$ (i.e., as $H_t \to 0$), the intercept on the ordinate axis $Y_0 = \lim_{X\to 0} Y$ and the limit slope $\mu_0 = \lim_{X\to 0} dY/dX$ are given by

$$Y_0 = K\left(1 + \frac{K'(1)}{1+KS_t}\right) \tag{9}$$

and

$$\mu_0 = K \frac{-1 - KS_t - K'(1)\left[1 + \dfrac{KS_t}{1 + KS_t + K'(1)} + \dfrac{KS_t}{(1 + KS_t)^2}\right] + K'(2)\dfrac{KS_t}{1 + KS_t}}{1 + KS_t\left(1 + \dfrac{K'(1)}{1 + KS_t + K'(1)}\right)}. \tag{10}$$

As $X \to 1$ (i.e., as $H_t \to \infty$), the asymptotic values are

$$Y_\infty = \lim_{X \to 1} Y = 0 \tag{11}$$

and

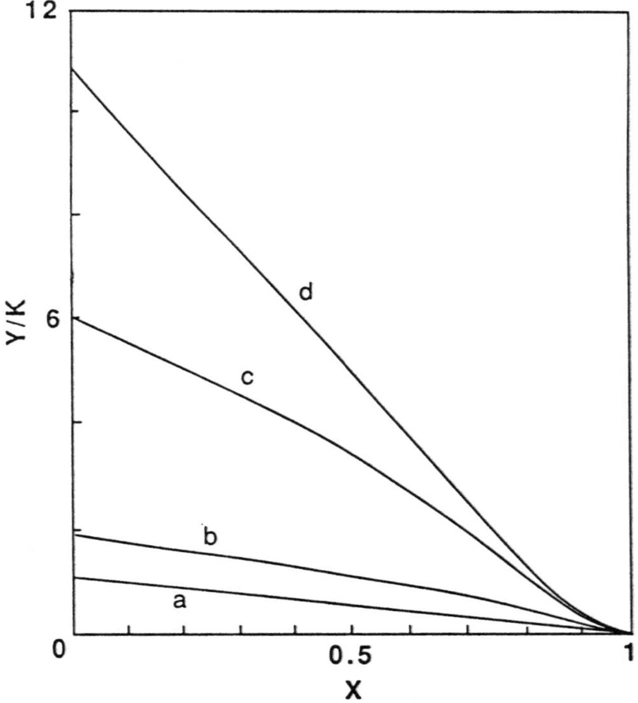

FIGURE 2 Scatchard plot of the binding curve when the value of the intrinsic ring closure constant K' is equal to 10, the KS_t product being equal to 10^2 (curve a), 10 (curve b), 1 (curve c), 10^{-2} (curve d).

$$\mu_\infty = \lim_{X \to 1} dY/dX = -K. \tag{12}$$

It can be seen that Y_0 depends only on the values of K, $K'(1)$ and S_t, while both sign and magnitude of μ_0 are also dependent on the value of $K'(2)$. The binding curve thus meets the ordinate axis at a value higher than that (equal to K) it would have in the absence of any ring formation. Such an increase of the intercept depends only on the ability of the system to form monogamous rings and is modulated by the value of the KS_t product. The limit slope of the binding curve is instead affected also by the formation of rings involving two antibody and two antigen molecules: thus, if formation of monogamous rings is somehow hampered, the binding curve can

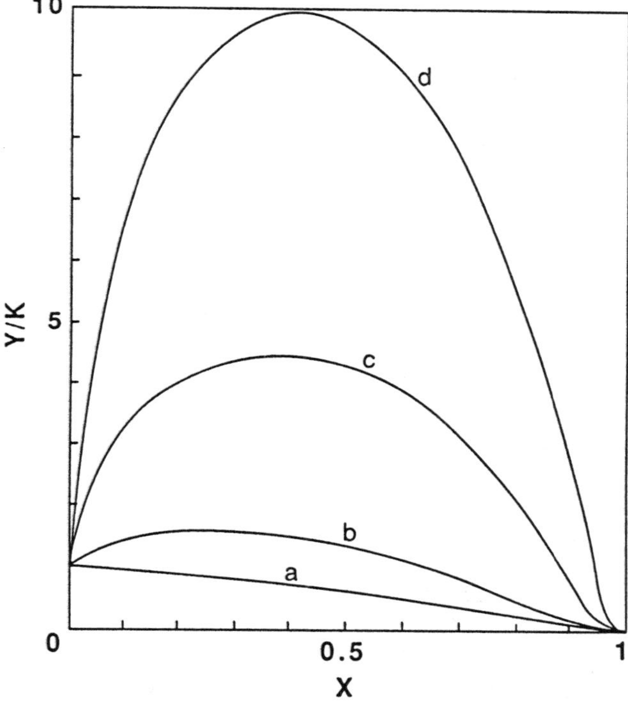

FIGURE 3 Dependence of the binding curve on the value of the intrinsic ring closure constant K', when formation of monogamous ring is totally prevented ($K'(1) = 0$) and the KS_t product is equal to 1. The values of K' are: 10 (curve a), 10^2 (curve b), 10^3 (curve c), $5 \cdot 10^3$ (curve d).

assume near the ordinate axis an upward trend which contrasts with the uniform negative slope (with value equal to $-K$) it would have in absence of ring formation. In the region of large excess of antigen, the binding curve is no more influenced by the possibility of ring formation, since all antibody sites are then occupied by epitopes belonging to different antigen molecules.

The simulation of the binding curves in the Scatchard plane was performed under the assumption of a random walk mechanism of ring closure. According to Jacobson and Stockmayer,[19] the values of $K'(n)$ should follow the law

$$K'(n) = K'n^{-3/2}, \qquad n = 1, 2, \ldots$$

where K' is to be considered as an "intrinsic" ring closure constant. The only exception to this assumption was the possibility that formation of monogamous rings be restricted by steric hindrance, $K'(1)$ thus having values lower than K'. In

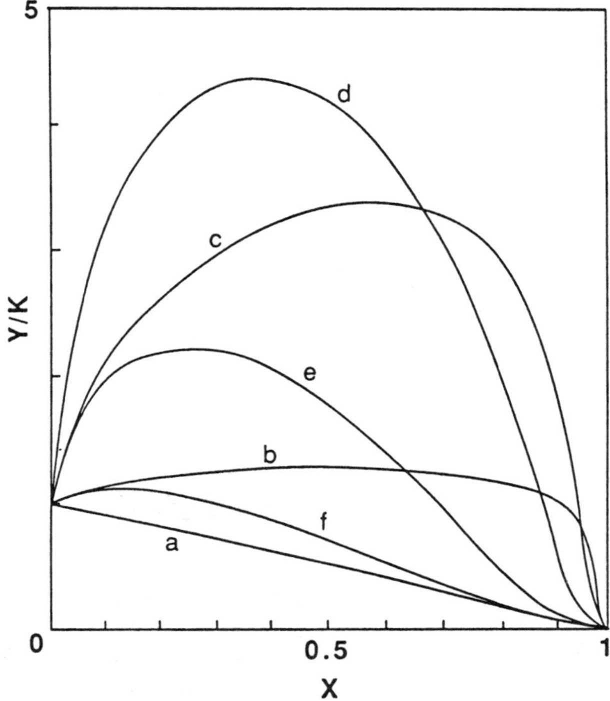

FIGURE 4 Dependence of the binding curve on the KS_t product, when K' is equal to 10^{-3}, except for monogamous rings which are totally prevented ($K'(1) = 0$). The values of KS_t are: 10^4 (curve a), 10^2 (curve b), 10 (curve c), 1 (curve d), 10^{-1} (curve e), 10^{-2} (curve f), 10^{-4} (curve a).

our simulations, the series in equation (8) was approximated by retaining all the terms with $n \leq 7$.

In the absence of any restriction on $K'(1)$, the representation of Y/K *vs.* X (such a further normalization of Y with respect to K allowing an easier comparison of curves at different values of the KS_t product) yields curves having a slight downward concavity in the central portion of the plot (Figure 2). This concavity vanishes as $KS_t \rightarrow 0$ or as $KS_t >> K'(1)$. As indicated by equation (9), the values of the KS_t product strongly affect the value of the intercept on the Y/K axis.

If the size and/or the flexibility of antibody molecules are such that their combining sites cannot accommodate both epitopes of an antigen molecule, formation of monogamous rings is impossible, or at least less probable than accounted for by the assumption of a pure random walk mechanism of ring closure. As shown by Figure 3, when $K'(1)$ is equal to zero the binding curve exhibits a downward concave shape, due to the impairment, in the region of large antibody excess, of the formation of linear chains having $n \geq 2$ and capable of undergoing ring closure. Such a shape resembles that occurring in the case of true positive cooperativity.[20] This effect depends sensibly on the value of the KS_t product, being maximum for $KS_t = 1$. The position of the curve's maximum is also affected by KS_t, being shifted toward low or high values of the abscissa as $KS_t < 1$ or $KS_t > 1$ respectively (Figure 4). It should, however, be noted that the otherwise downwardly concave curve reverses, as antibody sites tend to be saturated and then the number of chains susceptible of ring closure rapidly declines, to an opposite upward concavity, which differentiates it from similar curves originating from true cooperative interactions. In the Hill plot, this pseudo-cooperativity is revealed, provided that a correct estimate of S_t is available, by the presence of a region having slope lower than unity (Figure 5).

RING FORMATION IN TERNARY SYSTEMS: RATIONALE FOR AN IMMUNOASSAY

The previous model can be extended to include more complex systems, in which monogamous rings can be *a priori* impossible. In studying a radioimmunoassay procedure for human chorionic gonadotropin, Moyle et al.[8] have considered the case of an antigen carrying two different epitopes interacting with two divalent antibodies directed against either epitope. A similar ternary system had been investigated by Wolfsy et al.,[15] Wolfsy,[21] and by Perelson[22] as a possible basis for membrane receptor crosslinking in basophil degranulation. The availability of hybrid antibodies,[23] having the two binding sites of a same molecule directed against different epitopes, has prompted us to examine a system suggested by G. Görog as having a possible interest in the development of highly sensitive homogeneous enzyme immunoassays. In such a system, not formally different from those mentioned above, two different

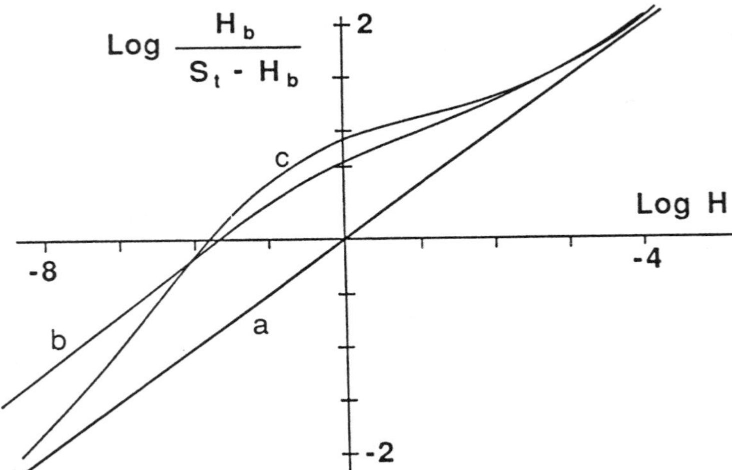

FIGURE 5 Hill plot of the binding curve in the following conditions: absence of ring formation (curve a), no restrictions on monogamous rings and $K' = 10$ (curve b), monogamous rings totally prevented ($K'(1) = 0$) and $K' = 10^3$ (curve c). In all cases the KS_t product is equal to 1.

antigen molecules, each carrying two identical epitopes of type A or type B, interact with bifunctional hybrid antibody molecules, the binding sites of which are directed one against epitope A and the other against epitope B. Our study aims to evaluate the level of binding of A epitopes as a function of the B antigen concentration, thus exploring the possibility of evaluating the amount of B antigen through measurements of bound A epitopes.

Assuming that all bimolecular reactions of a given epitope with the corresponding antibody combining sites are equivalent and independent, irrespective of the size of the immune complexes involved, let L_A be the concentration of free antigen molecules of type A, H_A be the concentration of free A epitopes, S_A be the concentration of free anti-A combining sites and let K_A be the intrinsic equilibrium constant for A bond formation in a bimolecular reaction. Let $c_A(n)$ be the concentration of linear chains containing n antibody molecules and terminating on one end with a free A epitope and on the other end with a free anti-A antibody site. Let L_B, H_B, S_B, K_B and $c_B(n)$ be the corresponding quantities of the B type. It can be noted that only the above defined linear chains are, in both cases, susceptible of ring closure and that they can exist only for even values of n. Let, moreover, $r(n)$ be the concentration of rings containing n antibody molecules.

At equilibrium, by mass conservation, we have

$$S_t = S_A + K_A S_A H_A + \sum_{n=2,4,\ldots} nr(n) \tag{13}$$

$$S_t = S_B + K_B S_B H_B + \sum_{n=2,4,\ldots} nr(n) \tag{14}$$

$$H_{A_t} = H_A + S_t - S_A \tag{15}$$

$$H_{B_t} = H_B + S_t - S_B, \tag{16}$$

where S_t is the concentration of total antibody binding sites of the anti-A type (and, due to the hybrid character of the antibody molecules, also of the anti-B type) and H_{A_t} and H_{B_t} are the concentrations of total A and B epitopes respectively.

Denoting by $K'_A(n)$ the ratio between the closure rate constant of the n-chains with ends of A type and the dissociation rate constant of any single A bond in the n-ring (and by $K'_B(n)$ the corresponding quantity for ring closure which involves B ends), the law of microscopic reversibility implies that

$$r(n) = \frac{K'_A(n)}{n} c_A(n) \tag{17}$$

$$r(n) = \frac{K'_B(n)}{n} c_B(n) \tag{18}$$

since, in each n-ring, n A-bonds and n B-bonds are present. We have moreover that, for $n = 2, 4, \ldots$,

$$c_A(n) = 2^n K_A^{n-1} K_B^n R^n L_A^{n/2} L_B^{n/2} \tag{19}$$

$$c_B(n) = 2^n K_A^n K_B^{n-1} R^n L_A^{n/2} L_B^{n/2}, \tag{20}$$

where R is the concentration of free antibody molecules. As a consequence, the following relation

$$\frac{K'_A(n)}{K_A} = \frac{K'_B(n)}{K_B}$$

holds for $n = 2, 4, \ldots$

The concentrations L_A, L_B and R can be computed in terms of concentrations of free epitopes and free combining sites following a probabilistic approach similar to that of section 2. It can be found that

$$L_A = \frac{H_A}{2(1 + K_A S_A)} \tag{21}$$

$$L_B = \frac{H_B}{2(1 + K_B S_B)} \tag{22}$$

and

$$R = \frac{S_A}{1 + K_B H_B}. \tag{23}$$

Inserting equations (17), (19) and (21-23) in equation (13), and subtracting equation (14) from equation (13), we obtain the equations

$$S_t = S_A(1+K_A H_A) + \sum_{n=2,4,\ldots} \frac{K'_A(n)}{K_A}\left[\frac{K_A^2 K_B^2 S_A^2 H_A H_B}{(1+K_B H_B)^2(1+K_A S_A)(1+K_B S_B)}\right]^{n/2} \tag{24}$$

$$H_{A_t} = H_A + S_t - S_A \tag{25}$$

$$H_{B_t} = H_B + S_t - S_B \tag{26}$$

$$\frac{S_B}{S_A} = \frac{1+K_A H_A}{1+K_B H_B} \tag{27}$$

which describe the equilibrium concentrations of free epitopes and free antibody combining sites.

In this system, the interaction of type A epitopes with anti-A combining sites can be expected to be enhanced as B-type interactions lead to ring formation. Preliminary experimental results by G. Görog (not shown) have indeed demonstrated that addition to the system of a suitable amount of B results, under well defined conditions, in a detectable enhancement of the binding of A. Our simulations have shown that the level of binding of A epitopes is increased only in a limited range of B antigen concentrations, approximately centered around the concentration of antibody sites. The extent of this increase is directly regulated by the magnitude of the ring closure constants (Figure 6), as well as by the value of the $K_B S_t$ product (data not shown). The level of binding over the background is moreover highly sensitive to the $K_A S_t$ product (Figure 7). The system needs, therefore, a careful optimization in order to be able to express its potentialities as a sensitive immunoassay.

CONCLUSIONS

In the interaction of divalent antigen with divalent antibody, formation of cyclic complexes results in an increase of the overall binding efficiency which is modulated by the total concentration of antibody sites. Moreover, the shape of the binding curves in the Scatchard representation is highly sensitive to restrictions on the occurrence of the smaller-sized rings.

The binding curves exhibit a region with a downward concavity, most marked when formation of monogamous rings is highly restricted, as if positive cooperativity between antibody sites existed. Nevertheless, experimentalists should be able to recognize the apparent character of this cooperativity because of the presence of an upward concavity in the antigen excess region. In the Hill plot, the logarithmic scale allows a better visualization of these opposite pseudo-cooperativities, although an error in the estimate of the total concentration of antibody sites may then lead to an erroneous evaluation of the experimental results. This complex shape is due to the fact that, at low antibody sites saturation, there is a low relative abundance of linear chains susceptible of ring closure and containing a number of antibody molecules greater than unity, while, at high antibody sites saturation, all linear chains which can undergo cyclization tends to vanish.

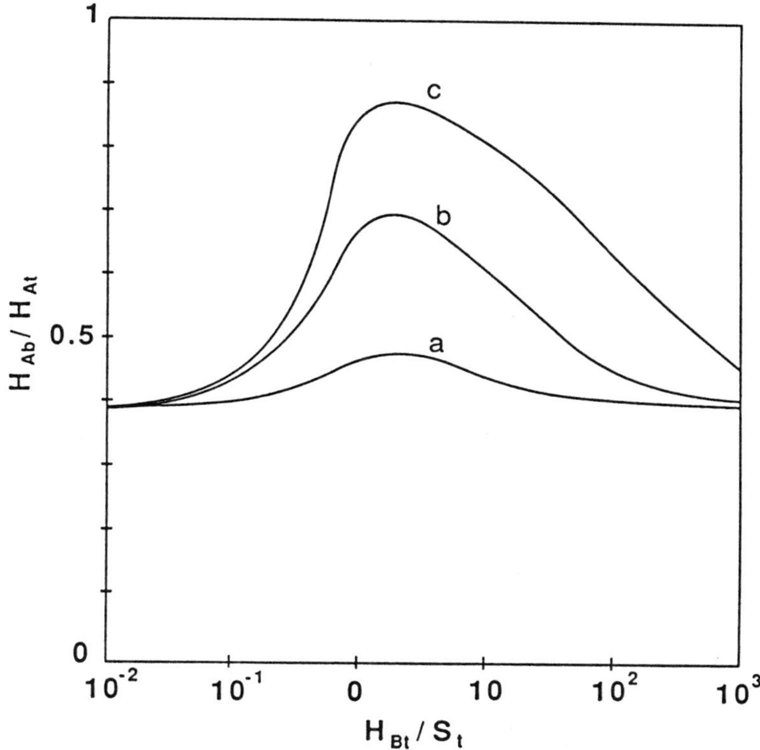

FIGURE 6 The fraction of bound A epitopes as a function of the ratio between B epitopes and antibody sites, for different values of the intrinsic ring closure constant. The dependence of ring closure on ring size is assumed according to Jacobson and Stockmayer.[19] Both bimolecular equilibrium constants K_A and K_B are equal to $1/S_t$, while the total concentration of the A epitopes is equal to S_t. The values of the ring closure constant $K'(2)$ are 10 (curve a), 10^2 (curve b), 10^3 (curve c).

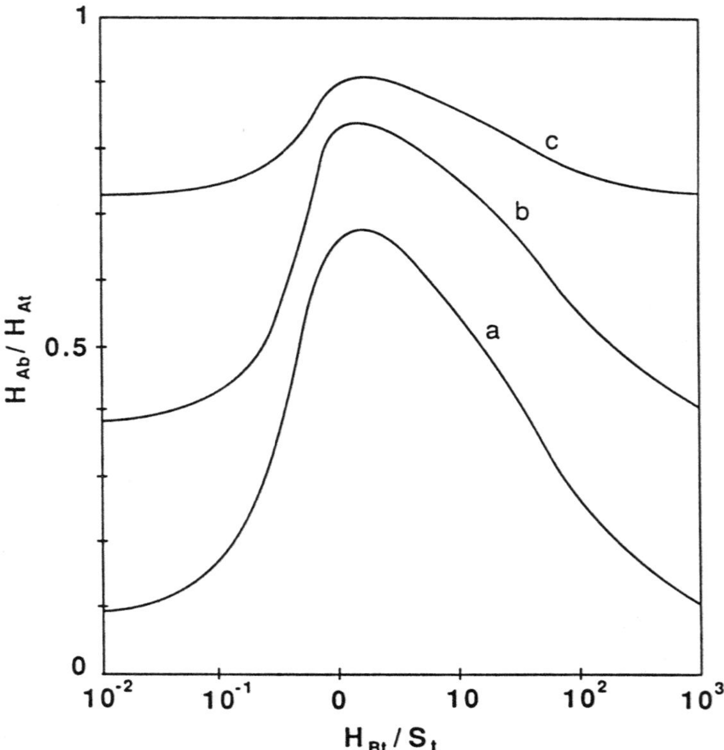

FIGURE 7 The fraction of bound A epitopes as a function of the ratio between B epitopes and antibody sites, for different values of the $K_A S_t$ product. The ring closure constant $K'(2)$ is equal to 10^3. The total concentration of A epitopes is equal to S_t. The bimolecular equilibrium constant K_B is equal to $1/S_t$, while K_A is equal to $0.1/S_t$ (curve a), $1/S_t$ (curve b), $10/S_t$ (curve c).

In more complex three-component systems, in which only rings containing an even number of antibody molecules are possible, the enhancement of binding due to ring closure can be utilized for the quantitation of one component. Since the interaction critically depends on the concentration of the various molecular species, maximal sensitivity requires however a careful optimization of the system parameters.

REFERENCES

1. Valentine, R. C., and N. M. Green (1967), "Electron Microscopy of an Antibody-Hapten Complex," *J. Molec. Biol.* **27**, 615–617.
2. Crothers, D. M., and H. Metzger (1972), "The Influence of Polyvalency on the Binding Properties of Antibodies," *Immunochemistry* **9**, 341–357.
3. Schumaker, V. N., G. Green and R. L. Wilder (1973), "A Theory of Bivalent Antibody-Bivalent Hapten Interactions," *Immunochemistry* **10**, 521–528.
4. DeLisi, C. (1976), *Antigen Antibody Interactions. Lecture Notes in Biomathematics* (Berlin: Springer-Verlag), Vol. 8.
5. Archer, B. G., and H. Krakauer (1977), "Thermodynamics of Antibody-Antigen Reactions. 2. The Binding of Bivalent Synthetic Random Coil Antigens to Antibodies having Different Antigen-Precipitating Properties," *Biochemistry* **16**, 618–627.
6. Dembo, M., and B. Goldstein (1978), "Theory of Equilibrium Binding of Symmetric Bivalent Haptens to the Cell Surface Antibody: Application to Histamine Release from Basophils," *J. Immunol.* **121**, 343–353.
7. Perelson, A. S., and C. DeLisi (1980), "Receptor Clustering on a Cell Surface. I. Theory of Receptor Cross-Linking by Ligands bearing Two Chemically Identical Functional Groups," *Math. Biosci.* **48**, 71–110.
8. Moyle, W. R., C. Lin, R. L. Corson and P. H. Ehrlich (1983), "Quantitative Explanation for Increased Affinity Shown by Mixtures of Monoclonal Antibodies: Importance of a Circular Complex," *Molec. Immunol.* **20**, 439–452.
9. Kuczek, T., and W. R. Moyle (1985), "Use of the Average Antibody-Antigen Bond Concept and Probability Theory to Simplify Modeling of Linear and Circular Antibody-Antigen Complex Formation," *J. Immunol. Meth.* **84**, 251–263.
10. Karush, F. (1976), 'Multivalent Binding and Functional Affinity," *Contemporary Topics in Molecular Immunology*, Eds. H.N. Eisen and R.A. Reisfeld (New York: Plenum Press), Vol. 5, pp. 217–228.
11. Karush, F., M. N. Chua and J. D. Rodwell (1979), "Interaction of a Bivalent Ligand with IgM Anti-Lactose Antibody," *Biochemistry* **18** , 2226–2232.
12. Holmes, N. J., and P. Parham (1983), "Enhancement of Monoclonal Antibodies against HLA-A2 is Due to Antibody Bivalency," *J. Biol. Chem.* **258**, 1580–1586.
13. Moyle, W. R., D. M. Anderson and P. H. Ehrlich (1983), "A Circular Antibody-Antigen Complex is Responsible for Increased Affinity Shown by Mixtures of Monoclonal Antibodies to Human Chorionic Gonadotropin," *J. Immunol.* **131**, 1900–1905.
14. Thompson, R. J., and A. P. Jackson (1984), "Cyclic Complexes and High-Avidity Antibodies," *Trends Biochem. Sci.* **9** , 1–3.
15. Wofsy, C., B. Goldstein and M. Dembo (1978), "Theory of Equilibrium Binding of Asymmetric Bivalent Haptens to Cell Surface Antibody: Application to Histamine Release from Basophils," *J. Immunol.* **121**, 593–601.

16. Gandolfi, A., and R. Strom (1987), "Ring Formation in Divalent Receptor–Divalent Ligand Interaction: Pseudo-Cooperative Shapes of Scatchard Plots," IASI-CNR Report R.184, submitted to *Math. Biosci.*.
17. Scatchard, G. (1949), "The Attraction of Proteins for Small Molecules and Ions," *Ann. N.Y. Acad. Sci.* **51**, 660–672.
18. Dahlquist, F. W. (1978), "The Meaning of Scatchard and Hill Plots," *Methods in Enzymology*, Eds. C. H. W. Hirs and S. N. Timasheff (New York: Academic Press), Vol. 48, pp.270–299.
19. Jacobson, H., and W. H. Stockmayer (1950), "Intramolecular Reactions in Polycondensations. I. The Theory of Linear Systems," *J. Chem. Phys.* **18**, 1600–1606.
20. Josè, M. V., and C. Larralde (1982), "Alternative Interpretation of Unusual Scatchard Plots: Contribution of Interactions and Heterogeneity," *Math. Biosci.* **58**, 159–170.
21. Wofsy, C. (1980), "Analysis of a Molecular Signal for Cell Function in Allergic Reactions," *Math. Biosci.* **49**, 69–86.
22. Perelson, A. S. (1980), "Receptor Clustering on a Cell Surface. II. Theory of Receptor Cross-Linking by Ligand Bearing Two Chemically Distinct Functional Groups," *Math. Biosci.* **49**, 87–110.
23. Lanzavecchia, A., and D. Scheidegger (1987), "The Use of Hybrid Hybridomas to Target Human Cytotoxic T Lymphocytes," *Eur. J. Immunol.* **17**, 105–109.

CHIN S. HSU
Associate Professor, Electrical and Computer Engineering Department, Washington State University, Pullman, WA 99164-2752

Estimation of Antibody Affinity via Reduced-Order Modeling

I. INTRODUCTION

It is well known that mathematical models of the immune system are very complicated. These immune models usually are time varying, nonlinear, non-robust and possibly high dimensional with time delays. Generally speaking, immune models involve feedback structures which interweave linear, bilinear, and nonlinear subsystems (Mohler et al., 1980; Hsu, 1978; Bell et al., 1978). Of particular concern in understanding immune system dynamics via mathematical models is the high dimensionality which renders mathematical analysis intractable. One approach to circumvent dimensionality problems is to generate reduced-order models (ROM) from high-order models (HOM), retaining as many attributes of the HOM as possible. Over the past two decades, various techniques for model order reduction have been proposed. Many of these reduction techniques are only applicable to linear systems (Jamshidi, 1983), though some results for reducing a special class of bilinear systems are now available (Hsu et al., 1985).

The objective of this paper is to introduce the notion of reduced-order modeling and to demonstrate its applicability in the estimation of antibody affinity.

It is known that antibodies produced during the immune response are heterogeneous with respect to their affinity toward the antigen. There are various methods for determining the affinity distribution from the experimental binding data. These methods include Fourier transform (Bowman and Aladjem, 1963), delta function (Erwin and Aladjem, 1976), Sips distribution (Kim et al., 1974), and Stieltjes transform (Bruni et al., 1983). A new method which is based upon the notion of reduced-order modeling will be introduced in the sequel. The organization of this paper is as follows: in Section II, we reformulate the problem of determining the affinity distribution into an equivalent problem of identifying a parametric model of a linear system. A numerical procedure to carry out the model identification (or realization) is presented along with computer-simulation results in Section III. The concept of "system inverse" is introduced in Section IV. Finally, in Section V, we discuss some possible topics for further immunology-theoretic research in the area of reduced-order modeling.

II. PROBLEM FORMULATION

The problem of determining the antibody affinity distribution from the binding data can be viewed as a problem of solving the following integral equation (Bruni, 1983)

$$R(H) = \int_0^\infty \frac{kP(k)}{k + \frac{1}{H}} dk \tag{1}$$

where $R(H)$ denotes the ratio between bound and total antibody sites for a given value of the free hapten concentration H in the titration assay, and $P(k)$ is the probability density function of the antibody affinity k. Depending upon the numerical method used, the affinity density may be unimodal or bimodal. Instead of using the variables defined in Eq. (1), a change of variable $y = 1/H$ yields

$$R(\frac{1}{y}) = \int_0^\infty \frac{kP(k)}{k + y} dk \tag{2}$$

which is, indeed, the Stieltjes transform. Based on Eq. (2), a comprehensive theory and numerical procedures were developed (Bruni et al., 1976; Bruni et al., 1983). To obtain more insightful and more efficient numerical procedures of affinity determination, we reformulate the problem in a system-theoretic setting. Let us make another change of variable in Eq. (2), i.e.,

$$y = e^{-t}, \quad k = e^{-\tau} \tag{3}$$

then Eq. (2) becomes

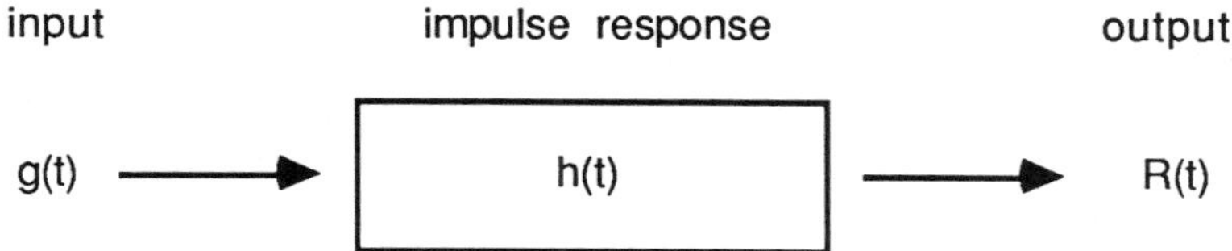

FIGURE 1 A linear model representation.

$$R(t) = \int_{-\infty}^{\infty} \frac{g(\tau)}{1 + e^{\tau - t}} d\tau = h(t) \star g(t) \tag{4}$$

where

$$g(t) = e^{-t} P(e^{-t})$$
$$h(t) = \frac{1}{1 + e^{-t}}$$

and "$\star$" stands for the convolution operator.

By examining Eq. (4), we can consider $h(t)$ as the impulse response function of a linear system with $g(t)$ and $R(t)$ as its input and output signals respectively. Figure 1 illustrates a diagrammatical relation among $h(t)$, $g(t)$ and $R(t)$.

The impulse response $h(t)$ specifies a linear system which has the following state-space representation

$$\begin{aligned} \dot{\underline{x}}(t) &= A\underline{x}(t) + \underline{b}g(t) \\ R(t) &= \underline{c}\,\underline{x}(t) \end{aligned} \tag{5}$$

where the state vector $\underline{x}(t)$ is an $n \times 1$ vector, the input vector $\underline{b}$ is $n \times 1$, the output vector $\underline{c}$ is $1 \times n$, and the system matrix A is $n \times n$. The dimension n is to be determined. From linear system theory (Chen, 1984), it is well known that any impulse response $h(t)$ can be expressed in terms of A, $\underline{b}$, and $\underline{c}$, namely,

$$h(t) = \underline{c}e^{At}\underline{b} \tag{6}$$

where the state-transion matrix e^{At} denotes the exponential of the matrix At.

Our objective is to determine $g(t)$ which, in turn, specifies $P(t)$, from the binding data $g(t)$. This can be accomplished by the following approach. Since $h(t)$ is known, we first obtain the system parameters A, $\underline{b}$, and $\underline{c}$ from $h(t)$. This step is called "system identification" or "state-space realization" in the system theory literature. Once we have A, $\underline{b}$, and $\underline{c}$, we can compute $g(t)$ from $R(t)$ via the system inverse to be discussed in Section IV. Now we are in a position to point out the essence of reduced-order modeling (ROM). The impulse response $h(t) = 1 - e^{-t} + e^{-2t} \ldots$

by itself is actually infinite-dimensional. However, as will be shown in the next section, $h(t)$ can be "approximated" by a finite-dimensional time-invariant system with $n < \infty$, where n will be the chosen dimension of a ROM. A numerical procedure to obtain A, $\underline{b}$, and $\underline{c}$ from $h(t)$ will be presented in the next section along with some computer simulation results. Here, we would like to mention that it is not coincidental that $h(t)$ (from the Stieltjes transform) happens to be in a very special form, such that it can be accurately approximated. There are other useful transforms, such as Abel transform and Fast Hankel transform, which can deal with the same procedure (Hansen, 1985; Hansen and Law, 1985). It should be noted here that the present work follows Dr. Hansen's research, though the numerical procedure of obtaining A, $\underline{b}$, and $\underline{c}$ is quite different. It should also be mentioned that, to this author's knowledge, the present work is the first attempt to bring the notion of reduced-order modeling in the context of control theory to modeling issues in theoretical immunology.

III. A REALIZATION PROCEDURE

In this section, we focus on the problem of finding A, $\underline{b}$ and $\underline{c}$ from the known impulse response $h(t)$. A numerical procedure will be developed along with computational results. Motivated by the recent availability of high-quality software, we chose the following numerical approach. A set of sampled data (time series) $h(kT)$ is first obtained by sampling the impulse response $h(t)$, where T is the sampling period. Recalling that $h(t) = \underline{c}e^{At}\underline{b}$, we have

$$h(kT) = \underline{c}e^{Akt}\underline{b} = \underline{c}(e^{AT})^k\underline{b} \tag{7}$$

or

$$h(kT) = \underline{c}A_d^k\underline{b} \tag{8}$$

where

$$A_d = e^{AT}$$

Suppose that we have 2p data points, then a $p+1 \times p$ Hankel data matrix H can be formulated as

$$H = \begin{pmatrix} h(1) & h(2) & \dots & h(p) \\ h(2) & h(3) & \dots & h(p+1) \\ h(3) & h(4) & \dots & h(p+2) \\ \vdots & \vdots & \ddots & \vdots \\ h(p+1) & h(p+2) & \dots & h(2p) \end{pmatrix}_{p+1,p} \tag{9}$$

Using Eq. (8), H can be rewritten as

$$H = \begin{pmatrix} \underline{c}A_d\underline{b} & \underline{c}A_d^2\underline{b} & \cdots \\ \underline{c}A_d^2\underline{b} & \underline{c}A_d^3\underline{b} & \cdots \\ \vdots & \vdots\ddots & \vdots \\ \cdots & \cdots & \underline{c}A_d^{2p}\underline{b} \end{pmatrix} \tag{10}$$

which, in turn, can be factored as

$$H = \begin{pmatrix} \underline{c} \\ \underline{c}A_d \\ \underline{c}A_d^2 \\ \vdots \\ \underline{c}A_d^p \end{pmatrix}_{p+1,n} [A_d\underline{b}, A_d^2\underline{b}, \ldots, A_d^p\underline{b}]_{n,p} \equiv OC, \tag{11}$$

where $n = \text{rank}\,(H)$ and the dimensions of the matrices O and C are indicated.

The numerical rank of a Hankel data matrix H is determined by its singular values (square roots of eigenvalues of HH'). By examining singular values of H, we are able to choose an integer n to be the dimension of a reduced-order model. In other words, n is the number of state variables which are "adequate" in describing the linear system specified by $h(t)$. Since the matrix H is given, factorization of H into a product of two matrices is always possible via singular value decomposition (Golub and Van Loan, 1983). After O and C are generated from the Hankel data matrix, the matrices A, $\underline{b}$ and $\underline{c}$ can be obtained as follows (Vaccaro, 1986):

$$\underline{c} = \text{1st row of O} \tag{12}$$

$$A_d\underline{b} = \text{1st column of C} \tag{13}$$

Define

$$O_1 = O \text{ without the last row}$$

$$O_2 = O \text{ without the first row}$$

then

$$O_2 = O_1 \bullet A_d \tag{14}$$

Solving the above equation yields

$$A_d = (O_1'\,O_1)^{-1}O_1'\,O_2 \tag{15}$$

It follows from Eq. (13) that

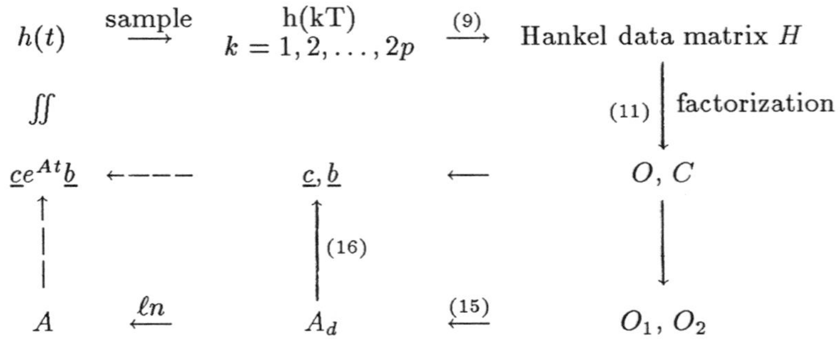

$$
\begin{array}{ccccc}
h(t) & \xrightarrow{\text{sample}} & \begin{array}{c}\text{h(kT)}\\ k=1,2,\ldots,2p\end{array} & \xrightarrow{(9)} & \text{Hankel data matrix } H \\
\iint & & & & \Big\downarrow \text{(11) factorization} \\
\underline{c}e^{At}\underline{b} & \longleftarrow - - - & \underline{c}, \underline{b} & \longleftarrow & O, C \\
\Big\uparrow & & \Big\uparrow (16) & & \Big\downarrow \\
A & \xleftarrow{\ell n} & A_d & \xleftarrow{(15)} & O_1, O_2
\end{array}
$$

FIGURE 2 A Realization of $h(t)$.

$$\underline{b} = A_d^{-1}\,(\text{1st column of C}) \tag{16}$$

Finally, we recall the relationship $A_d = e^{AT}$ and obtain A from A_d by

$$A = \frac{1}{T}\ell n(A_d) \tag{17}$$

where ℓn denotes the natural log of a matrix.

In Figure 2, we summarize the proposed numerical procedure of computing A, $\underline{b}$ and $\underline{c}$ from the given $h(t)$.

Some remarks are in order. The numerical procedure may seem to be sophisticated. On the contrary, it is not. By taking advantage of numerical software, such as MATLAB, it is straightforward to carry out the realization algorithm on a personal computer (IBM-PC preferred). There are advantages of using the above realization approach over other methods. First, the proposed method can be extended to process unequally spaced (nonuniform sampling) experimental data. Second, if the experimental data is noise contaminated, the well-known Kalman filtering techniques can be called upon to estimate the true signal (Hansen and Law, 1985). Prior to the presentation of a numerical example, we mention that Eq. (5) can also be rewritten in terms of the variable y instead of t,

$$
\begin{aligned}
\frac{dx(y)}{dy} &= -\frac{1}{y}A\underline{x}(y) + \underline{b}yP(y) \\
R\left(\frac{1}{y}\right) &= \underline{c}\,\underline{x}(y)
\end{aligned}
\tag{18}
$$

Reduced-order models of dimension 5 and 6 obtained from the realization approach depicted in Figure 2 are presented below. For the sake of saving space, other numerical data are omitted here.

$n = 5$

$$A = \begin{pmatrix} 0.0989 & -0.2550 & -0.0127 & -0.0158 & -0.0018 \\ 0.2841 & -0.4869 & -0.6131 & -0.0840 & -0.0393 \\ -0.0130 & 0.6691 & -1.1251 & -0.9567 & -0.1646 \\ 0.0180 & -0.0859 & 1.0396 & -1.7800 & -1.3701 \\ -0.0020 & 0.0435 & -0.1712 & 1.4706 & -2.7554 \end{pmatrix}$$

$$\underline{b} = \begin{pmatrix} 0.8263 \\ -0.3801 \\ 0.0783 \\ -0.0110 \\ 0.0011 \end{pmatrix}, \quad \underline{c} = (0.7688 \quad 0.3716 \quad 0.0784 \quad 0.0114 \quad 0.0012)$$

$n = 6$

$$A = \begin{pmatrix} 0.0989 & -0.2550 & -0.0127 & -0.0159 & -0.0016 & -0.0012 \\ 0.2841 & -0.4870 & -0.6131 & -0.0847 & -0.0364 & -0.0091 \\ -0.0130 & 0.6689 & -1.1243 & -0.9601 & -0.1505 & -0.0661 \\ 0.0181 & -0.0866 & 1.0435 & -1.7954 & -1.2956 & -0.2551 \\ -0.0018 & 0.0410 & -0.1567 & 1.4120 & -2.4902 & -1.7508 \\ 0.0014 & -0.0102 & 0.0745 & -0.2719 & 1.8957 & -3.6767 \end{pmatrix}$$

$$\underline{b} = \begin{pmatrix} 0.8263 \\ -0.3801 \\ 0.0783 \\ -0.0110 \\ 0.0011 \\ -0.0001 \end{pmatrix}, \quad \underline{c} = (0.7688 \quad 0.3716 \quad 0.0784 \quad 0.0114 \quad 0.0012 \quad 0.0001)$$

Figure 3 illustrates the approximation of $h(t)$ by its reduced-order models $n = 2$, 4, 5, 6, and 7. It is seen that reduced-order models of order $n = 6$ and $n = 7$ closely approximate the given $h(t)$.

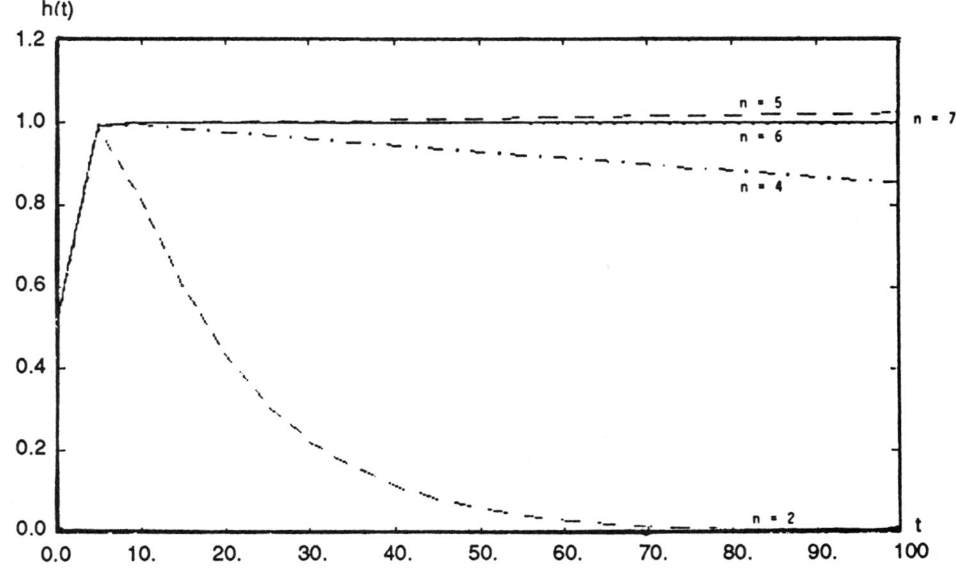

FIGURE 3 Computer Simulation Results.

IV. SYSTEM INVERSE

In the previous section, we have established a procedure to compute A, $\underline{b}$ and $\underline{c}$ from $h(t)$. This enables us to compute $R(t)$ from $g(t)$ via Eq. (5). Actually, our ultimate objective is to do the reverse. That is, we aim at determining $g(t)$,

which is related to the affinity density $P(k)$, provided that $R(t)$, the binding data, is available. In order to do so, we call upon the notion of "system inverse" (Brockett, 1965) which is a linear system reversing the input and output variables of a given linear system. To construct the system inverse of Eq. (5), the following result due to Brockett can be used.

THEOREM 4.1. Consider a single-inpu, single-output, completely controllable, linear system,

$$\begin{aligned} \dot{\underline{x}}(t) &= A\underline{x}(t) + \underline{b}u(t); \quad \underline{x}(0) = \underline{x}_0,\ \underline{x}(t) \in R^n \\ y(t) &= \underline{c}\,\underline{x}(t) \end{aligned} \tag{19}$$

If the relative order α of Eq. (19) is finite, i.e., $\alpha < \infty$, then there is a unique inverse system which is also completely controllable and is defined by,

$$\begin{aligned} \dot{\hat{\underline{x}}}(t) &= \left[A - \left(\frac{\underline{b}\underline{c}A^{\alpha}}{\underline{c}A^{\alpha-1}\underline{b}}\right)\right]\hat{\underline{x}}(t) + \left(\frac{1}{\underline{c}A^{\alpha-1}\underline{b}}\right)\underline{b}\hat{u}(t), \quad \underline{x}(0) = \underline{x}_0 \\ \hat{y}(t) &= -\left(\underline{c}A^{\alpha-1}\underline{b}\right)\hat{\underline{x}}(t) + \left(\frac{1}{\underline{c}A^{\alpha-1}\underline{b}}\right)\hat{u}(t) \end{aligned} \tag{20}$$

Let $\hat{u}(t) = y^{(\alpha)}(t)$, then $\hat{y}(t) = u(t)$.

The relative order α used above is referred to as the difference between degrees of numerator and denominator polynomials of the transfer function of Eq. (19), i.e., the Laplace transform of the impulse response $h(t)$. A system inverse takes the derivatives of the output, $y^{(\alpha)}(t)$, as its input and reproduces the input of the original system. Depending upon the relative order α of a given system, $y(t)$ has to be differentiated α times. This may not be numerically acceptable, though various numerical approximations have been devised to avoid differentiations. Employing the terminology from linear system theory, the relative order α is, indeed, the difference of the numbers of poles (eigenvalues of A) and zeros. The reduced-order models presented in the previous section have $\alpha = 1$, due to the following calculations of poles and zeros:

$$\begin{aligned} &n = 5 \\ &\qquad \text{poles}: 0.0003, -1.1473 \pm j0.1604, -1.8771 \pm j1.4615 \\ &\qquad \text{zeros}: -1.4191 \pm j0.9547, -1.8552 \pm j1.3200 \\ &n = 6 \\ &\qquad \text{poles}: 0, -0.9822, -1.7649 \pm j0.5547, -2.4814 \pm j1.9387 \\ &\qquad \text{zeros}: -1.3288 \pm j1..0569, -2.2944, -2.5113 \pm j1.9428 \end{aligned}$$

Hence, in order to use the system inverse to produce $g(t)$ given $R(t)$, $R(t)$ needs to be differentiated only once.

V. FURTHER RESEARCH

Some preliminary results have been presented to highlight the applicability of reduced-order modeling concepts known in linear system theory to the determination of antibody affinity. Undoubtedly, there are many other system concepts which can be exploited to benefit the study of various topics in theoretical immunology. The complexity of the immune dynamics will never vanish. To delineate complexity issues, techniques of reduced-order modeling are believed to be an effective tool.

We conclude this paper by suggesting some topics for further research:

a. The modeling of immune systems or subsystems requires data processing of experimental data contaminated with noise. Well-known theories, such as digital signal processing and Kalman filtering, can be effective tools in processing immunological data or in estimating parameters of immune models.

b. Due to the fact that an immune system contains bilinear and nonlinear components (subsystems), the techniques of reduced-order modeling as described in this paper need to be extended beyond the class of linear time-invariant models.

c. The overall immune system is a highly structured, nonlinear, multi-level feedback-control system. The theory of decentralized control for interconnected large-scale systems (Jamshidi, 1983) can be fully used to identify the role each feedback connection plays, both in the healthy and pathological environments.

REFERENCES

1. Bell, G. I., A. S. Perelson, and G. H. Pimbley, Jr. (1978), *Theoretical Immunology* (New York: Marcel Dekker).
2. Bowan, J. D., and F. Aladjem (1963), "A Method for the Determination of Heterogeneity of Antibodies," *J. Theoret. Biol.* **4**, 242–253.
3. Brockett, R. W. (1965), "Poles, Zeros, and Feedback: State Space Interpretation," *IEEE Trans. Automatic Control* **AC-10**, 129–135.
4. Bruni, C., A. Germani, G. Koch, and R. Strom (1976), "Derivation of Antibody Distribution from Experimental Binding Data," *J. Theoret. Biol.* **61**, 143–170.
5. Bruni, C., A. Gandolfi, and A. Germani (1983), "Theoretical Investigation and Numerical Procedure for the Estimation of Antibody Affinity Distribution from Binding Data," *Math. Modeling in Immunology and Medicine*, Eds. G. I. Marchuk and L. N. Belykh (New York: North Holland Publishing Company, IFIP).
6. Chen, C. T. (1984), *Linear System Theory and Design* (New York: Holt, Rinehart and Winston).
7. Erwin, P. M., and F. Aladjem (1976), "The Heterogeneity of Antibodies with Respect to Equilibrium Constants, Calculation of a New Method using Delta Functions, and Analysis of the Results," *Immunochemistry* **13**, 873–883.
8. Golub, G. H., and C. F. Van Loan (1983), *Matrix Computations* (Baltimore: The Johns Hopkins Press).
9. Hansen, E. W. (1985), "Fast Hankel Transform Algorithm," *IEEE Trans. Acoustics, Speech, and Signal Processing* **ASSP-33, No. 3**, 666–671.
10. Hansen, E. W., and P. L. Law (1985), "Recursive Methods for Computing the Abel Transform and its Inverse," *J. Opt. Soci. Am. A.* **Vol. 2, No. 4**, 510–520.
11. Hsu, C. S. (1978), *Bilinear Control Processes with Applications to Immunology, Unpublished Ph.D. Dissertation, Oregon State University, September 1978.*
12. Hsu, C. S., U. B. Desai and C. A. Crawley (1985), "Realization and Approximation of Discrete Bilinear Systems," *Applied Digital Control*, Ed. S. G. Tzafestas (New York: North Holland Publishing Company).
13. Jamshidi, M. (1983), *Large-Scale Systems: Modeling and Control* (New York: North Holland Publishing Company).
14. Kim, Y. T., T. P. Weblin, and G. W. Siskind (1974), "Distribution of Antibody-Binding Affinity, III. Detection of Low Affinity Antibody in the Presence of High Affinity Antibody," *J. Immunol.* **112**, 2002–2012.
15. Mohler, R. R., C. Bruni, and A. Gandolfi (1980), "A Systems Approach to Immunology," *IEEE Proceedings* **Vol. 68, No. 8**, 964–990.
16. Vaccaro, R. J. (1986), "Finite-Data Algorithms for Approximate Stochastic Realization," *Modelling and Application of Stochastic Processes*, Ed. U. B. Desai (Boston: Kluwer Academic Publishers), 105–122.

ANDREW WOHLGEMUTH
Department of Mathematics, University of Maine, Orono, Maine 04469

Symbolic Interpretation of Data and the Definition of Factors in Immunogenetics

ABSTRACT

Contemporary immunogenetics uses a notation that does not allow for the complexity found in biological systems. All attempts at a more general approach to interpreting immunogenetic data have been shown, in one respect, to be mathematically equivalent. The new, more general symbolism admits more interpretations of data than can be handled—even with a computer. The intractable computational questions that have been asked historically are replaced by new questions that are more meaningful biologically and more tractable computationally.

As an application, data on HLA is interpreted in terms of the new symbolism. The results suggest a genetic model that differs from the currently accepted model. The new model is shown to be more economical in its interpretation of statistical and family (recombination) data.

THEORY

There have been several independently developed approaches to the symbolic interpretation of immunogenetic data. These symbolic approaches or models all attempt to interpret data in terms of abstract entities we shall call labels. The labels of interest may be genes, antigens, specificities, factors, epitopes, etc. On an abstract level, there is a common ground on which all these models are equivalent in spite of very different ideas used as starting points. This is a strong argument for the correctness of "the" model. The symbolic approach in general use in immunogenetics is, however, much more restricted than this model. When our more general symbolism is used to reinterpret certain data, alternate genetic hypotheses are suggested.

Cotterman (1953) defined a (one-locus) *phenotype system* as a partition (into phenotypes) of the set of all subsets of size one or two of a set of labels—which here represent alleles. He defined *phenotype systems* (Cotterman, 1969) which form one manifestation of our general model. Using the notation of Markowsky (1983), a *factor-union representation* of a phenotype system F consists of a set X of factors and an assignment f of subsets of X to the alleles (labels) of F with the following two properties:

a. If ab and cd are two genotypes in the same phenotype, then $f(a) \cup f(b) = f(c) \cup f(d)$.

b. If P and Q are distinct phenotypes, then $f(P) \neq f(Q)$.

The elements of X can be any properties whatever. The word "factor" is used in deference to blood group workers.

Other forms of the model are all concerned with interpreting cell/antibody reaction data in the form of a zero/one matrix. We will relate factor-union systems to these models after examples of the various forms of the model are given.

LABELED REACTION MATRICES

(Wohlgemuth 1976, 1978, 1979a)

In the following (hypothetical) matrix R, rows represent cells being tested, columns represent reagents containing antibodies, and a *one* appears in a given row and column if the corresponding cell has an antigen recognized by some antibody in the reagent corresponding to the column.

$$\begin{array}{cc} & \text{Reagents} \\ & R:\ \ 1\ \ 2\ \ 3\ \ 4 \\ \text{cells} & \begin{array}{c} 1 \\ 2 \\ 3 \end{array} \begin{pmatrix} 1 & 0 & 0 & 0 \\ 1 & 1 & 1 & 0 \\ 1 & 1 & 1 & 1 \end{pmatrix} \end{array} \tag{1}$$

A hypothetical reaction matrix

A labeled reaction matrix is an interpretation of R of the following form. Rows are each labeled with Latin letters representing antigens. Columns are each labeled with Greek letters representing antibodies. A *one* appears in a given row and column if some labeling antigen and antibody react. This ability to react is denoted by adjacency (connected by an edge) in a bipartite graph. The following gives one interpretation of R as a labeled reaction matrix.

$$\begin{array}{c} R: \\ a \\ ac \\ b \end{array} \begin{array}{c} \begin{array}{cccc} \alpha\beta & \beta\gamma & \gamma & \beta \end{array} \\ \begin{pmatrix} 1 & 0 & 0 & 0 \\ 1 & 1 & 1 & 0 \\ 1 & 1 & 1 & 1 \end{pmatrix} \end{array} \qquad \begin{array}{ccc} \alpha & \beta & \gamma \\ a & b & c \end{array} \tag{2}$$

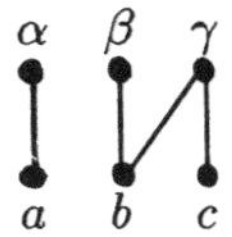

A labeled reaction matrix.

The fact that γ reacts with both b and c is called "crossreactivity." Crossreactivity is a well-known phenomenon in immunology and depends on the properties of the antibody V region (see Hoffmann, 1987, with references). The nomenclature and symbolism in current use in immunogenetics, however, do not allow for crossreactivity. This means that the models presented here are capable of "uncovering" real-world relationships hidden by the current nomenclature.

The interpretation of R above can be given completely in matrix notation.

$$\underset{\text{matrix})}{R\text{ (reaction}} \qquad \underset{\text{matrix})}{A\text{ (labeling}} \qquad \underset{\text{matrix})}{B\text{ (definition}} \qquad \underset{\text{matrix})}{C\text{ (labeling}}$$

$$\begin{array}{c} \\ 1 \\ 2 \\ 3 \end{array}\begin{array}{c} \begin{array}{cccc} 1 & 2 & 3 & 4 \end{array} \\ \begin{pmatrix} 1 & 0 & 0 & 0 \\ 1 & 1 & 1 & 0 \\ 1 & 1 & 1 & 1 \end{pmatrix} \end{array} = \begin{array}{c} \\ 1 \\ 2 \\ 3 \end{array}\begin{array}{c} \begin{array}{ccc} a & b & c \end{array} \\ \begin{pmatrix} 1 & 0 & 0 \\ 1 & 0 & 1 \\ 0 & 1 & 0 \end{pmatrix} \end{array} \times \begin{array}{c} \\ a \\ b \\ c \end{array}\begin{array}{c} \begin{array}{ccc} \alpha & \beta & \gamma \end{array} \\ \begin{pmatrix} 1 & 0 & 0 \\ 0 & 1 & 1 \\ 0 & 0 & 1 \end{pmatrix} \end{array} \times \begin{array}{c} \\ \alpha \\ \beta \\ \gamma \end{array}\begin{array}{c} \begin{array}{cccc} 1 & 2 & 3 & 4 \end{array} \\ \begin{pmatrix} 1 & 0 & 0 & 0 \\ 1 & 1 & 0 & 1 \\ 0 & 1 & 1 & 0 \end{pmatrix} \end{array} \tag{3}$$

The matrix form of a labeled reaction matrix.

Thus, $R = A \times B \times C$ is the Boolean matrix production of A, B and C. A gives the labels for the rows of R and C gives the labels for the columns of R. B gives the same information as the bipartite graph and defines antigens by specifying the antibodies with which they reaction and *vice versa*. The product is associative so that factors can be grouped together as desired. For example, $B \times C$ is a (labeling) matrix that labels the columns (reagents) of R with antigens. A given antigen labels a column in $B \times C$ if and only if that antigen recognizes (by B) some antibody in the reagent (given by C).

$$D = B \times C: \quad \begin{array}{c} \\ a \\ b \\ c \end{array} \begin{array}{c} \begin{array}{cccc} 1 & 2 & 3 & 4 \end{array} \\ \begin{pmatrix} 1 & 0 & 0 & 0 \\ 1 & 1 & 1 & 1 \\ 0 & 1 & 1 & 0 \end{pmatrix} \end{array} \tag{4}$$

The matrix assigning antigens to reagents.

Thus, $R = A \times (B \times C)$.

$$R: \begin{array}{c} \\ 1 \\ 2 \\ 3 \end{array} \begin{array}{c} \begin{array}{cccc} 1 & 2 & 3 & 4 \end{array} \\ \begin{pmatrix} 1 & 0 & 0 & 0 \\ 1 & 1 & 1 & 0 \\ 1 & 1 & 1 & 1 \end{pmatrix} \end{array} = A: \begin{array}{c} \\ 1 \\ 2 \\ 3 \end{array} \begin{array}{c} \begin{array}{ccc} a & b & c \end{array} \\ \begin{pmatrix} 1 & 0 & 0 \\ 1 & 0 & 1 \\ 0 & 1 & 0 \end{pmatrix} \end{array} \times D: \begin{array}{c} \\ a \\ b \\ c \end{array} \begin{array}{c} \begin{array}{cccc} 1 & 2 & 3 & 4 \end{array} \\ \begin{pmatrix} 1 & 0 & 0 & 0 \\ 1 & 1 & 1 & 1 \\ 0 & 1 & 1 & 0 \end{pmatrix} \end{array} \tag{5}$$

An interpretation of R by one set of labels.

This factorization of R into two factors A and $B \times C$ interprets R in terms of a single set of labels. The labels may represent genes, alleles, antigens, specificities, antibodies, etc. Unfortunately, the word "factor" is also applied to the labels, as in "blood group factors" or "factor analysis." The phrase "set of factors" is always used in the sense of "set of labels."

COMPLEX-COMPLEX CODE

(Hirschfeld 1965, 1972, 1975)

Another interpretation of zero/one matrices was given by Hirschfeld with his *codes*. According to the *simple-complex* code, each column of R is labeled with a single letter. Rows are labeled with all letters necessary to account for the reaction pattern seen. According to the *complex-complex* code, both rows and columns may have multiple labels. The simple-complex labeling of R and one example of a complex-complex labeling are given by:

Simple-Complex

	anti–a	anti–b	anti–c	anti–d
a	1	0	0	0
abc	1	1	1	0
abcd	1	1	1	1

Complex-Complex

	anti–ab	anti–bc	anti–bc	anti–b
a	1	0	0	0
ac	1	1	1	0
b	1	1	1	1

(6)

Hirschfeld's codes.

In both codes, there is a *one* in a given row and column of R if and only if the row and column share a letter label. Note that a complex-complex interpretation of R is equivalent to a factorization of R into two factors (Wohlgemuth, 1978). The complex-complex interpretation shown is equivalent to factorization $R = A \times D$ given above.

Contemporary immunogenetics uses a notation equivalent to the simple–complex code. That is, reagents used in practice are, in effect, labeled with a single identifier—supposedly the identifier of the "single thing" recognized by the reagent. This notation oversimplifies things immunologically. The difficulty with using the complex-complex code, on the other hand, is that there are an enormous number of possible interpretations of a matrix. For example, Denniston (1976) gives fifteen distinct interpretations of the matrix:

$$\begin{pmatrix} 1 & 0 \\ 1 & 1 \end{pmatrix}$$

By considering a solid block of ones and the definition of specificity cover below, it is clear that in general the number of possible interpretations grows exponentially as a function of the size of the reaction matrix.

SPECIFICITY COVER

(Nau, Markowsky, Woodbury and Amos, 1978)

A third method of interpreting a zero/one matrix was given by Nau et al. in terms of a *specificity cover.* This was shown in Wohlgemuth (1979b) to be again equivalent to a two-fold matrix factorization. The specificity cover equivalent to the factorization $R = A \times D$ is given by:

$$\begin{array}{c} \\ 1 \\ 2 \\ 3 \end{array} \begin{array}{c} \begin{array}{cccc} 1 & 2 & 3 & 4 \end{array} \\ \begin{pmatrix} \boxed{1} & 0 & 0 & 0 \\ \boxed{1a} & \boxed{1 \quad 1c} & & 0 \\ \boxed{1} & \boxed{1 \quad 1 \quad 1b} & & \end{pmatrix} \end{array} \tag{7}$$

A specificity cover interpretation of R.

A specificity cover is an assignment of (not necessarily contiguous) blocks of solid ones that completely cover the ones in the matrix. A block or "specificity" is a Cartesian product of a subset of rows with a subset of columns. In the matrix above, block a (specificity a) is the product of rows 1 and 2 with column 1; block b, row 1 with all columns; and block c, row 2 with columns 2 and 3.

Thus, all approaches to interpreting a zero/one reaction matrix (in terms of a single set of labels) which are more general than the simple-complex code are equivalent to a Boolean factorization of the given matrix. Denniston's count of 15 interpretations for the given 2x2 matrix indicates that for any matrix representing data for a real immunogenetic system, there are an enormous number for factorizations. Nau (1976) showed that for a given matrix and a given natural number k, the question of whether the matrix has a specificity cover with k specificities is NP-complete, roughly, computationally intractable.

We are, therefore, faced with the following problem: the customary notation for doing serology (equivalent to the simple-complex code) does not adequately reflect the complexity of the real world and is, therefore, inappropriate. The immunological oversimplification may lead to an artifactual genetic overcomplication (preserving the actual complexity in nature). In Wohlgemuth (1978), it was shown that *linkage disequilibrium* and the cis-trans effect can be such artifacts. On the other hand, the only alternative notation (equivalent to the complex-complex code) admits more interpretations than can be handled.

Mathematical (Wohlgemuth, 1978) and computer approaches (Nau and Woodbury, 1977; Wohlgemuth, 1987) based on heuristic principles have produced interesting symbolic interpretations from real-world data. For example, the labeling of the Ag blood group system given by Wohlgemuth (1978) is shown in Table 1.

Note that the new labels are indicative of a six-allele, one-locus system. In fact, the assignment of the new labels to the rows of the matrix forms a complete

TABLE 1 Standard and New Labels of the Common Ag Phenotypes

Standard Labels	Row No.	Reaction Matrix	New Labels
xy$a_1$dcgth	1	1 1 1 1 1 1 1 0 1	DF
xy$a_1$dcgt	2	1 1 1 1 1 1 1 0 0	AD
xy$a_1$dgt	3	1 1 1 1 0 1 1 0 0	BD
xy$a_1$gtz	4	1 1 1 0 0 1 1 1 0	CD
xy$a_1$gt	5	1 1 1 0 0 1 1 0 0	DE
x$a_1$gt	6	1 0 1 0 0 1 1 0 0	D
y$a_1$dcgtzh	7	0 1 1 1 1 1 1 1 1	CF
y$a_1$dcgtz	8	0 1 1 1 1 1 1 1 0	AC
y$a_1$dcgth	9	0 1 1 1 1 1 1 0 1	EF
y$a_1$dcgt	10	0 1 1 1 1 1 1 0 0	AE
y$a_1$dgtz	11	0 1 1 1 0 1 1 1 0	BC
y$a_1$dgt	12	0 1 1 1 0 1 1 0 0	BE
y$a_1$gtz	13	0 1 1 0 0 1 1 1 0	CE
y$a_1$gz	14	0 1 1 0 0 1 0 1 0	C
ydcgth	15	0 1 0 1 1 1 1 0 1	BF
ydcgt	16	0 1 0 1 1 1 1 0 0	AB
ydcth	17	0 1 0 1 1 0 1 0 1	F or AF
gdct	18	0 1 0 1 1 0 1 0 0	A
ydgt	19	0 1 0 1 0 1 1 0 0	B
y$a_1$gt	20	0 1 1 0 0 1 1 0 0	E

factor-union system. The rows of new labels form the phenotype categories in the phenotype system. Row 17 has two genotypes, A and AF. Note that the lack of distinction between A and AF can be interpreted genetically (dominance) or immunologically (crossreactivity). Cotterman (1953) points out that dominance cannot be distinguished from a "blindness" in observing factors. The "factors" in this factor-union system are the positive reactions indicated by ones in the matrix. Observe, for example, that the ones in the row labeled by BC are the union of the ones in the rows labeled by B and C. In a Boolean matrix factorization $R = A \times D$, the rows of R are linear combinations of the rows of D. (A specifies the combinations.) This linear combination is merely the union of the one-entries in the row vectors. Thus, the matrix product once again describes the symbolic biological model.

The NP-complete problem of Nau (1976) is connected with the question of interpreting a given zero/one matrix in terms of a minimum number of labels (specificities). Although this may be of peripheral interest to the biologist, it is not the question one ought to ask. For example, after the matrix for the common Ag phenotypes has been interpreted by six labels forming a complete factor-union system

TABLE 2 HLA Phenotypes with Standard and New Labels

Phenotype Labeling: Standard 3-Locus						Phenotype Labeling: Suggested 2-Locus			
locus A		locus B		locus C		locus 2		locus 1	
A1	A25	B44	B8			172	173	171	219
A2	A32	B7	B51	C7		174	176	177	178
A3	A24	B15		C3	C4	185	254	186	205
A31	A3	B18	B8			181	185	206	219
A2		B44	B8			174		171	219
A1	A31	B8	B39			172	181	219	226
A25	A31	B44	B51			173	181	171	178
A1		B8				172		219	
A1		B18	B8			172		206	219
A31	A3	B18	B49			181	185	206	221
A32	A31	B51	B35	C4		176	181	178	193
A3		B7				185		175	
A31	A3	B35	B62	C3		181	185	192	223
A25	A2	B7	B18			173	174	175	206
A32	A2	B44	B39	C2		176	241	171	226
A2		B7				174		175	
A2	A3	B18	B17			174	185	200	218
A32	A3	B7	B35	C4		176	185	175	193

TABLE 3 Reagents Recognizing the New Labels

Locus 1		Locus 2	
antigen	recognizing reagents	antigen	recognizing reagents
	anti-		anti-
171	B44	174	A2
206	B18	176	A32
177	B7, C7	181	A31
178	B51	173	A25
193	B35, C4	213	A24
175	B7	185	A3
186	B15	172	A1
205	C3	254	A24, C4
219	B8	241	A2, C2
223	B62,C3		
192	B35		
218	B17		
226	B39		
221	B49		

(one-locus six-allele), does one really care whether there is some complicated interpretation in terms of five labels? (Four labels would not suffice since $2^4 = 16 < 20$.) This motivates the following questions:

1. Can a given zero/one matrix be interpreted as a factor-union representation of a complete phenotype system?
2. If not, then how close to a complete system can we come?

We are, therefore, not interested in the most mathematically efficient interpretations, but in the biologically meaningful ones.

Note that Markowsky (1983) has answered the question dual to (1), namely, "When does a phenotype system have a factor-union representation?" (That is, given the labels, find the matrix.)

APPLICATION

The computer program of Wohlgemuth (1987) was applied to data on the HLA class I antigens. The results are given in Wohlgemuth and Dubey (1987a) and suggest a two-locus instead of the usual three-locus model. Analysis of the labels (numbers 171 through 226) generated by the computer, shows these are forced into two distinct sets which we interpret as corresponding to two distinct loci. Figures 6 and 5 from Wohlgemuth and Dubey (1987a) give first the comparison of the standard labeling with the new labeling (for those phenotypes uniquely labeled with the new labels; see Table 2).

Second, these labels (now called antigens) are given with those reagents which recognize each (Table 3).

It is seen that no reagents recognizing supposed A-locus alleles recognize locus 1 antigens. No reagents recognizing B-locus alleles recognize locus 2 antigens. Thus, the computer program has rediscovered the A and B loci for class I HLA. But this has happened in such a way that the C-locus is no longer necessary—at least for our phenotypes.

It is interesting to note the connection of the new symbolism with linkage disequilibrium. According to the table above, antigen 205 is recognized by reagent anti-C3. To remember this association, we will denote 205 by $\overline{C3}$. Antigen 223 is recognized by anti-C3 and anti-B62. Denote 223 by $\overline{B62C3}$. Anti-C3 then crossreacts. That is, it recognizes $\overline{C3}$ and $\overline{B62C3}$ and should be denoted by anti-$\overline{C3}$, $\overline{B62C3}$. The 1980 International Histocompatibility Workshop listing of haplotype frequencies gives the frequency of B62 and C3 together (p(B62 & C3)) as 0.0417. The marginal frequencies of B62 and C3 are

$$\begin{aligned} p(B62) &= 0.0477 \\ p(C3) &= 0.1208 \end{aligned}$$

Since B62 occurs in the standard model without C3 ($477 - 417 = 60$ times per ten thousand population), there must be an antigen other than $\overline{B62C3}$ and $\overline{C3}$. Let us call this $\overline{B62}$.

In the standard model

$$\begin{aligned} p(B62) \times p(C3) &= 0.477 \times .1208 = .00576 \\ \text{and} \quad p(B62\ \&\ C3) &= 0.417 (= 7.23 \times .00576) \end{aligned}$$

The combination B62 & C3 occurs 7.23 times more often than expected. The difference between the numbers above is taken as a measure of linkage disequilibrium:

$$\begin{aligned} D &= p(B62\&C3) - p(B62) \times p(C3) \\ D &= 0.365 = \text{disequilibrium constant.} \end{aligned}$$

In the new model, $\overline{B62C3}$, $\overline{B62}$, and $\overline{C3}$ are alleles at the same locus. Gene frequencies for these can be calculated from the equations:

TABLE 4 Recombination between Locus One and Two that Looks Like Recombination between A and C in the Standard Model

Standard and alternate genotype for family WER 02			
Standard Genotype	Standard Labels	New Labels	Alternate Genotype
Paternal:			
(a) A29,-,B44	A2,A29,C2,B44,B51	241,A29,171,178	(1) A29,171
(b) A2,C2,B51			(2) 241,178
Maternal:			
(c) A23,-,B44	A3,A23,C4,B35,B44	185,A23,171,193	(3) 185,193
(d) A3,C4,B35			(4) A23,171
Offspring:			
(a) A29,-,B44	A3,A29,C4,B35,B44	185,A29,171,193	(1) A29,171
(d) A3,C4,B35			(3) 185,193
(b) A2,C2,B51	A2,A23,C3,B44,B51	241,A23,171,178	(2) 241,178
(c) A23,-,B44			(4) A23,171
(a) A29,-,B44	A23,A29,C4,B35,B44	A23,A29,171,193	(1) A29,171
(c/d) A23/C4,B35			(4/3) A23/193

$$p(\overline{B62}) + p(\overline{B62B3}) = p(B62) = .0477$$
$$p(\overline{C3}) + p(\overline{B62C3}) = p(C3) = .1208$$
$$p(\overline{B62C3}) = p(B62 \ \& \ C3) = .0417$$

So that

$$p(\overline{B62}) = .0060$$
$$p(\overline{C3}) = .0791$$
$$p(\overline{B62C3}) = .0417$$

Thus, what appears in the new model as three alleles with the calculated frequencies appears in the standard model as two genes at different loci in disequilibrium.

It has been thought that the recombination observed in family studies has demonstrated the separation of the three loci, A, B, and C. Tables 7 and 8 from Wohlgemuth and Dubey (1987b) show what has been taken as recombination between A and C and also between B and C in the standard model. A comparison

TABLE 5 Recombination between Locus One and Two that Looks Like Recombination between B and C in the Standard Model

Standard and alternate genotypes for family JEA 02			
Standard Genotype	Standard Labels	New Labels	Alternate Genotype
Paternal:			
(a) A2,C2,B27	A2,A32,C2,B27,B44	175,241,B27,171	(1) 176,171
(b) A32,-,B44			(2) 241,B27
Maternal:			
(c) A24,C4,B35	A24,A31,C4,B35,B51	181,254,192,178	(3) 254,192
(d) A31,C4,B51			(4) 181,178
Offspring:			
(a) A2,C2,B27	A2,A24,C2,C4,B27,B35	241,254,192,B27	(2) 241,B27
(c) A24,C4,B35			(3) 254,192
(b) A32,-,B44	A24,A32,C4,B35,B44	176,254,171,192	(2) 176,171
(c) A24,C4,B35			(3) 254,192
(a/b) A2,C2/B44	A2,A24,C2,C4,B35,B44	241,254,171,192	(2/1) 241/171
(c) A24,C4,B35			(3) 254,192

with the labeling of the phenotypes in the new symbolism shows that recombination between just two loci suffices to explain the family data in terms of the more general symbolism (see Tables 4 and 5).

In Table 5, the child with genotype (a/b) (c) exhibits recombination of the paternal chromosomes, crossover apparently occurring between the B and C loci (A2,C2 on (a) recombining with B44 on (b)). According to the new labeling, there is one antigen, 241 (from Table 3), that accounts for the positive reactions with anti-A2 and anti-C2. In the new model, crossover occurs between locus one and two. Table 4 is similarly interpreted.

CONCLUSION

The notation used in immunogenetics has the effect of symbolically interpreting data. Contemporary notation, however, is not general enough to represent the complexity found in biological systems. Alternate methods of symbolic interpretation can suggest alternate genetic models for the same data. Having two distinct interpretations for the same data allows us to see the effect of using a notation that does not sufficiently allow for immunological complexity. The result is that systems appear more complex genetically. Thus what appear as genetic laws for our system may be symbolic artifacts.

The development of general methods for symbolically interpreting data has proceeded along three lines: mathematical analysis of models incorporating new symbolism; computer implementation; and application of the new methods to data from real immunogenetic systems. This paper suggests that computation ought to relate data to a descriptive model of Cotterman rather than merely to the number of symbols needed mathematically to interpret the data. The new problems will be more meaningful biologically and more tractable computationally. This paper also continues to examine an application of the new methodology to the HLA system. It shows that what appears to be linkage disequilibrium in the inadequate customary notation may disappear completely in the new symbolism. Thus, the customary notation may lead to artifactual interpretations of data.

REFERENCES

1. Cotterman, C. W. (1953), "Regular Two-Allele and Three-Allele Phenotype Systems," *Amer. J. Human Genet.* **5**, 193–235.
2. Cotterman, C. W. (1969), "Factor-Union Phenotype Systems," *Computer Applications in Genetics*, Ed. N. E. Morton (Honolulu: University of Hawaii Press).
3. Denniston, C. (1976), "A Note on Serological Interpretations," *Anim. Blood Grps. Biochem. Genet.* **7**, 101–108.
4. Hirschfeld, J. (1965), "Serologic Codes: Interpretation of Immunogenetic Systems," *Science* **48**, 968–971.
5. Hirschfeld, J. (1962), "Immunogenetic Models," *Nature* **239**, 385–386.
6. Hirschfeld, J. (1975), "Introduction to a Conceptual Framework in Serology," *Prog. Allergy* **19**, 275–312.
7. Hoffman, G. W. (1987), "On Helpers, Suppressors and I-J: Is There a Centre Pole in the Network of V Regions," *The Semiotics of Cellular Communication in the Immune System*, Eds. E. Sercarz, F. Celada, N. A. Mitchison, and T. Tada (New York: Springer Verlag), in press.
8. Markowsky, G. (1983), "Necessary and Sufficient Conditions for a Phenotype System to Have a Factor-Union Representation," *Math. Biosciences* **66**, 115–128.
9. Nau, D. (1976), *Specificity Covering: Immunological and Other Applications, Computational Complexity and Other Mathematical Properties, and a Computer Program, A.M. Thesis, Technical Report CS—176-7, Computer Science Dept., Duke University, Durham, N.C.*
10. Nau, D., and M. Woodbury (1977), "A Command Processor for the Determination of Specificities from Matrices of Reactions between Blood Cells and Antisera," *Computers and Biomed. Res.* **10**, 259–269.
11. Nau, D., G. Markowsky, M. Woodbury, and D. B. Amos (1978), "A Mathematical Analysis of Human Leukocyte Antigen Serology," *Math. Biosciences* **40**, 243–270.
12. Wohlgemuth, A. (1976), "A Histocompatibility Model," *Notices Am. Math. Soc.* **23**, A442–443.
13. Wohlgemuth, A. (1978), "Abstract Immunogenetic Systems," *J. Theor. Biol.* **73**, 469–508.
14. Wohlgemuth, A. (1979a), "Labeled Reaction Matrices, A Histocompatibility Model," *Discrete Math.* **26**, 285–292.
15. Wohlgemuth, A. (1979b), "Modeling Immunogenetic Specificities," *Math. Biosciences* **45**, 175–177.
16. Wohlgemuth, A. (1987), "An Interactive Program for Determining Tentative Gene Assignments from Immunological Data," *Computers and Biomed. Res.* **20**, 76–84.
17. Wohlgemuth, A., and D. P. Dubey (1987a), "The Impact of Symbolism on Immunogenetics: An Application to HLA," *J. Theor. Biol.* **126**, 149–165.

18. Wohlgemuth, A., and D. P. Dubey (1987b), "Symbolic Reinterpretation of HLA Gene Products—Impact on Interpretation of HLA at the Molecular Level," *Computer Applications in the Biological Sciences* **3–3**, 233–238.

Immune Suppression and Self-Tolerance

Theoretical Immunology, Part Two, SFI Studies in the Sciences of Complexity
Ed. A. S. Perelson, Addison-Wesley Publishing Company, 1988

JANICE NORTH, N. RANDALL CHU, AGNES CHAN, J. KEVIN STEELE,*
RAKESH SINGHAI, ANTHEA TENCH STRAMMERS, and JULIA G. LEVY**
Department of Microbiology, University of British Columbia, Vancouver, British Columbia, Canada, V6T 1W5, *Department of Pathology, Harvard Medical School, Boston, and **Tufts Medical School, Boston

Idiotypic Control of the Immune Response to Ferredoxin

IDIOTYPIC INTERACTIONS AT THE T CELL LEVEL

Most of the elements of any control system look inwards at one another—in the immune response, most lymphocytes interact with each other rather than with the external environment. Thus, the role of idiotype-bearing (id^+) and anti-idiotypic (id^-) T cells in the regulation of the immune response has been the subject of numerous investigations. In most instances, more than a single T cell phenotype has been implicated in the generation of suppression.[1–3] A currently dominant paradigm is that there is a suppressor pathway or "cascade" consisting of cells usually denoted as Ts1, Ts2 and Ts3.[4] The model, as described, depicts the unidirectional activation of these cells with Ts1 activating Ts2 and Ts2 activating Ts3. The model usually does not accommodate the possibility of stimulation occurring in the reverse direction, although it has been suggested that suppression could be activated in this way.[5] In contrast to this cascade model, which invokes asymmetric (unidirectional) interactions, a symmetric model for immune regulation has been described[6–8] in which one would predict mutual stimulation between T cells that recognize each other, such as between Ts1 and Ts2. When symmetric interactions are predicted between idiotypically related T cell subsets, it could also be predicted that varying dynamics between the interacting subsets could result in either suppression or stimulation of a given immune response.

THE FERREDOXIN SYSTEM

Work in this laboratory has been directed to understanding the immune response to the ferredoxin (Fd) molecule. Fd is a simple antigen of bacterial origin that consists of 55 amino-acid residues. There are only two antigenic determinants in the molecule (as assessed by antibody production), one of which is located within the NH_2-terminal heptapeptide (the N determinant) and the other within the COOH-terminal pentapeptide (the C determinant).[9,10] Responsiveness to Fd is under control of I-A region gene products of the murine major histocompatibility complex.[11] Mice of the H-2^k haplotype are high responders and, after initial immunization, produce antibody predominantly to the C determinant (this prevalence of anti-C diminishes as animals become hyperimmunized). Mice of the H-2^d haplotype are absolute non-responders to Fd,[11] although a variety of protocols can be used to convert H-2^d animals to the responder status.[12] A number of monoclonal antibodies with specificities for the N or C determinants have been raised to further study immunoregulation in the Fd system. One such monoclonal, FdB2, directed to the C determinant, was found to have interesting properties of considerable relevance to idiotypic interactions in controlling the response to Fd. Using a rabbit anti-FdB2 idiotype antiserum, it was found that the FdB2 id was only rarely expressed (and then only at very low levels) as antibody in the serum of Fd-immune H-2^k mice, regardless of their Igh allotype. Therefore, it did not represent a commonly expressed id at the B-cell level. When either the id (FdB2) or the anti-id was administered to Fd-immune or nonimmune B10.BR (H-2^k) mice, the subsequent response to Fd was significantly enhanced. However, expression of the FdB2 idiotype was still not detectable in the resulting antisera. Similar results were observed in H-2^k mice of allotypes other than Igh^b, indicating that this effect was not linked to the immunoglobulin genes. Adoptive transfer experiments in which nylon wool-purified T cells were treated with either id or anti-id plus complement prior to injection into irradiated syngeneic recipients clearly showed that the effects observed were attributable to T cells in each case. Such immune manipulation had no effect on the expression of the FdB2 id as antibody remained negligible even after treatment. Thus, it was suggested that there existed, in H-2^k mice, a set of interacting T cells bearing the FdB2 id and its reciprocal anti-id which regulated the immune response to Fd.[12] In a parallel set of experiments, it was determined that adoptive transfer of id or anti-id plus complement treated syngeneic T cells into irradiated non-responder H-2^d mice resulted in conversion of recipient animals to the responder status.

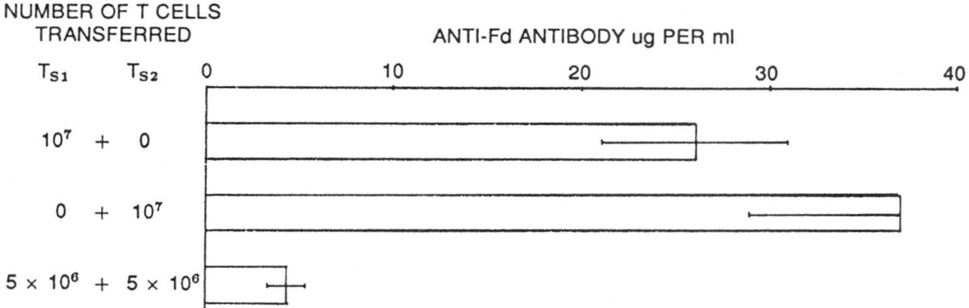

FIGURE 1 Effect of Id^+ and Anti-Id^+ T Cells in Co-Adoptive Transfer. Reconstitution of Unresponsiveness with id^+-depleted and id^--depleted T cells. Selectively depleted T cells from B10.D2 mice were mixed with non-immune B cells and adoptively transferred to two groups of 12 syngeneic irradiated recipients. A third group of mice was treated with a mixture of an equal number (5×10^6) of each depleted T-cell population. Mice were immunized with Fd and KLH. The 21-day primary anti-Fd response is shown. There were no differences in the anti-KLH responses of the three groups (data not shown).

CONNECTIVITY BETWEEN ID^+ AND ID^- CELLS

The finding that FdB2 id^+ T cells and their respective id^- counterparts controlled the magnitude of immune responses to Fd in mice of various haplotypes, was strongly suggestive of network interaction. An experiment was designed to test this hypothesis. Splenic T cells from B10.D2 non-responder mice were treated with id or anti-id plus complement, yielding populations A (depleted of id^- cells) and B (depleted of id^+ cells). 10^7 cells from populations A, B or A+B were adoptively transferred to syngeneic sublethally (500 rads) irradiated mice. Animals were subsequently challenged with Fd in 50% complete Freund's adjuvant and their ability to mount a response to Fd was assessed by measurement of serum antibody levels. The results of such an experiment are shown in Figure 1, in which it can be seen that mice receiving cells which had been depleted of either anti-idiotypic (Ts id^-) or id^+ (Ts id^+) cells had converted to the Fd responder status. However, animals which had received a mixture of populations A+B maintained the non-responder status. It was felt that these results established connectivity between the id^+ and id^- populations, and implied that these populations constituted part of a suppressor network which, in non-responder mice, abrogated the response to Fd, and, in high responder H-2^k animals, down regulated it. Furthermore, these results together with those mentioned above strongly suggested that this idiotypic expression was linked to neither the MHC nor the Igh locus.[13]

ISOLATION OF T CELL HYBRIDOMAS OF THE Ts1 PROTOTYPE

Some time ago, we were able to produce a B-cell hybridoma which secreted a monoclonal antibody (B16G) which appeared to be capable of binding to a population of T suppressor cells (TsC) and the suppressive molecules which they secrete. B16G appears to bind to an "invariant" epitope on T suppressor molecules since it is capable of reacting with polyclonal-suppressive molecules isolated from lysates of mouse splenocytes.[14] Thus, B16G was used to identify T cell hybridomas capable of secreting molecules with which it reacted. T cell hybridomas were produced by fusing splenocytes from B10.D2 mice which had been primed with Fd to BW5147

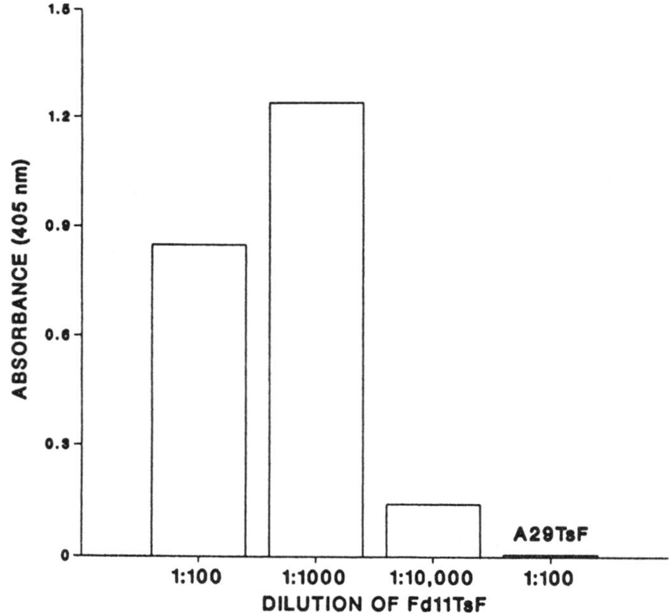

FIGURE 2 ELISA results of culture supernatants from Fd11 or a control T cell hybridoma (A29). ELISA plates were coated with Fd at 5.0 μg/ml. Culture supernatants were added as indicated. Plates were developed with alkaline phosphatase-labeled, rabbit, anti-mouse Ig after they had been reacted with the monoclonal antibody B16G.

cells. Fusion products were grown in HAT medium and culture supernatants were tested in a sandwich ELISA for molecules which bound to Fd (which constituted the solid phase of the test) and reacted with B16G. One strongly reactive hybridoma, Fd11, was identified and cloned by this assay. Specificity of reactivity of the "factor" secreted by Fd11 is shown in Figure 2, in which it is compared to a comparable preparation of an equivalent TsF, A29, which has no specificity for Fd.

Studied were undertaken to determine if this putative factor for Fd had any effect *in vivo*. Culture supernatants from overgrown cultures of Fd11, or the control A10 hybridoma were affinity-enriched over Sepharose 4-B-B16G immunoadsorbent columns. Because the Fd11 hybridoma had been raised by fusion of cells from

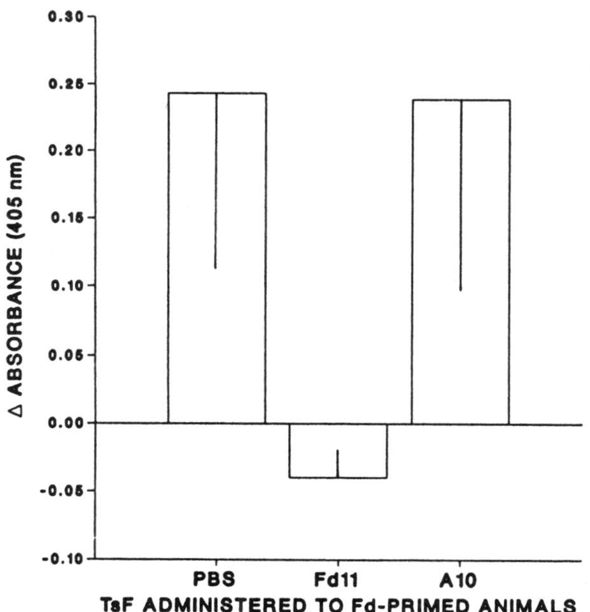

FIGURE 3 The influence of the Fd11 or a control TsF (A10) on the secondary response of B10.D2 mice to Fd. Thirty mice were converted to responder status and primed with Fd. On day 21 following priming, individual animals were bled and their anti-Fd titers recorded. Individual mice were ranked according to titer and divided into three equivalent groups (according to titers). On day 28, all animals were challenged with a secondary injection of Fd in complete Freund's adjuvant. At the same time, one group received PBS intravenously (i.v.), one received 20 μg of affinity-purified Fd11, and the third group received 20 μg of affinity-purified control TsF (A10). All animals were bled on day 35. Anti-Fd individual by ELISA and the net difference between the two levels are shown in the figure. Animals receiving Fd11 all showed a decrease in titer whereas all animals in the other two groups showed a titer increase.

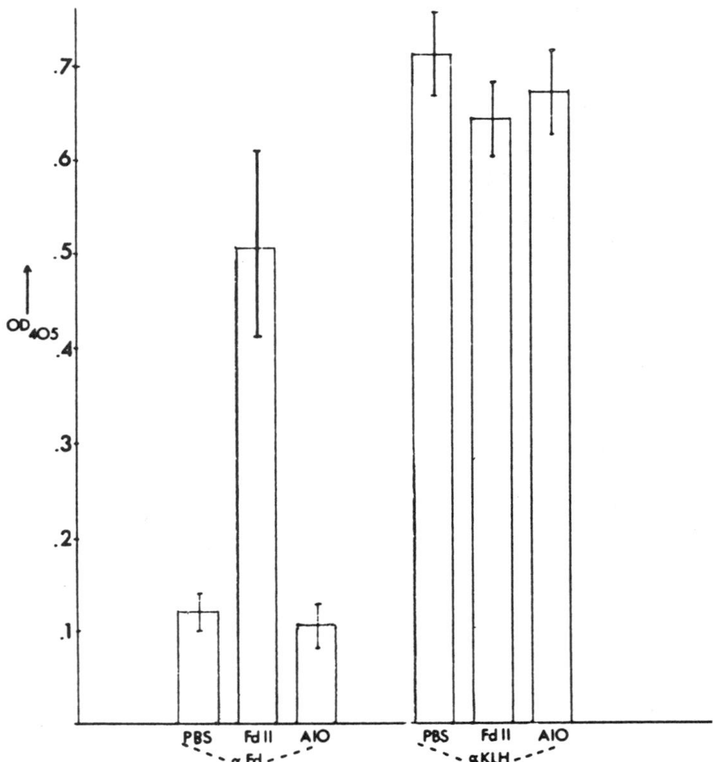

FIGURE 4 The effect of Fd11 administered to B10.D2 mice on their subsequent response to Fd. Groups of ten mice were injected intravenously with 20μg of affinity-purified Fd11 or control TsF (A10), or PBS. Ten days later, all animals were immunized with both Fd and KLH in CFA. They received a second antigenic stimulus 28 days later and were bled 7 days above. Only those mice receiving Fd11 show a significant anti-Fd response.

B10.D2 (non-responder) mice with the BW5147 cell line, it was desirable to carry out initial experiments on the biological activity of Fd11 "TsF" in B10.D2 animals. Therefore, it was essential to use mice which had been converted from the non-responder status. Converted animals usually give a very weak primary response to Fd, but produce a substantial secondary response. For this reason, B10.D2-converted and Fd-primed animals were used in the initial experiment. Animals received 20 μg of affinity-enriched Fd11, A10 (a control TsF), or PBS intravenously on the same day that they received a subcutaneous boost with Fd in 50% complete Freund's adjuvant. Animals were bled 7 days later and their serum anti-Fd titers assessed by ELISA. The results (Figure 3) are shown as net differences in ELISA titers from sera taken from individual animals prior to treatment and 7 days post treatment, and clearly demonstrate that animals receiving Fd11 showed essentiallyeceiving A10 or PBS had comparable secondary responses. Further detail on these experiments are given elsewhere.[15]

These experiments showed that culture supernatants containing Fd11 material eluted from B16G columns appeared to be able to suppress the Fd response

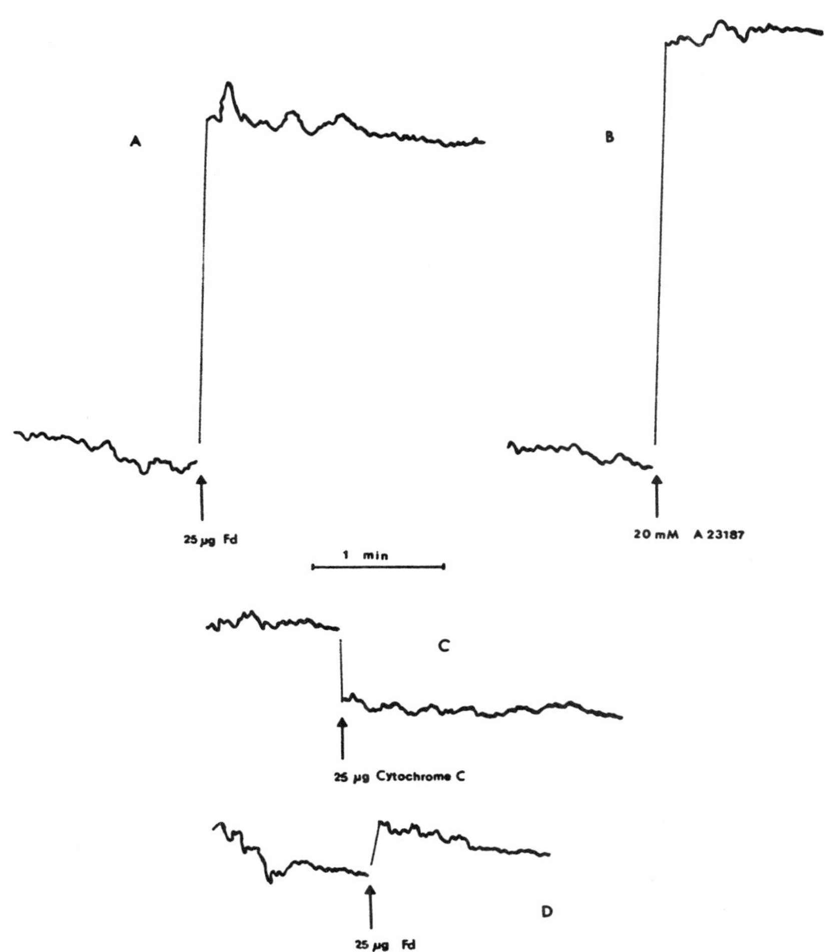

FIGURE 5 The Fd11 hybrid responds specifically to antigen as demonstrated by calcium flux using Quin 2/AM. The Fd11 hybrid was loaded with the Ca^{2+}-selective fluorescent indicator, Quin 2/AM. Exposure of cells to Fd (A) or the calcium ionophore, A23187 (B) caused enhanced fluorescence of Quin 2, indicating calcium flux. Comparable changes in fluorescence did not occur when the Fd11 hybrid was exposed to the irrelevant antigen, cytochrome c (C) or when the A10 hybrid was exposed to Fd (D).

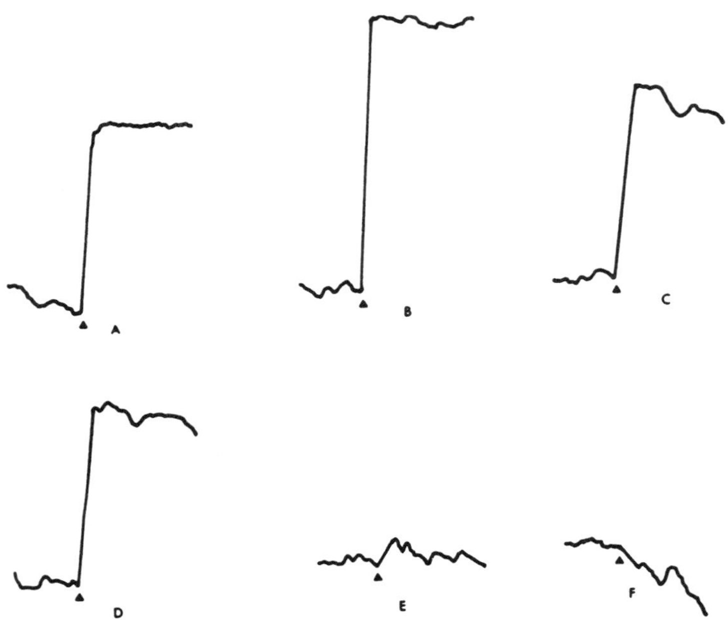

FIGURE 6 Calcium flux of Fd11 or A10 using Quin 2/AM. The Fd11 and A10 hybridomas were loaded with Quin 2/AM and treated as follows: (A) Fd11 cells treated with 50 μg Fd; (B) Fd11 cells preincubated with 50 μg of B16G, washed, and stimulated with 50 μg Fd; (C) Fd11 cells preincubated with 50 μg of control MAb, washed and stimulated with 50 μg of Fd; (D) Fd11 cells stimulated simultaneously with 50 μg of B16G and 50 μg of Fd; (E) A10 cells treated with 50 μg Fd; (F) A10 cells preincubated with 50μg of B16G, washed, and stimulated with 50 μg Fd.

in an antigen-specific manner. A further experiment was carried out to ascertain the effect of Fd11 "TsF" when administered prior to antigenic stimulus. It was thought that, if idiotypic interactions regulated the anti-Fd response, then prior administration of id$^+$-bearing molecules could have the effect of modifying the immune status of animals when antigen was introduced. An experiment was set up in which non-responder DBA/2 mice were injected intravenously with 20 μg of B16G affinity-enriched Fd11 or A10, or with PBS. They were rested for 10 days after which they were challenged with 20 μg of Fd in 50% complete Freund's adjuvant. Animals were bled on day 21, rechallenged with Fd on day 28 and bled again on day 35. Their serum levels of anti-Fd antibody were measured both day 21 and day 35. In both instances, the results were the same, and the results of the day-35 bleed are shown in Figure 4 in which it can be seen that the animals which had been treated with Fd11 showed a strong anti-Fd response, whereas animals in the other two groups maintained the non-responder status. These results are compatible with idiotypic control of the response to Fd.

ANTIGEN BINDING OF Fd11

The original studies carried out with the FdB2 monoclonal and its anti-idiotypic antibody indicated that the T cells involved in the idiotypic control of the immune responses to Fd were capable of recognizing antigen. In order to examine whether Fd11 cells had analogous antigen-binding molecules on their cell surface, Ca^{++} channeling experiments were undertaken. Somewhat surprisingly, it was found that

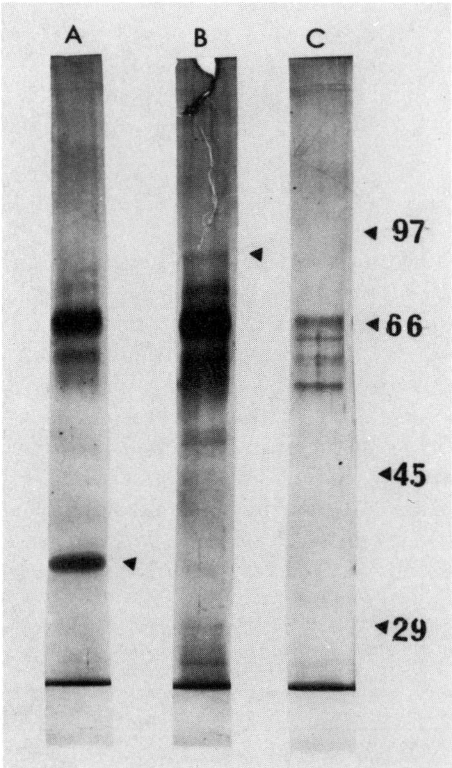

FIGURE 7 SDS-PAGE appearance of Fd11 TsF eluted from Fd immunoadsorbent columns; lane A, Fd11 material eluted at pH 4.3; lane B, Fd11 material eluted (after pH 4.3 elution) at pH 1.0; lane C, BW5147 culture supernatant eluted from Fd columns at pH 1.0.

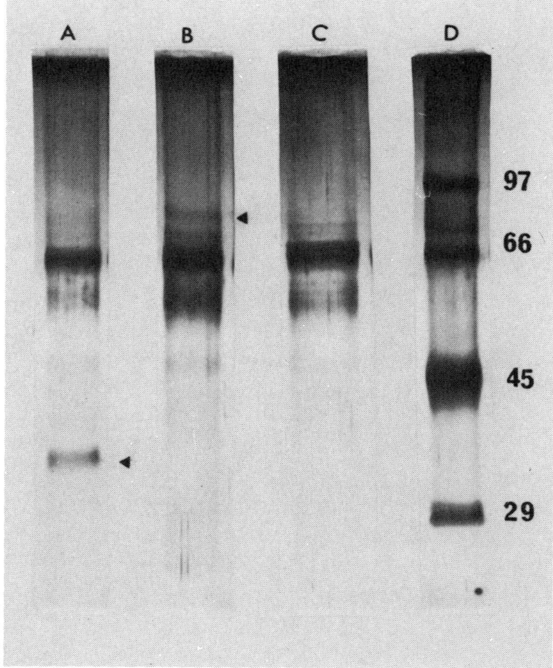

FIGURE 8 SDS-PAGE appearance of Fd11 TsF eluted from B16G immunoadsorbent columns; lane A, Fd11 culture supernatant eluted at pH 4.3; lane B, Fd11 material eluted (after pH 4.3 elution) at pH 1.0; lane C, BW5147 culture supernatant eluted at pH 1.0.

Fd11 cells responded to Fd in a specific and significant manner (Figure 5) in the absence of any antigen-presenting cells. These results implied the presence, on the surface of Fd11 cells, of molecules capable of binding antigen, and responding to it, which may differ from the conventional T cell receptor, which binds and responds to processed antigen presented in the context of self-MHC antigens. Attempts to demonstrate Ca^{++} flux using anti-idiotypic reagents or B16G, however, were not successful. An interesting additional observation to these antigen-binding studies was that, when Fd11 cells were pre-incubated with the B16G monoclonal, and subsequently stimulated with Fd, the level of Ca^{++} channeling was enhanced over that seen with Fd alone, or Fd stimulation following pretreatment with an irrelevant monoclonal antibody (Figure 5). These observations are, to our knowledge, unprecedented in T cell hybridomas.[15]

BIOCHEMICAL CHARACTERIZATION OF Fd11 TsF

In order to determine the nature of the Fd11 TsF material, Fd11 culture supernatants were affinity purified over B16G immunoadsorbent columns, eluted using either pH 4.3 or pH 1.0 buffers, run on SDS-PAGE (sodium dodecylsulphate polyacrylamide gel electrophoresis) and silver stained. These experiments demonstrated the present of a number of apparently specific molecular entities. They included bands with molecular weights of 140, 90, 80, 45, 30 and 16 kD. Ratios of relative concentrations of these bands varied form one preparation to another. Similar gel patterns were observed if materials were passed over immunoadsorbent columns to which Fd had been covalently bound.[15] Predominant bands were those at 90, 80 and 30 kD. It was observed that the lower molecular weight materials (30 kD) eluted more easily, at pH 4.3, than did higher molecular weight entities (90 and 80 kD), which required pH 1.0 (0.1 N HCl) for elution (Figures 7 and 8).

Because of the complexity of the gel patterns, there was some uncertainty regarding the biological relevance of the various bands visualized on SDS-PAGE. In order to attempt to answer the question as to their relevance, preparative gels of Fd-affinity purified Fd11 material were run, individual predominant bands (80 kD and 30 kD) were cut out and eluted and used to raise specific antisera in rabbits.

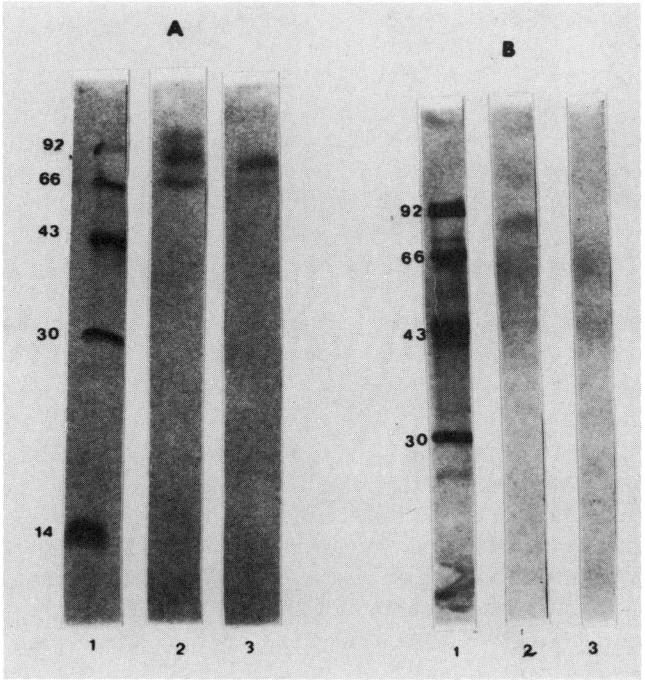

FIGURE 9 Western blot analysis of TsF using rabbit antibody raised against Fd11 80 kD protein. Gel A: lane 1, low molecular weight standards; lane 2, affinity-purified Fd11 TsF; lane 3, medium control. Gel B: lane 1, low molecular weight standards; lane 2, affinity-purified A10 TsF; lane 3, medium control.

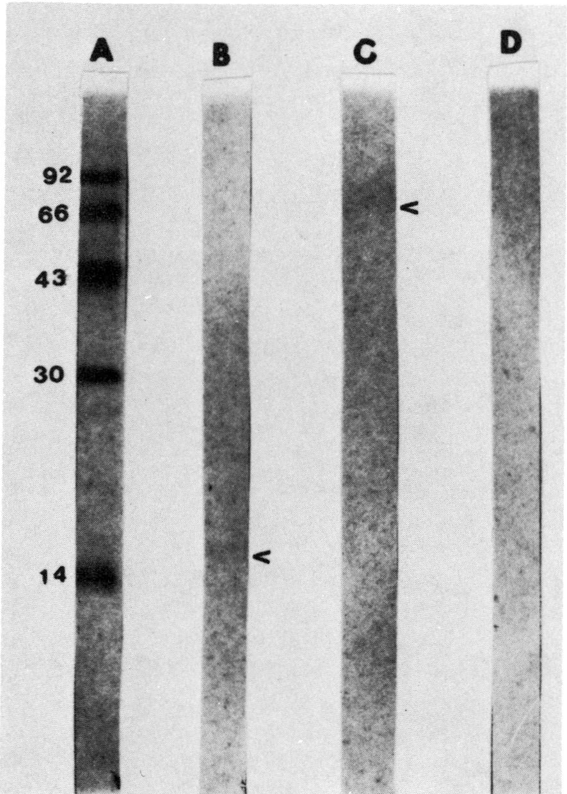

FIGURE 10 Western blot analysis of TsF using rabbit antisera raised to the Fd11 30 kD protein. Lane A, low molecular weight standards; lane B, affinity-purified Fd11 TsF; lane C, affinity-purified A10 TsF; lane B, medium control.

Antisera raised in this manner were used to examine, by Western blot analysis, and by flow cytometry, the nature of the proteins against which they were raised.

Using Western blot analyses, it became clear that the anti-p80 and anti-p30 antisera reacted with proteins that were produced by the T cell hybridomas tested, and that they were interrelated (Figures 9 and 10). When these antisera were tested for their ability to bind to the surface of the Fd11 T cell hybridoma, it became clear that both were reactive to molecules expressed on the surface of these cells. Further experiments showed that both antisera bound to the surface of the A10 Ts hybridoma (although not as intensely as they did to Fd11), but not to the fusion partner BW5147. Representative results are shown in Figure 11. These results support the contention that Fd11 cells bear on their surfaces, molecules which bind nominal antigen and are capable of modulating the immune response to that antigen.

DISCUSSION AND CONCLUSIONS

In a general study of idiotypic dominance using B-cell hybridomas secreting antibodies specific for the Fd molecule, we somewhat inadvertently found a monoclonal antibody which bore an idiotype apparently expressed in a dominant manner on Fd-specific regulatory T cells. Further experimentation showed that these id^+ T cells and their id^- T cell counterparts controlled the non-responder status of $H\text{-}2^d$ mice to Fd. Adoptive transfer experiments established connectivity between the id^+ and id^- T cell populations in this system.

In an attempt to clearly define the interactive T cell populations in this system, we undertook to raise T cell hybridomas, reactive with Fd and capable of down regulating the anti-Fd response. The Fd11 T cell hybridoma has demonstrated the

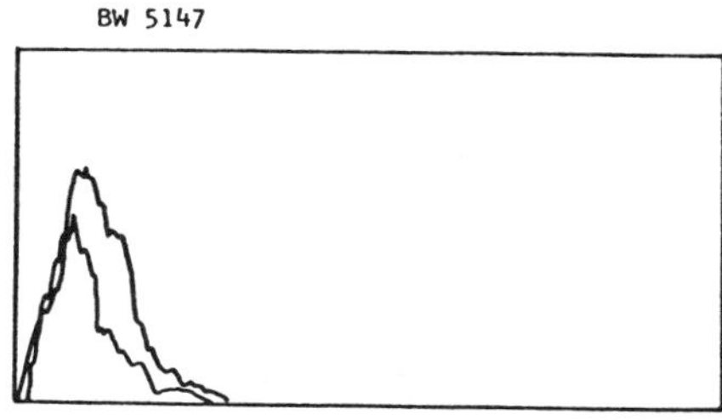

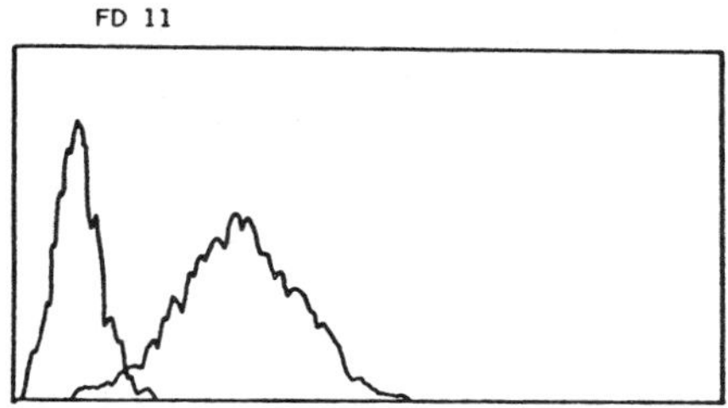

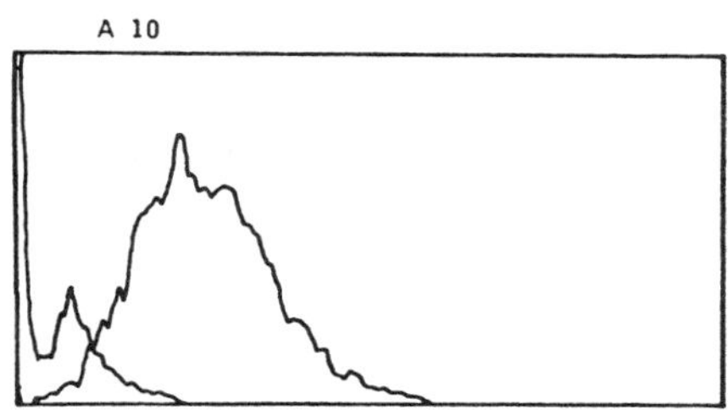

FIGURE 11 Fluorescence-activated cell sorter analysis of BW5147, Fd11 and A10 using rabbit antisera raised to the 30 kD protein from Fd11. Staining was done with FITC-labeled goat anti-rabbit Ig. Specific anti-p30 data were compared to cells treated with normal rabbit serum under identical conditions as controls.

required characteristics. The soluble material secreted by this hybridoma was shown to bind specifically to Fd as well as to an anti-TsF monoclonal B16G. Further, Fd11 TsF was shown to down regulate the secondary antibody response to Fd by B10.D2 mice when administered concurrently with the secondary antigenic stimulus. The possibility that the Fd11 TsF was acting via idiotypic interactions was supported by the observations that if Fd11 material was injected into Fd non-responder animals 10 days prior to antigenic stimulus, the animals were converted to the responder status. These results suggest that the Fd11 material (id^+) when administered in large amounts causes an imbalance in the connectivity between id^+ and id^- populations such that the connectivity between id^+ and id^- T cells is no longer effective when antigen is administered, thus abrogating the down regulation normally effected by this idiotypic regulatory network. The mechanism for this "paralysis" of suppression is not understood.

There are still a number of questions to answer. We do not know at this time whether the Fd11 hybridoma and its TsF product share idiotypes with the FdB2 monoclonal antibody. This question will be addressed and answered in order to determine whether th FdB2 and Fd11 patterns are parallel or identical.

The experimental data emanating from the Fd11 hybridoma and its ability to react with Fd as shown by Ca^{++} channeling raise further interesting and testable questions. We have shown that Fd11 apparently reacts in a specific manner to external stimulation with nominal antigen (i.e., Fd) in the absence of antigen-presenting cells. These observations imply that Fd11 can bind antigen specifically in the absence of presentation in the context of syngeneic class II antigens. Preliminary experiments with Fd11 indicate that they do not produce mRNA for the β chain of the T cell receptor, as assessed by Northern analysis. The results, to date, imply the existence of an alternative antigen-binding complex to the conventional T cell receptor, on at least some T cells. This possibility has been suggested by others and has been supported by some experimental observations.[16]

Our flow cytometry results, using antisera raised in rabbits immunized with PAGE-purified proteins, again supports the contention that Fd11 cells bear on their surfaces molecules which are identifiable with antigen-binding molecules secreted from the cells. Perhaps most convincing is our observation (data not shown) that murine T cells also react strongly (on analysis by flow cytometry) with the anti-p80 and anti-p30 antisera. Our data indicate that about 10% of murine T cells bear the "TsF" markers, and by double-labeling experiments we have shown that both Lyt 2^+ and $L3T4^+$ cells express this in a ratio of 2:1, respectively. These findings indicate the presence of a distinct population of T cells which bear a unique marker as defined by our antisera and which could well identify a new unique family of antigen-reactive molecules. Their relationship to idiotypic control of the immune response remains to be clarified.

REFERENCES

1. Weinberg, J. Z., R. N. Germain, B. Benacerraf, and M. E. Dorf (1980), "Role of Idiotypes in the Suppressor Pathway," *J. Exp. Med.* **152**, 161.
2. Hirai, Y., and A. Nisonoff (1980), "Selective Suppression of the Major Idiotypic Component of an Anti-Hapten Response by Soluble T-Cell-Derived Factors with Idiotypic or Anti-Idiotypic Receptors," *J. Exp. Med.* **151**, 1213.
3. Sy, M.-S., M. H. Dietz, R. N. Germain, B. Benacerraf, and M. I. Greene (1980), "Antigen- and Receptor-Drive Regulation Mechanisms. IV. Idiotype-Bearing I-J$^+$ Suppressor T Cell Factors Induce Second-order Suppressor T Cells which Express Anti-Idiotypic Receptors," *J. Exp. Med.* **151**, 1183.
4. Greene, M. I., M. J. Nelles, M.-S. Sy, and A. Nisonoff (1982), "Regulation of Immunity to Azobenzenearsonate Hapten," *Adv. Immunol.* **32**, 253.
5. Gershon, R. K., and H. Cantor (1980), in *Strategies of Immune Regulation*, Eds. E. E. Sercarz and A. J. Cunningham (New York: Academic Press), 43.
6. Hoffmann, G. W. (1980), "On Network Theory and H-2 Restriction," *Cont. Top. Immunobiol.* **11**, 185.
7. Gunther, N., and G. W. Hoffmann (1982), "Qualitative Dynamics of a Network Model of Regulation of the Immune System: A Rationale for the IgM to IgG Switch," *J. Theoret. Biol.* **94**, 815.
8. Herzenberg, L. A., S. J. Black, and L. A. Herzenberg (1980), "Regulatory Circuits and Antibody Responses," *Eur. J. Immunol.* **10**, 1.
9. Kelly, B., J. G. Levy, and D. Hull (1973), "Cellular and Humoral Immune Responses in Guinea Pigs and Rabbits to Chemically Defined Synthetic Peptides," *Eur. J. Immunol.* **3**, 574.
10. Sikora, L. K. J., M. Weaver, and J. G. Levy (1982), "The Use of Unideterminant Fragments of Ferredoxin in the Genetic Mapping of Determinant Specificity of the Immune Response," *Mol. Immunol.* **19**, 693.
11. Sikora, L. K. J., and J. G. Levy (1980), "Genetic Control of the Immune Response to Ferrodoxin: Linkage and Mapping of T Cell Proliferation and Antibody Production Genes to the MHC of Mice," *J. Immunol.* **124**, 2615.
12. Singhai, R., M. Weaver, L. Sikora, and J. G. Levy (1984), "Evidence for the Presence of Idiotype-Bearing Regulatory T Cells in which Idiotype Expression is Not Linked to Either Igh Alleles or the MHC," *Immunology* **51**, 743.
13. Singhai, R., G. W. Hoffmann, and J. G. Levy (1985), "Abrogation and Reconstitution of Nonresponsiveness: a Correlation with High Network Connectivity," *Eur. J. Immunol.* **15**, 526.
14. Maier, T., A. Tench Stammers, and J. G. Levy (1983), "Characterization of a Monoclonal Antibody Directed to a T Cell Suppressor Factor," *J. Immunol.* **131**, 1843.
15. Steele, J. K., N. R. Chu, A. Chan, J. North, and J. G. Levy (1987), "Isolation of an Antigen-Specific T Suppressor Factor that Suppresses the *In Vivo* Response of DBA/2 Mice to Ferredoxin," *J. Immunol.*, in press.

16. Chan, A., A. T. Stammers, J. North, J. K. Steele, N. R. Chu, and J. G. Levy (1987), "Characterization of Antigen-Binding Molecules from T-Suppressor Hybridomas," *Int. Rev. Immunol.*, in press.

MASARU TANIGUCHI, KENJI IMAI AND HARUHIKO KOSEKI
Department of Immunology, School of Medicine, Chiba University, Chiba, Japan

Predominant Use of the Particular V Gene in the KLH-Specific Suppressor T Cell Family

INTRODUCTION

Jerne proposed his idiotype network hypothesis[1] to explain how the immune system generates and maintains enormous diversity of antigen-specific lymphocytes. However, there are several questions to be asked if idiotype networks function as the main communication systems in immune responses. For example, most studies on the idiotype network have been done in particular animal models with specific haptens, such as 4-hydroxy-3-nitro-phenyl (NP) in C57BL/6, azobenzenearsonate (ABA) in A/J, phosphorylcholine (PC) in BALB/c mice, etc. These haptens only stimulate limited germ-line gene-encoded clones in these animals (reviewed in [2]). Thus, particular idiotype and anti-idiotype sets are only allowed to expand. If this is the case, the idiotype network system can not be generalized. It is necessary to demonstrate how idiotype networks function in conventional antigen systems with multiple epitopes in order to generalize their biological significance.

Secondly, it is interesting to know how idiotype networks function in regulatory cell interactions, because enormous negative energy would be necessary to accomplish the effective regulation of immune responses, if the regulatory T cells, like B cells, see every epitope and possess a huge repertoire corresponding to the various epitopes on conventional antigen. Therefore, a more economical system than that of B cells can be expected to exist at the T-cell level to circumvent the above problems. In this paper, we focus on the above-mentioned subjects in the idiotype network hypothesis.

REPERTOIRE LIMITATION OF SUPPRESSOR T CELLS EVEN IN THE CONVENTIONAL ANTIGEN SYSTEM WITH MULTIPLE EPITOPES

Since Köhler and Milstein developed the cell hybridization technique,[3] it has become possible to study the network hypothesis at a T-cell level. By the technique, three sets of suppressor T cell (Ts) hybridomas that regulate immune responses have been established by fusion of BW5147 thymoma (H-2^a) and keyhole limpet hemocyanin (KLH)-primed C57BL/6 (H-2^b) splenic T cells.[4] The first type of Ts is idiotype (Id) positive, binds to KLH and suppresses antibody responses in a KLH/H-2^b-specific manner. These are 34S-18, 34S-704, 34S-11, 9F181a, 9F8C2, and 9F8F3. The second type is complementary to the first one. Thus, the Ts bears anti-Id receptor (34S-281 and 1L-5), and suppresses antibody responses in KLH/H-2^b restricted fashion like the first one. The third set of Ts has KLH-binding activity, but mediates antigen-nonspecific suppressor function (34S-44). Interestingly, the anti-Id Ts hybridoma (34S-281) expresses the structure mimicking the epitope on KLH which is preferentially seen by Ts and also by the particular anti-KLH antibody with KLH/H-2^b-specific suppressor activity.[5] The structure on the anti-Id Ts hybridoma (34S-281) is called the regulatory internal image of KLH (Figure 1).

By this regulatory internal image, Ts and B cells recognizing the suppressor epitope can only communicate with each other. Therefore, the internal image works as a communication device and the anti-Id Ts receives the signal from antibody and transmits to the effector phase of suppression. Based on the investigation of the size of the repertoire of suppressor antibodies, it is possible to speculate about those of anti-Id Ts and Id Ts. In fact, we successfully established 4 monoclonal anti-KLH antibodies with KLH/H-2^b-specific suppressor activities from 120 anti-KLH antibodies.[5] All suppressor antibodies mediate functions via Lyt-2^+ suppressor T cells, because the suppressor activities of these antibodies were completely abrogated by treatment of responding cells with anti-Lyt-2 and complement. Among 4 suppressor antibodies, two anti-KLH recognizing different epitopes can react with the particular anti-Id Ts hybridoma 34S-281. The other two antibodies with KLH/H-2^b-specific suppressor activities do not react with the hybridoma, indicating that at least two or three types of anti-Id Ts expressing distinct internal images of suppressor KLH epitopes are present in the KLH-primed population. Although we could not precisely estimate the size of the Ts repertoire, the repertoires of anti-Id and their complementary Id Ts seem to be restricted even in the KLH-primed population. This observation is consistent with other antigen systems with multi-antigenic epitopes; for example, in the hen-egg lysozyme system, the amino acid

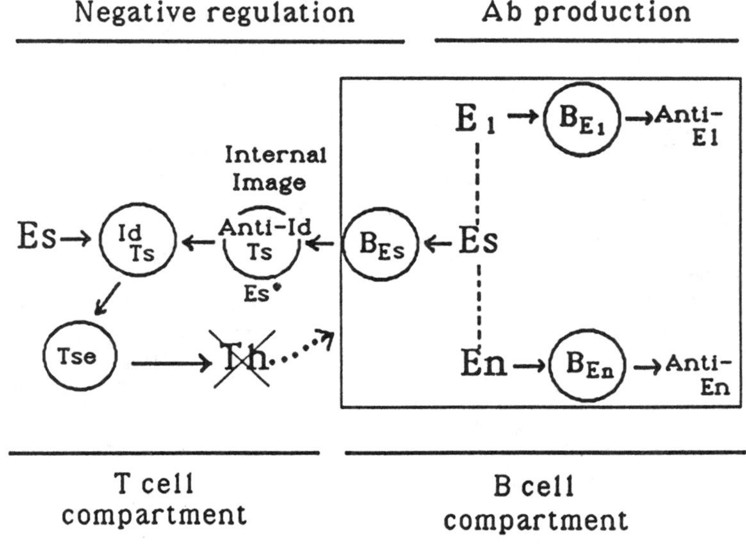

FIGURE 1 Regulatory T cell interactions in the conventional antigen system. E_{1-n}: antigenic epitopes. Es: suppressor epitope. Es′: internal image of suppressor epitope. Tse: effector Ts. B_{E1}: B cells recognizing epitope E_1. Id Ts: Ts recognizing epitope. Anti-Id Ts: Ts with complementary to Id Ts.

peptides 13–35 and 74–96 of hen-egg lysozyme-induced specific helper T cells but not Ts, and the reverse was true for the amino acids 1–17.[6] Therefore, it is suggested that T cells see an antigen by pattern recognition.

The mode of this T-cell antigen recognition is similar to that of the central nervous system when recognizing an object (Figure 2). For example, if we casually look at the shape of the heart, the major viewpoints are those of the V-shaped parts of the heart but not other parts.[7] Therefore, the most typical structure of the heart for visual sensory neurons is the V-shaped configuration. In the similar manner, T cells see only the restricted part of the antigen. T cells and neurons, thus, understand the whole structure of the object by recognizing the most characteristic parts. The mode of this T-cell recognition seems to be quite economical, especially in regulatory cell interactions.

MOLECULAR BASIS OF THE REPERTOIRE LIMITATION OF KLH-SPECIFIC Ts

As mentioned in the previous section, the repertoire of Ts specific for a conventional antigen with multiple epitopes seems to be limited. This limitation of Ts repertoire appears to be advantageous in the effective regulation of immune responses. In other words, a single antigen-specific suppressor signal leads to regulate the whole repertoire of effector lymphocytes against a specific antigen.

The repertoire limitation of Ts has been demonstrated at a molecular level. In this experiment, we isolated the Vα gene segment encoding T cell antigen receptor from the KLH-specific Ts hybridoma 34S-281 lacking β- and γ-chain gene expression and investigated the use of this particular Vα gene segment in the three types of Ts hybridomas (Id Ts, anti-Id Ts and antigen-non-specific Ts with KLH-binding activity) described in the previous section. First, we investigated the sequence of the α-chain cDNA of C57BL/6 Ts origin obtained from the anti-Id Ts hybridoma,

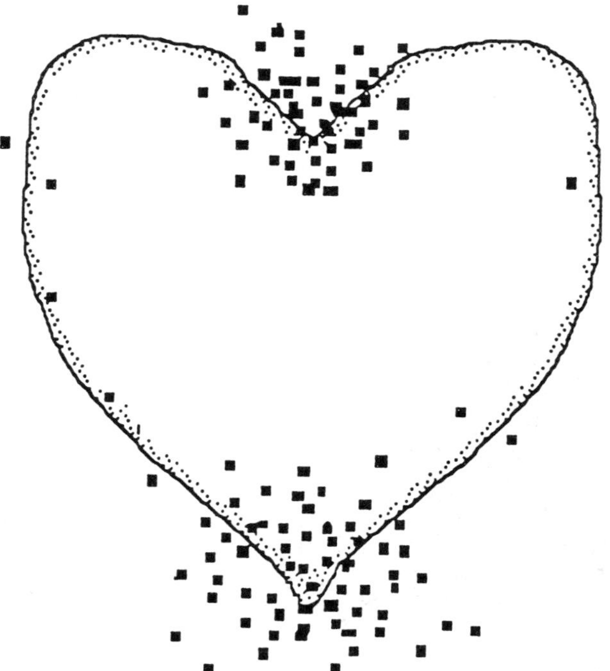

FIGURE 2 Viewpoints of the shape of heart.

34S-281.[8] The Ts α-chain cDNA clone (TsRα281) has an ATG start codon for a long open-reading frame with the V, J and C consensus sequences of the conventional α-chain. The deduced amino-acid sequence of the clone has a unique feature showing that the Vα281 segment possesses an additional cystein residue at amino-acid position 19, and only 34% homology to other Vα gene families is detected by computer analysis. Therefore, the Vα281 belongs to a new Vα family.

By using the Vα281 probe, Northern and Southern blot analyses were carried out on RNAs and DNAs from the aforementioned three types of Ts hybridomas including 8 individual clones. All Ts hybridomas tested showed the same rearrangement patterns of the Vα281 in Southern blot analysis. The DNA rearrangements detected in Ts hybridomas all seemed to be active because the Vα281 probe detected the 1.7 kb mRNA by Northern blot analysis, strongly suggesting that KLH-Ts hybridomas express a functional message of the α-chain encoded by the identical V genes. This also implies that the Vα281 may be responsible for generation of the KLH-specificity of the Ts-antigen receptor.

The above findings were also confirmed at the level of DNA sequence. We cloned and sequenced genomic DNAs from two KLH-Ts (34S-18 and 34S-704) and one anti-Id Ts (34S-281) hybridomas, and they were compared with the sequence of germline DNA from C57BL/6 kidney. Surprisingly, the entire sequences of the Vα region from two KLH-Ts and their anti-Id hybridomas are identical, and are found to be

TABLE 1 Preferential Use of the Particular Vα and Jα gene segments in KLH-Ts Hybridomas

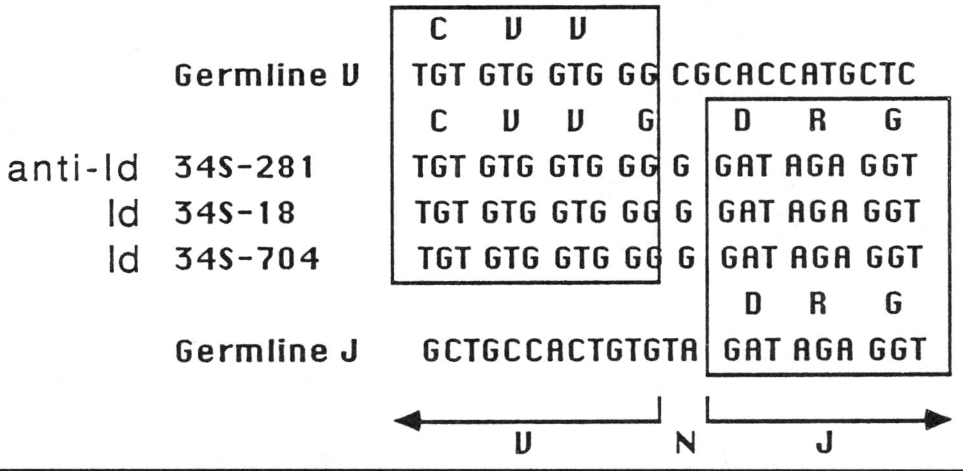

encoded by the germ line V and J region segments without any mutations except for one nucleotide (G) insertion in the V-J junctional region (Table 1). Therefore, KLH-Ts hybridomas use the identical Vα281 gene which is encoded by a germ-line gene. Moreover, the same nucleotide substitution in the junctional region was represented in two complementary receptors, namely Id and anti-Id receptors. Thus, the complementary receptors share the one chain.

Although all KLH-Ts hybridomas use the same Vα281 and the particular Jα genes, these Ts hybridomas were found to be independent because all hybridomas displayed different chromosome patterns. Furthermore, chromosome patterns of each hybridoma clone at 4-year intervals (1979 and 1983) were identical. This indicates that the different chromosome patterns in the individual clone did not occur during culturing the cells for a long time *in vitro*. Although the other chain associated with the α-chain of the Ts receptor has not been identified, this predominant use of one particular Vα gene apparently supports the previous idea that the repertoire of Ts specific for KLH is very limited.

INTERPRETATION

The possible mechanisms for this restricted usage of the particular Vα and Jα genes have not yet been identified. The repertoire limitation and the predominant use of the particular V-J sets do not seem to be due to the restricted number of the germ-line Vα and Jα sets, because a number of Vα and Jα segments have so far been detected. This suggests that the predominant use of the Vα281 gene in the KLH system may reflect its selection during antigen priming. Ts recognizing the particular KLH epitope (suppressor epitope) may have a selective growth advantage over Ts recognizing another epitope. Thus, the immune system provides Ts with a particular receptor specific for one antigenic epitope.

Most intriguing is the finding that two complementary Ts hybridomas, namely Id and anti-Id Ts, use the same V gene composed of identical V and J germ-line gene segments with identical joints. The molecules recognizing and recognized share at least one-half of the structure. The undefined chain should be responsible for construction of both Id and anti-Id specificities.

There is a similar example in anti-NP antibodies and their anti-Id antibodies as reported by Sablitzky and Rajewsky.[9] The V_H genes used in anti-Id antibodies are strongly homologous to the germ-line V_H186.2 gene encoding the V_H regions of the anti-NP antibody, implying that the V_H regions of the anti-Id and anti-NP antibodies are derived from the same germ-line V_H gene. Moreover, both anti-NP and their anti-Id antibodies use parts of, or the same D_HFl16.1 and J_H segments. Thus, Id and anti-Id specificities may be accounted for by somatic mutation in the V region of the heavy chain as mentioned by Sablitzky and Rajewsky.[9] However, as the V_κ light chain is always used by anti-Id antibodies, whereas the V_{λ_1} light chain is always associated with anti-NP antibodies, it is highly probable that the light chain is largely responsible for generation of Id and anti-Id specificities. The Id and anti-Id specificities on Ts can also be explained in a similar manner.

Simple antigen-selection mechanisms may not account for the fact that the two complementary Ts share one receptor chain composed of identical V and J segments with identical V-J joining. Some other mechanisms may operate in Ts. The Id and its complementary anti-Id Ts may be derived from the same progeny which had already expressed the α-chain with the particular V region (for example, Vα281 in the KLH system). The second (σ)-chain gene rearrangement will occur later and construct a functional Ts receptor with a pre-existing α-chain. Ts with a receptor composed of a particular combination of α- and σ-chains are preferentially selected by antigenic stimulation.

If the β-chain is used as a second chain or if the β-chain is subsequently rearranged after the α-chain gene rearrangements, the cell will be destined to be a helper or a cytotoxic T lymphocyte. When precursor cells with rearranged α-chains select the σ-chain as a second chain rather than β-chains, they may become Ts with Id and anti-Id receptor. Taking our results collectively, the idiotypic complementarity is highly genetically restricted, and the idiotypic interactions occur at the level of the germ-line-encoded Ts repertoire. This restricted use of the particular germ-line V gene may limit the repertoire of Ts to a small size even against the conventional antigen with multiple epitopes.

ACKNOWLEDGMENTS

We wish to express our appreciations to Dr. M. Tagawa for generous discussion. This work was partly supported by grants-in-aid for Scientific Research from the Ministry of Education, Science and Culture, Japan. We thank Ms. Rieko Shimomura and Ms. Sachie Kurata for preparation of the manuscript.

REFERENCES

1. Jerne, N. K. (1974), "Towards a Network Theory of the Immune Response," *Ann. Immunol. (Paris)* **125C**, 373.
2. Mäkelä, O., and K. Karjalainen (1977),"Idiotypes of T and B cells," *Immunol. Rev.* **34**, 119.
3. Köhler, G. and C. Milstein (1975), "Continuous Cultures of Fused Cells Secreting Antibody of Predicted Specificity," *Nature* **256**, 495.
4. Taniguchi, M., T. Saito, I. Takei, M. Kanno, T. Tokuhisa, and H. Tomioka (1982), "Suppressor T-cell Hybridomas and Their Soluble Products," *Lymphokines* **5**, 77.
5. Sumida, T., and M. Taniguchi (1985), "Novel Mechanisms of Specific Suppression of Anti-Hapten Antibody Response Mediated by Monoclonal Anti-Carrier Antibody," *J. Immunol.* **134**, 3675.
6. Benjamin, D. C., J. A. Berzofsky, I. J. East et al. (1984), "The Antigenic Structure of Proteins: A Reappraisal," *Ann. Rev. Immunol.* **2**, 67.
7. Zusne, L, and K. M. Michels (1964), "Nonrepresentational Shapes and Eye Movements," *Percept. Motor Skills* (Purdue University Press), 18 (1), 11.
8. Imai, K., M. Kanno, H. Kimoto, K. Shigemoto, S. Yamamoto, and M. Taniguchi (1986), "Sequence and Expression of Transcript of the T-cell Antigen Receptor α-Chain Gene in a Functional, Antigen-Specific Suppressor T-Cell Hybridoma," *Proc. Natl. Acad. Sci. USA* **83**, 8708.
9. Sablitzky, F., and K. Rajewsky (1984), "Molecular Bases of an Isogenic Anti-Idiotypic Response," *EMBO J.* **3**, 3005.

A. MALLEY and S.M. SHIIGI
Oregon Regional Primate Research Center, Beaverton, OR

The Role of I-J$^+$ Macrophages in the Induction of Antigen B-Specific T-Suppressor Cells

INTRODUCTION

In previous studies, we have demonstrated that both timothy grass pollen antigen B-specific (AgB-specific) anti-idiotypic (anti-Id) antibody and AgB-specific T suppressor factors cultured with normal spleen cells in mini-Marbrook chambers will produce significant levels of AgB-specific T-suppressor (T_s) cells.[1,2] Attempts to induce AgB-specific T_s cells *in vitro* using the $F(Ab)_2$ fragment of anti-Id were unsuccessful ,[3,4] and others have similarly found that the $F(Ab)_2$ fragments of anti-Id to be unable to induce suppressor cells.[5–8] These studies suggested that a cell having Fc receptors might be involved in the induction of T_s cells. Earlier, we reported that both normal B cells (anti-Thy and complement-treated spleen cells) and bone-marrow-derived macrophages (BMDM) were able to "present" anti-Id to target T cells and induce significant levels of T_s cells.[4] Analysis of the two cell populations by cell surface antigens and a nonspecific esterase staining method[9] that detects macrophages indicated that the B cells contained a 1 to 2% contamination of macrophages while the macrophage population was uncontaminated.

In this report, we shall show that successful induction of AgB-specific T_s cells requires as few as 10^3 histocompatible and I-J$^+$ BMDM. Flow cytometric analysis of BMDM with a polyclonal anti-I-J^k (3R anti-5R IgM antibody), monoclonal anti-I-A^k, and a monoclonal anti-I-J^k (clone JK10-23) not only confirmed the presence of I-J antigens on the surface of BMDM, but also demonstrated that the appearance of I-J on the cell surface began after the bone marrow cultures were devoid of contaminating cells and paralleled the development of macrophages. In addition, NP-40 extracts of BMDM immunoprecipitated with either monoclonal anti-I-A^k or anti-I-J^k antibodies, and analyzed by SDS-PAGE and Western Blots indicated that the anti-I-A^k antibody showed reactivity with a 30 and 33Kd protein bands while the anti-I-J^k antibody showed reactivity only with a 30Kd protein band.

RESULTS

THE REQUIREMENT OF MACROPHAGES IN T_s CELL INDUCTION

Previously we reported that T cells from normal spleens enriched by panning on anti-Ig coated petri dishes cultured *in vitro* with anti-Id do not produce AgB-specific T_s cells.[2,4,12] However, the addition of either 10^6 BMDM or 5×10^6 normal B cells would permit reconstitution of these cultures, and significant levels of AgB-specific T_s cells were induced. Upon closer examination of the macrophages and B cell populations, we found the B cells contained a 1 to 2% macrophage contamination. On the other hand, the BMDM were free of any other cell contamination as determined by flow cytometry (Ig^-, $L3T4^-$, Lyt 2^-, Thy^-, and Mac 1^+). Therefore, we chose to use BMDM in all subsequent experiments to avoid the potential confusion of mixed cell populations in the interpretations of our experiments.

Table 1 shows the effect of BMDM in anti-Id induction of AgB-specific T_s cells. Mice given only a primary and secondary immunization with WST give normal secondary IgG (27.5 micrograms/ml) and IgE (12 to 18 units) responses (Grp 1). Mice treated with cells harvested from cultures of normal spleen cells and anti-Id had significantly suppressed IgE responses (>70%), but control levels of IgG (Grp 3). In contrast, mice receiving cells from cultures of normal spleen cells and normal rabbit IgG (nRGG) gave control level IgG and IgE responses (Grp 2). Mice receiving cells from cultures of anti-Id and enriched T cells give control level IgG and IgE responses (Grp 4). The addition of 500 to 1000 histocompatible BMDM to cultures of enriched T cells and anti-Id reconstituted the ability of these cultures to produce AgB-specific T_s cells (Grp 5). The addition of 10^3 to 10^6 histoincompatible (H-2^b or H-2^d) BMDM to cultures of enriched T cells and anti-Id failed to produce T_s cells (Grp 6).

TABLE 1 The Role of Macrophages in Anti-Idiotypic Antibody Induction of AgB-Specific T-Suppressor Cells

GROUP	*IN VITRO* CULTURES[2]	NO. OF RECIPIENTS[3]	IMMUNE RESPONSE[1] IgE (units) Day 7	IgG (ug/ml) Day 7
1	none[4]	30	12–18	27.5 ±.3
2	nIgG + Normal spleen	15	12–17	27.0 ±.2
3	anti-Id + Normal spleen	30	3–5	25.5 ±.2
4	anti-Id + T-enriched	15	13–18	26.5 ±.1
5	anti-Id + T-enriched + 10^3 H-$2^{k/b}$ macrophages	30	3–5	25.5 ±.4
6	anti-Id + T-enriched + 10^3 H-2^d macrophages	12	14–17	26.0 ±.5

[1] IgE and IgG responses were measured in duplicate by ELISA. One unit of IgE represents a PCA titer of 1:100 of a standard sera.

[2] 25ug of affinity purified normal rabbit IgG (nIgG) or affinity purified anti-timothy IgE antibody (anti-Id) was added to either 10^7 normal spleen cells or 5×10^6 T cells enriched by panning or anti-mouse Ig-coated petri dishes in mini-Marbrook chambers and cultured for 4 days at 37°C. Bone-marrow-derived macrophages from H-2^d (Balb/c) or H-$2^{k/b}$ (CBA/JxC57BL/6) F_1 mice were grown in L-cell-conditioned media for at least 6 days and were added as a source of macrophages in some cultures.

[3] Recipient mice [CBA/JxC57BL/6 F_1] , 3-5 animals/group, were primed 20 days earlier with 10ug of WST adsorbed on alum, injected intravenously with either saline (Grp 1) or 4×10^6 cells harvested from mini-Marbrook chambers, and given a secondary boost with the same dose of antigen within 24 hours of cell transfer.

[4] Represents control mice given only a primary and secondary antigen challenge.

SPECIFICITY OF THE MACROPHAGE ASSOCIATED ANTI-I-J REAGENT

Earlier, we reported the preparation of a polyclonal anti-I-J^k antibody using BMDM from B10.A(5R) mice to immunize B10.A(3R) mice.[15] The resulting antisera was shown to be functionally specific for H-2^k macrophages and block the primary sheep red blood cell response of various H-2^k mice, but had no effect on the response of C57BL/6, B10.A(3R), and Balb/c mice. Additional specificity of the macrophages associated anti-I-J^k IgM antibody was shown by flow cytometry (Figure 1).[16] Treatment of BMDM from CBA/J mice with the IgM fraction of B10.A(3R) serum bind less than 1% of the BMDM (Figure 1B), and is not significantly different from the autofluorescence control (Figure 1A). BMDM treated with the IgM fraction of polyclonal anti-I-J^k antibody (3R anti-5R) showed that 85% of these cells stained positively when treated with FITC-F$(Ab)_2$-anti-mouse IgM (heavy chain specific) (Figure 1C). These same cells show greater than 99% staining with FITC-anti-Mac 1 antibody (data not shown). The haplotype specificity of the polyclonal anti-I-J^k IgM antibody was demonstrated by the failure of this reagent to bind BMDM from either C57BL/6 or Balb/c mice (Figure 1D). In addition, the polyclonal anti-I-J^k does not bind to BW5147 or HT_2 T cell lines, normal enriched T cells, a number of monoclonal B cell lines or a P388D1 macrophage cell line. The polyclonal anti-I-J^k antibody does bind weakly (20 to 30% positively stained cells) to some I-J^+ AgB-specific T_s cells, suggesting that the I-J epitope on macrophages cross-reacts weakly with the I-J epitiope on T_s cells.

EFFECT OF TEMPERATURE ON PULSING MACROPHAGES WITH ANTI-ID

Our data suggests that I-J^+ macrophages are necessary for the *in vitro* induction of AgB-specific T_s cells. Previously we have reported that pre-treatment of BMDM with anti-I-J^k antibody and complement rendered the remaining BMDM (after washing) unable to reconstitute cultures of enriched T cells and anti-Id to produce AgB-specific T_s cells.[12] On the other hand, pretreatment of BMDM with anti-I-A^k and complement did not alter the ability of the remaining macrophages to reconstitute our *in vitro* induction of AgB-specific T_s cells. In an attempt to further understand the mechanism of T_s cell induction, we attempted to determine if we could pulse our macrophage population with anti-Id, wash away the excess anti-Id, and add a fixed number of pulsed macrophages to our enriched T cells and retain the ability to induce AgB-specific T_s cells. Table 2 examines the effect of temperature on pulse-labeling BMDM and their ability to induce T_s cells. Animals given only a primary and secondary antigen challenge (Grp 1) gave the identical IgE and IgG response seen in mice receiving cells harvested from cultures of enriched T cells and BMDM pulsed with either nRGG (Grps 3 and 5) or anti-Id pulsed BMDM at 4°C (Grp 2). On the other hand, recipient mice treated with cells harvested from cultures of enriched T cells and BMDM pulsed with anti-Id at 37°C (Grp 4) showed normal IgG responses, but an 86% suppression of the AgB-specific IgE response.

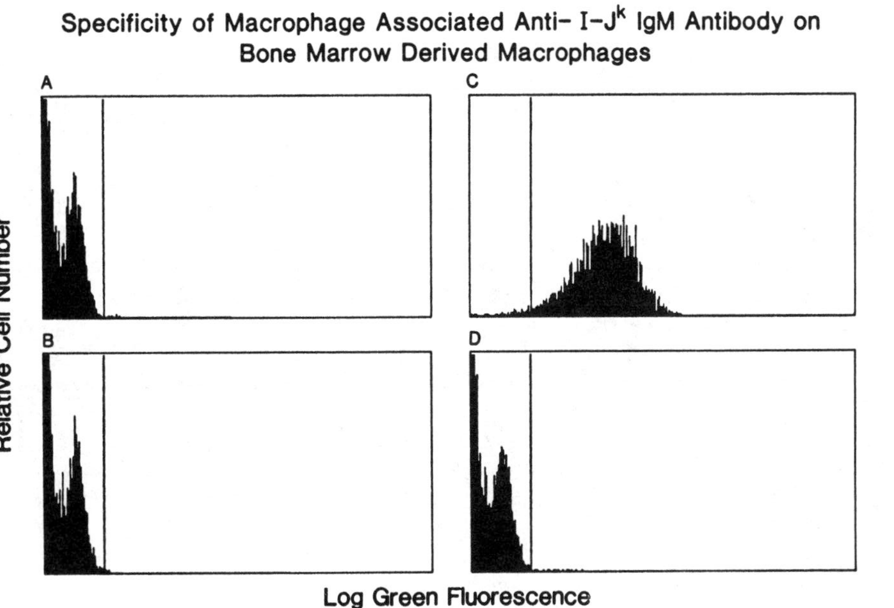

FIGURE 1 Specifically of Macrophage-Associated Anti-I-J^{k} IgM Antibody on Bone-Marrow-Derived Macrophages. CBA/J, Balb/c, and C57BL/6 bone-marrow-derived macrophages (BMDM) were cultured in L-cell-conditioned medium (LCM) for 7 days. A. Autofluorescence control - 5×10^5 H-2^k BMDM; B. 5×10^5 H-2^k BMDM treated with 30ug of 3R IgM fraction; C. 5×10^5 H-2^k BMDM treated with 5 ug of anti-I-J^{k} fraction. D. 5×10^5 H-2^b or H-2^d BMDM treated with 5 ug of anti-I-J^{k} IgM fraction. All cell mixtures were stained by the addition of 20μl of a 1:40 dilution FITC-F(Ab)$_2$-goat anti-mouse IgM (Tago, Inc., Burlingame, CA), and analyzed in a Coulter EPICS C cell sorter.

TABLE 2 Effect of Temperature on Ability of Pulsed Macrophages to Induce Antigen-Specific T Suppressor Cells *In Vitro*

Group	Treatment[2]	Cultures[3]	No. of recs.[4]	Immune response[1] IgE (units)	IgG ug/ml
1	None[5]	None	6	8.2	28.1
2	anti-Id + 10^4 macrophages 4°C 1 hr, wash 3 times	5 x 10^6 T cells + 10^3 pulsed macrophages	8	8	29.2
3	nIgG +10^4 macrophages 4°C 1 hr, wash 3 times	5 x 10^6 T cells +10^3 pulsed macrophages	8	8.3	28.1
4	anti-Id +10^4 macrophages 37°C 1 hr, wash 3 times	5 x 10^6 T cells +10^3 pulsed macrophages	8	1.1	27.5
5	nIgG +10^4	5 x 10^6 T cells +10^3	8	8.25	28.7

[1] IgG response was compared to a standard curve of monoclonal AgB-specific IgG run simultaneously with unknown serum samples. The IgE response was compared to a standard AgB-specific IgE serum. One unit of activity represents a PCA titer of 1:100. Serum IgG and IgE titers represent maximum responses seen in recipient mice. Maximum IgE and IgG titers occurred on day 7 and day 10, respectively, after secondary antigen challenge.

[2] 25ug of affinity purified rabbit IgG (nIgG) or anti-ID was incubated with 10^4 bone-marrow-derived macrophages from (CBA/J x C57BL/6)F_1 mice for 1 hour at either 4°C or 37°C. The cells were washed 3 times in cold PBS-azide and resuspended in PBS.

[3] Cultures of 5 x 10^6 T cells enriched by panning on anti-Ig-coated petri dishes and 10^3 pulsed macrophages were maintained in mini-Marbrook chambers for 4 days at 37°C in 10% CO_2. The harvested cells were washed 2 times in PBS, and 3 x 10^6 cells were injected intravenously into recipient mice 20 days after they were primed with 10ug of WST adsorbed to 1mg of alum.

[4] Recipient mice (CBA/J x C57BL/6) F_1, 3-4 animals/groups, were primed earlier with 10ug of WST adsorbed on alum and given a secondary boost with the same antigen dose within 24 hours at cell transfer.

[5] Recipient mice were given only a primary and secondary WST in alum antigen challenge.

T CELL REQUIREMENTS FOR THE INDUCTION OF AgB-SPECIFIC T_s CELLS

It appears that macrophages may "present" anti-Id to an appropriate target T cell to initiate the formation of T_s cells. In an effort to obtain sub-populations of normal T cells, T cells enriched on anti-Ig coated petri dishes were further purified by panning on anti-Lyt 2 coated petri dishes.[17] The non-adherent T cells were enriched for $L3T4^+$ $Lyt\ 23^-$ T cells (suppressor inducer), and the adherent T cells were enriched for $L3T4^-$ $Lyt23^+$ and $L3T4^+$ $Lyt23^+$ T cells. The adherent cells were further purified by panning on anti-L3T4 coated petri dishes into $L3T4^-$ $Lyt23^+$ (non-adherent) and $L3T4^+$ $Lyt\ 23^+$ T cells. Table 3 shows the effect of the addition of $L3T4^+$ $Lyt\ 2^-$ or $L3T4^+$ $Lyt23^+$ T cells to anti-Id pulsed BMDM upon the induction of AgB-specific T_s cells. As previously reported,[12] anti-Id-pulsed BMDM cultured with enriched T cells at 37°C show a significant induction of AgB-specific T_s cells (Grp 2). On the other hand, anti-Id pulsed BMDM cultured with either $L3T4^+$ $Lyt\ 2^-$ T cells or $L3T4^+$ $Lyt\ 23^+$ T cells do not produce AgB-specific T_s *in vitro*, and recipient mice made control level responses of antibody. However, the addition of both $L3T4^+$ $Lyt\ 2^-$ and $L3T4^+$ $Lyt23^+$ T cells to anti-Id-pulsed BMDM resulted in the formation of significant levels of T_s cells, and a >75% suppression in the level of AgB-specific IgE. The number of $L3T4^+$ $Lyt2^-$ and $L3T4^+$ $Lyt23^+$ T cells that appeared optimal was 10^5 cells, but lower levels (10^4 to 5×10^4 cells) of these T cells resulted in the induction of a smaller level of AgB-specific T_s cells.

THE BINDING OF MONOCLONAL ANTI-I-J^k ANTIBODY TO BMDM

Earlier we reported on the failure of pretreatment of BMDM with anti-F_c receptor antibody to block the binding of our IgM polyclonal anti-I-J^k antibody on these cells.[16] In addition, these studies showed that when BMDM were pretreated with 200 micrograms of the anti-F_c receptor antibody that less than 5% of the cells were stained when they were treated with FITC-IgG_{2a} antibody (Figure 2B). When BMDM were pretreated with anti-F_c receptor antibody and then treated with monoclonal anti-I-J^k (clone JK10-23) antibody, a significant number of these cells were positively stained (Figure 2C and 2D). The cells stained positively with the monoclonal anti-I-J^k antibody varied whether the BMDM had been in culture for either 7 (40% positive, Figure 2C) or 14 days (90% positive, Figure 2D).

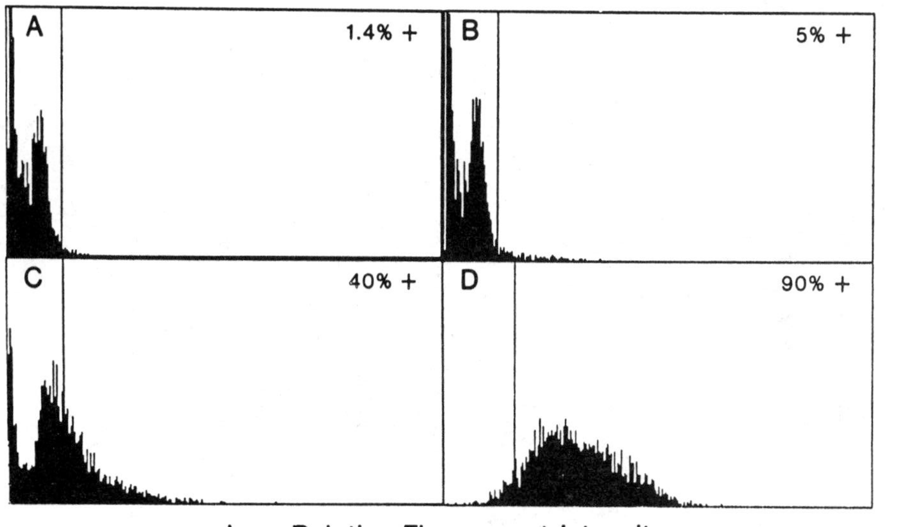

FIGURE 2 Pretreatment of $H-2^k$ Bone-Marrow-Derived Macrophages (BMDM) with Monoclonal Anti-F_c Receptor Antibody Prior to the Addition of FITC-Monoclonal Anti-I-J^k Antibody. A. Auto control 7 day CBA/J BMDM (5×10^5); B. 7-day CBA/J BMDM (5×10^5) were treated with 200 ug of anti-F_c receptor antibody for 10 minutes at 4° prior to the addition of 10 ug of FITC-murine IgG_{2a} myeloma protein; C. 7 day CBA/J BMDM (5×10^5) were treated with 200 ug of anti-F_c receptor antibody for 10 minutes at 4°C prior to the addition of 20 ug of FITC-anti-I-J^k antibody; D. 14 day CBA/J BMDM (5×10^5) were treated with 200 ug of anti-F_c receptor antibody for 10 minutes at 4°C prior to the addition of 20 ug of FITC-anti-I-J^k antibody.

TABLE 3 T Cell Requirement for the *In Vitro* Induction of Antigen B-Specific T_s Cells

Group	Enriched T Cells[1]	anti-Id-pulsed macrophases[2]	L3T4+Lyt2− T Cells[3]	L3T4+Lyt23+ T Cells[4]	Number of Recipients	AgB-IgE Responses	% Suppres
(units)							
1	+	-	-	-	8	16–18	0
2	+	+	-	-	8	1–3	81
3	-	+	+	-	8	16–18	0
4	-	+	-	+	8	16–18	0
5	-	+	+	+	8	2–4	>75

1 5×10^6 normal enriched T cells obtained by panning on anti-Ig couted petri dishes.

2 10^3 anti-Id pulsed BMDM.

3 10^5 L3T4+Lyt2− T cells obtained by panning on anti-Lyt coated petri dishes.

4 10^5 L3T4+Lyt23+ T cells obtained by panning T cells adherent on anti-Lyt 2 coated petri dishes and further purity them on anti-L3T4 coated petri dishes.

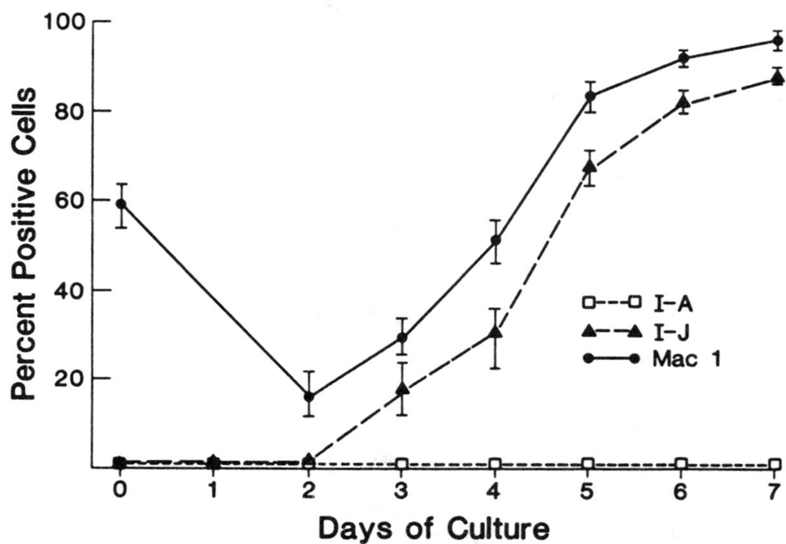

FIGURE 3 Expression of I-A, I-J, and Mac-1 on Bone-Marrow-Derived Macrophages. Bone marrow cells collected from CBA/J femurs were washed with balanced salt solution and resuspended in PBS-FCS at a final concentration of 5×10^6 cells/ml. A portion of these cells (5×10^5) were analyzed for I-A, I-J, and Mac-1 cell surface determinants by flow cytometry (day 0). The remaining cells were placed in culture containing IMDM media, LCM, 10% fetal calf serum, and 5% horse serum for 14 days. At various times after the bone marrow cells were maintained in culture, an aliquot of these cells were reexamined by flow cytometry to determine the expression of I-A, I-J, and Mac-1 cell surface determinants.

WHEN DOES THE I-J EPITOPE APPEAR ON BMDM?

The expression of I-A, I-J, and Mac 1 determinants on the surface of macrophages during the development of BMDM is shown in Figure 3. Bone marrow cells collected from the femur, washed in PBS-FCS, and stained immediately with reagents to detect cell surface I-A, I-J, and Mac 1 showed that these cells lack detectable I-A and I-J determinants, but about 60% of the cells express the Mac 1 determinant. Once the bone marrow cells have been placed in culture with L-cell-conditioned medium, many of the cells die during the first two days of culture, and the number of Mac 1^+ cells decreases rapidly to about 20%; there are still no detectable I-A and I-J positive cells. Three days after the bone marrow cells are placed in culture, the number of Mac 1 positive cells steadily increase, up to greater than 99% as the total number of macrophages grows in these cultures. During this same period of time (days 3 to 7) there is a parallel increase in the I-J+ macrophages up to 90%. The BMDM grown in L-cell-conditioned medium did not express detectable cell surface I-A during the entire period that these cells were examined (up to day 14).

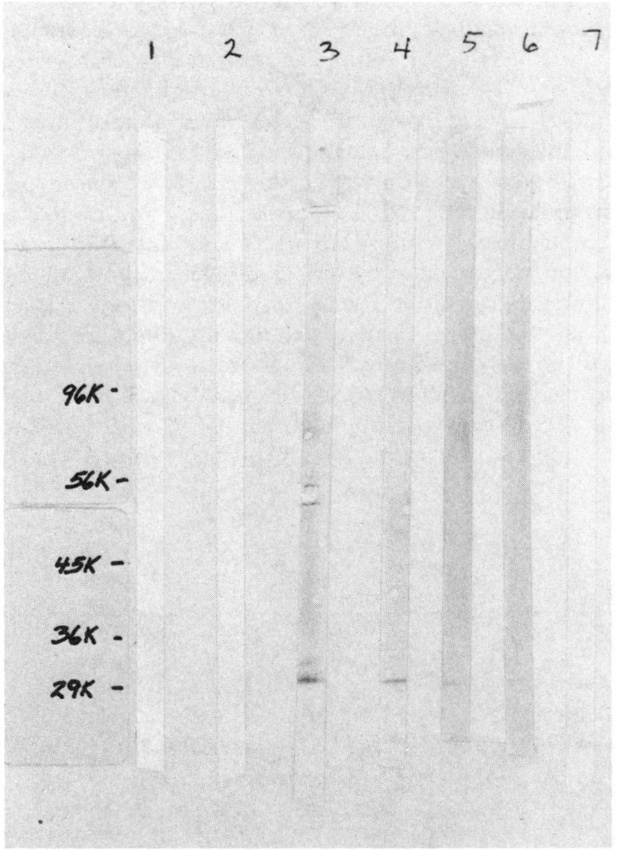

FIGURE 4 Western blot analysis of monoclonal anti-I-A, anti-I-J, and anti-I-E immunoprecipitated NP-40 BMDM extract. Lane 1 represents molecular weight standards; Lane 2 represents material immunoprecipitated with normal mouse IgG and stained with anti-I-A^k; Lane 3 represents material immunoprecipitated with anti-I-A^k and stained with the same antibody; Lane 4 represents material immunoprecipitated with anti-I-A^k and stained with anti-I-J^k antibody; Lanes 5 and 6 represent material immunoprecipitated with anti-I-J^k and stained with the same antibody; Lane 7 represents material immunoprecipitated with anti-I-E^k and stained with the same antibody.

DOES THE I-J PROTEIN EXIST ON BMDM?

The presence of an I-J epitope on BMDM is clear. However, I-J epitopes on T_s cells has been reported by many investigators, but attempts to isolate the I-J protein or mRNA from I-J^+ T_s cells have been unsuccessful. In fact, this failure to either isolate the I-J protein or identify a gene associated with the formation of I-J has led to the "I-J Paradox."[22] A NP-40 extract (30 microliter aliquots) of BMDM maintained in culture for 7 days was treated with 10 microgram amounts of nMGG, monoclonal anti-I-A^k, monoclonal anti-I-J^k, or monoclonal anti-I-E^k for 30 minutes at 4°C before these mixtures were treated with an excess of a Protein A-bacterial adsorbent. The samples eluted from the bacterial adsorbent were run on a 10% SDS-PAGE gel and transferred to a nitrocellulose filter (Western Blot). Individual 4-mm-wide strips were reacted with either monoclonal anti-I-A^k anti-I-J^k, or anti-I-E^k antibody. Figure 4 shows the results of this experiment. Lane 1 represents protein molecular-weight standards. Lane 2 and Lane 7 represent NP-40 extracts immunoprecipitated with either nMGG or anti-I-E^k and incubated with anti-I-A^k (Lane 2) antibody or anti-I-E^k (Lane 7) antibody. Lane 3 shows that NP-40 extracts immunoprecipitated with anti-I-A^k antibody and stained with the same antibody shows two protein bands; one band at 30Kd, and the other at 33Kd. Lane 4 represents the NP-40 extract immunoprecipitated with anti-I-A^k antibody but stained with anti-I-J^k antibody, and shows only a 30Kd protein band. Lanes 5 and 6 represent NP-40 extract immunoprecipitated with anti-I-J^k antibody and stained with the same antibody and these strips show only a weakly stained 30Kd band.

DISCUSSION

In vitro cultures of normal spleen cells with anti-Id in mini-Marbrook chambers induce significant levels of T_s cells that suppress AgB-specific IgE, but not the IgG response (Table 1). Cultures of enriched T cells with anti-Id *in vitro* do no produce T_s cells, but the addition of as few as 10^3 histocompatible BMDM to these cultures restores their ability to produce significant levels of T_s cells. Histoincompatible BMDM are unable to restore cultures of enriched T cells and anti-Id allowing them to make T_s cells. Thus, the interaction of the macrophage and T cell populations involved in T_s cell induction are MHC-restricted. The T_s cells produced by anti-Id are isotype specific, and they do not reduce primary or secondary IgG or IgE responses of ovalbumin, Ascaris suum, or Keyhole Limpet hemocyanin (data not shown).[13] Macrophages are not only required for antigen or anti-Id induction of T_s cells, but, as we have shown, they also are required for the induction of T-suppressor factor induced T_s cells.[12]

We have previously described the preparation of a IgM anti-I-J^k antibody and shown the functional specificity of this antibody.[15] Flow cytometric analysis of the anti-I-J^k antibody shows that it binds specifically with H-$2^{k/b}$ or H-2^k BMDM, but does not bind with H-2^b or H-2^d BMDM, BW5147 or HT_2 T cell lines, normal T or B cells, or the P388D1 macrophage cell line (Figure 1).[16] Treatment of BMDM with anti-I-J^k antibody and complement prevents the remaining viable cells from restoring cultures of enriched T cells and anti-Id producing T_s cells.[12] On the other hand, treatment of BMDM with anti-I-A^k antibody and complement did not effect the ability of BMDM to restore cultures of enriched T cells and anti-Id from producing T_s cells.[12] Further evidence for the presence of I-J on BMDM was presented in Figure 2. BMDM maintained in culture for 7 to 14 days show a significant binding (40 to 90%) with monoclonal anti-I-J^k antibody. On the other hand, these same cells show no significant binding (<5%) with either FITC-monoclonal anti-I-A^k, -monoclonal anti-I-E^k, or -IgG_{2a} myeloma protein (Figure 2B).[16]

Although the participation of macrophages was initially not considered for the induction of T_s cells,[23,24] evidence has accumulated from several laboratories indicating a requirement of macrophages at several stages in the induction of T_s cells.[25–30] Recently, Nakamura et al.,[31] using a cloned population of I-J^+ macrophages, also demonstrated their requirement in the induction of T_s cells *in vitro*. These observations raise a question regarding the relationship between the I-J seen on macrophages and the I-J expressed on T_s cells or T suppressor factors. The polyclonal anti-I-J^k antibody made against the I-J on BMDM reacts strongly with a variety of H-2^k BMDM, but weakly with some I-J^+ AgB-specific T_s hybrids, and not at all with other I-J^+ AgB-specific T_s hybrids. These observations suggest that the I-J expressed on macrophages is not identical, but cross-reactive with the I-J expressed on T cells. Nakamura et al.[32] and Murphy et al.[33] has also presented evidence suggesting this difference in the I-J expressed on macrophages and T cells, and the serological heterogeneity in I-J determinants associated with different T cell regulatory factors was discussed by Flood et al.[34] Although the monoclonal anti-I-J^k (clone JK10-23) was made against I-J expressed on T_s cells, this antibody also bind to an I-J epitope in common with the I-J epitope expressed on BMDM (Figure 2), and the level of expression of this common epitope increases significantly as these cells are maintained in culture beyond 7 days (Figure 2). Further work will be necessary to resolve the shared and unique I-J epitopes on T_s cells and BMDM.

Other investigators have made similar observations regarding the involvement of I-J^+ macrophages in the induction of T_s cells,[26–28,35] and some have suggested that macrophages having both I-A and I-J cell surface determinants were needed for the induction of T_s cells.[26] The analysis of BMDM by flow cytometry using reagents specific for I-A, I-J, and Mac 1 permits us to determine both the frequency and intensity of these determinants. Mac 1 is found on both macrophages and granulocytes. The Mac 1 positive cells in initial bone marrow suspensions are predominantly due to granulocytes as only 3-6% of the cells are macrophages at this time.[18] As the granulocytes die, there is a decrease in the fraction of Mac 1 positive cells (through day 2) (Figure 3). As the macrophage precursors proliferate exponentially, they become the predominant cell type, and by day 7 greater than 98% of the cells are Mac 1 positive macrophages. The increase in I-J positive cells parallels the increase in macrophages (Figure 3). Since the expression of I-J on BMDM does not occur until these cultures are virtually 100% macrophages, the BMDM must be able to synthesize their own I-J protein. While expression of I-J on BMDM is unambiguous in this study, the I-A expression appears to be less clear. Several investigators have reported the presence of I-A determinants on BMDM,[36–40] but they used anti-I-A antibodies which were IgG_{2a}. BMDM bind monomeric IgG_{2a} protein via their F_c receptors and, while specificity controls were performed, the possible binding of the anti-I-A antibody to F_c receptors could not be completely excluded. In the experiments reported here, we circumvented this problem by using an anti-I-A^k reagent that was an IgM antibody or by blocking F_c receptors with anti-F_c receptor antibody prior to staining with FITC-anti-I-A^k antibody (IgG_{2a}).(Figure 2).[16] Using these reagents, we cannot demonstrate constitutive I-A expression on BMDM.

Lee and Paraskevas et al.[41] have shown that T cells can acquire I antigens released from macrophages during antigen presentation. Several investigators have suggested that the role of I-J might be to focus antigen or T suppressor factors to their appropriate target cells by interaction with I-J binding sites. The data presented in Table 2 showing that anti-Id pulse labeled BMDM cultured at 37°C with enriched T cells produce significant levels of AgB-specific T_s cells.is consistent with the macrophage "presenting" anti-Id to the appropriate target T cell. Table 3 demonstrates that both a L3T4$^+$ Lyt 2$^-$ and a L3T4$^+$ Lyt 23$^+$ T cells are required to interact with anti-Id pulsed BMDM to induce significant levels of T_s cells *in vitro*. Recently, Zupko et al.[42] used a preparation of monoclonal anti-I-J to prepare an anti-idiotypic antibody that specifically blocked T_s cell function without effecting T helper cell activation. These investigators used their anti-idiotypic antibody to identify a receptor for the I-J determinant that is important in interactions between T_s cells and possibly macrophages and B cells. The availability of I-J$^+$ BMDM may permit one to further define the requirements for presentation and processing of antigen to enriched AgB-specific T_s cells or T_s cell hybrids.

Preliminary studies of NP-40 extracts of BMDM indicated significant levels of protein with molecular weights of 30 and 33 Kd. Other I-region proteins exhibit molecular weights similar to those seen in our BMDM extracts. Immunoprecipitation analysis of the BMDM extract with monoclonal anti-I-A^k, anti-I-E^k, and anti-I-J^k antibodies has shown that the anti-I-A^k and anti-I-J^k antibodies will immunoprecipitate both 30 and 33 Kd protein bands. Transfer of the immunoprecipitated protein from SDS-PAGE gels to nitrocellulose paper and staining the transferred protein with the above monoclonal antibodies resulted in the following (Figure 4):

1. The monoclonal anti-I-A^k antibody reacted with both the 30Kd (beta chain) and 33Kd (alpha chain) bands (Figure 4, Lane 4). The same immunoprecipitated material stained with anti-I-J^k antibody reacted only with the 30Kd (beta chain) band (Figure 4, Lane 4). This suggests that the I-J protein expressed on BMDM may represent an I-A molecule which has a modification of its beta chain permitting the expression of the I-J epitope.
2. In contrast, the material immunoprecipitated with monoclonal anti-I-J^k antibody and stained with this antibody (Figure 4, Lanes 5 and 6) reacted only with the 30Kd or beta chain band.

One explanation of the above results is that I-A may be produced in BMDM, but due to glycosylation or some other post-translational modification the protein that reaches the surface of BMDM expresses only the I-J determinant. Precedent for a glycosylation event altering the function of a protein from a helper function to a suppressor function is seen among the IgE-binding factors reported by Ishizaka and his associates.[43] Alternatively, the expression of I-J may be due to altered splicing of an I-A gene transcript. At this time both possibilities exist, and we do not have evidence favoring one alternative or the other. Future efforts to differentiate between these alternatives are now in progress.

MATERIAL AND METHODS

The animals used in our experiments, antigens,[10] immunization route and schedules,[11,13] affinity adsorbents,[13] enrichment of T cells,[12] *in vitro* culture methods,[2,15,18] enzyme-linked immunosorbent assay (ELISA),[1] and flow cytometric methods[16,19] have been previously described in detail elsewhere.

ANTIBODIES

Normal rabbit IgG (nRGG) and normal mouse IgG (nMGG) was prepared by passing normal serum over a Sepharose-Protein A adsorbent (Pharmacia, Inc., Piscataway, N.J.), and the bound IgG was eluted with 0.05 M glycine-HCl buffer, pH 3. Rabbit anti-Timothy IgE (anti-Id) was prepared as described elsewhere.[13,14] Briefly, the anti-Timothy IgE was precipitated with 33% saturated ammonium sulfate, and passed over Sepharose-DNP IgE, -normal mouse serum, -AgB, -mouse IgG, -mouse light chains, and -fetal calf serum adsorbents to remove any contaminating antibodies.[12]

The preparation of polyclonal anti-I-J^k antibody was accomplished by immunizing B10.A(3R) mice with 5 to 10×10^6 B10.A(5R) BMDM emulsified in Freunds' complete adjuvant as previously described.[15,16]

Monoclonal anti-I-A antibodies were obtained from three separately cloned cell lines (HB15, IgM anti-I-A^k; HB38, IgM anti-I-A^b; and TIB 93, IgG_{2b} anti-I-A^k) from the American Type Culture Collection (ATCC, Rockville, MD). Monoclonal anti-I-J^k (clone JK10-23, IgG_{2a}) was a generous gift from Dr. Y. Asano (University of Tokyo). The fluorescent labeled anti-Mac 1 (FITC-anti-Mac 1) antibody was obtained from Dr. Carleton Stewart. The monoclonal anti-I-E^k was purchased from Accurate Chem. Co. (Hicksville, N.Y.).

The IgG fractions of anti-I-A^k (TIB 93), anti-I-E^k, and anti-I-J^k (clone JK10-23) were obtained by elution from a Sepharose-Protein A adsorbent or from a Affigel-mouse IgG (gamma chain specific) adsorbent. The IgM fraction of polyclonal anti-I-J^k antibody, and monoclonal anti-I-A^k or monoclonal anti-I-A^b antibody was obtained by passage over a Sepharose-Protein A adsorbent, and further passage of the non-adherent fraction over a Affigel-mouse IgM adsorbent. The bound IgM fractions were eluted with 0.05 M glycine-HCl buffer pH 3.[16] Anti-F_c receptor antibody was obtained from a cloned cell line (2.4G 2) generously provided by Dr. Unkeless (Rockefeller University), and was purified over a anti-Rat IgG adsorbent. FITC labeled goat anti-IgM $F(Ab)_2$ fragment, and the FITC-labeled goat-anti-mouse IgG (H and L) was obtained from TAGO, Inc., Burlingame, CA. The IgG_{2a} myeloma protein used for blocking experiments was a gift from Dr. Marvin Rittenberg (Oregon Health Sciences Univ., Portland, OR), and the FITC-labeled IgG_{2a} was prepared as previously described.[16] Monoclonal cell lines producing antibodies directed against T cell antigens Lyt 1(ATCC, TIB 104), Lyt 2 (ATCC, TIB105), and L3T4 (ATCC, TIB 207) were obtained from ATCC, and were grown in culture with the media specified by ATCC. The resulting antibodies were purified by ammonium sulfate precipitation and affinity methods.

EXTRACTION OF I-J$^+$ BONE-MARROW-DERIVED MACROPHAGES

Bone marrow cells were maintained in culture for 7 days, and the harvested cells were washed twice in balanced salt solution, once in PBS, and resuspended in 3 ml of PBS. The cells were counted by hemocytometer, and centrifuged at 500 x g for 10 minutes at 4°C. The supernatant was removed, and 3 ml of a NP-40 Protease inhibitor cocktail was added to give a final concentration of 5×10^7 BMDM cells/ml of detergent. The Protease inhibitor cocktail contained 3mM phenylmethylsulfonylfluoride, 60mM iodoacetamide, 3 micro-grams Leupeptin, 2.1 International Units Aprotinin, and 0.5% NP-40 in 0.01 M Tris-HCl buffer pH 7.5 containing 0.14 M NaCl and 1.5 mM $MgCl_2$. The cell suspension was vortexed at high speed for 1 minute, and placed on a rocking platform at 4° for 20 minutes. The suspension was centrifuged at 28,000 x g for 30 minutes, and the supernatant was filtered through a 0.22 micron filter and stored at -20°C. Removal of the NP-40 detergent was accomplished by passage of the BMDM extract over Bio-Rad SM2 beads (10mmx 60mm) column, and 75 ml of effluent was concentrated by negative pressure to a final volume of 1.5 ml. This sample was stored at -20°C until used.

IMMUNOPRECIPITATION

BMDM extract (30 microliters) was incubated for 30 minutes with 10 micrograms of normal mouse IgG, monoclonal anti-I-A^k (Clone TIB 93), or monoclonal anti-I-J^k (clone JK10-23) antibody for 30 minutes at 4°C. These mixtures were treated with 150 micro-liters of a Protein A bacterial adsorbent (Miles Scientific, Naperville, IL), and these suspensions were mixed for 30 minutes at 4°C. These suspensions were centrifuged for 5 minutes in a microcentrifuge to remove the supernatants. The remaining adsorbent was transferred to a 15 ml centrifuge tube, and the adsorbent was washed three times with a large excess of 0.63 M phosphate buffer pH 7.6 containing 0.2 M NaCl, 0.02% sodium azide, 0.05% NP-40 and 2mM methionine. The final pellet was transferred to a microcentrifuge tube (1.5 ml capacity), and centrifuged at 1000 x g to remove the excess supernatant. The individuals adsorbents were treated with 50 microliters of elution buffer (0.63 M Tris buffer, pH 6.8 containing 5% beta-mercaptoethanol, 2% SDS, and 5% NP-40), and vortexed at maximum speed for 1 minute. The suspensions were centrifuged at 1000 x g, and the supernatants carefully removed and stored at -20°C.

Western blotting was performed essentially according to Towbin et al.[20] Initially, polyacrylamide gel electrophoresis (PAGE) was performed according to the procedure of Laemmli.[21] Immunoprecipitated samples were diluted with sample buffer containing 5% beta-mercaptoethanol followed by boiling for 10 minutes. The separated samples were then transferred to nitrocellulose in a Bio-Rad Trans Blot apparatus in 0.025 M Tris-0.192 M glycine buffer at pH 8.3 with 20% methanol. The nitrocellulose paper was blocked with 3% gelatin in the Tris-glycine buffer at 37°C for 1 hour. Individual strips (4mm wide) were treated with either monoclonal anti-I-A^k or monoclonal anti-I-J^k antibody (10 micrograms/3 ml) diluted with the Tris-glycine buffer containing 1% gelatin and 0.5% Tween 20. The strips were washed three times with the Tris-glycine buffer containing both gelatin and Tween 20, and twice with the buffer without the Tween 20. A conjugate of goat anti-mouse IgG-horse radish peroxidate (Dynatech) antibody diluted 1:500 in the buffer containing 1% gelatin was applied. After washing the strips to remove unbound antibody, the strips were placed in a horse radish peroxidate substrate (Bio-Rad Labs) for 10 minutes, and finally washed with water.

ACKNOWLEDGMENTS

The work reported in this article, publication number 1518 of the Oregon Regional Primate Res. Ctr., was supported by U.S. Public Health Service Grants AI-18624, AI-19532, and RR00163, the Collins Medical Trust, and the Medical Research Foundation of Oregon.

REFERENCES

1. Malley, A. and D.W. Dresser (1983), "Regulation of Timothy Grass Pollen IgE Antibody Formation: In Vitro Induction of Suppressor T Cells by Soluble T Suppressor Factors," *Immunology* **49**, 463.
2. Malley, A. and D.W. Dresser (1983), "Anti-Idiotypic Regulation of the Formation of IgE Antibody to Timothy Grass Pollen. II. In Vitro Induction of Suppressor T Cells in Mini-Marbrook Cultures," *Immunology* **48**, 93.
3. Malley, A. and D.W. Dresser (1982), "Anti-Idiotypic Regulation of Timothy Grass Pollen IgE Antibody Formation. I. In Vitro Induction of Suppressor T Cells," *Immunology* **46**, 653.
4. Malley, A. , L. Bradley, and S. Shiigi (1983), "Anti-Idiotypic Antibody Regulation of Timothy Grass Pollen IgE Responses," *Intracellular Communication in Leucocyte Function*, Eds. J.W. Parker and R.L. O'Brien (Chichester, England: John Wiley and Sons, Ltd.), p. 623.
5. Fitch, F.W. (1975), "Selective Suppression of Immune Responses. Regulation of Anti-Body Formation and Cell-Mediated Immunity by Antibody," *Prog. Allergy* **19**, 195.
6. Forni, L. and B. Pernis (1975), "Interactions between F_c Receptors and Membrane Immunoglobulins on B Lymphocytes," *Membrane Receptors of Lymphocytes*, Eds. M. Seligman, J.L. Preud'Home and F.M. Kourilinsky (New York: American Elsevier).
7. Moretta, L. M.C. Mingari, and C.A. Romangi (1978), "Loss of F_c Receptors for IgG from Human T Lymphocytes Exposed to IgG Immune Complexes," *Nature (London)* **272**, 618.
8. Pierce, S.K. and N.R. Klinman (1977), "Antibody-Specific Immunoregulation," *J. Exp. Med.* **146**, 509.
9. Koski, I.R., D.G. Poplack, and R.M. Blaese (1976) "A Nonspecific Esterase Stain for the Identification of Monocytes and Macrophages," *In Vitro Methods in Cell-Mediated Tumor Immunity*, Eds. B.R.Bloom and J.R. David (New York: Academic Press), p.359,.
10. Malley, A. and R.J. Harris, Jr. (1967), "Biological Properties of a Nonprecipitating Antigen from Timothy Pollen Extracts," *J. Immunol.* **99**, 825.
11. Malley, A., D. Begley, and A. Forsham (1979), "Chemical Modification of Timothy Grass Pollen Antigen B. I. Loss of Antibody Binding Properties," *Mole. Immunol.* **16**, 929.
12. Malky, A., L. M. Bradley, and S. M. Shiigi (1987), "The Role of Macrophages in Anti-Idiotypic Antibody and T Suppressor Factor Induction of Timothy Grass Pollen Antigen Specific T Suppressor Cells," *J. Immunol.* **139**, 1046.
13. Malley, A., C.J. Brandt, and L.B. Deppe (1982), "Preparation and Characterization of the Anti-idiotypic Properties of Rabbit Anti-Timothy Antigen B Helper Factor and Anti-Mouse Timothy IgE Antisera," *Immunology* **45**, 217.
14. Campbell, D.H., J.S. Garvey, N.E. Cremer, and D.H. Sussdorf (1962), in *Methods in Immunology* (New York: W.A. Benjamin, Inc.), p. 118.

15. Bradley, L.M., S.M. Shiigi, and A. Malley (1986), "Anti-I-J Allosera Elicited by Immunization of B10.A(3R) (I-Jb Mice with Bone-Marrow-Derived Macrophages from B10.A(5R) (I-J^k Mice," *Immunology* **57**, 443.
16. Malley, A. C.C. Stewart, S.J. Stewart, L. Waldbeser, L.M. Bradley, and S.M. Shiigi, "A Flow Cytometric Analysis of I-J Expression on Murine Bone-Marrow-Derived Macrophages," *J. Immunol.* (in press).
17. Wyocki, L.J., and V.L. Sato (1978), "'Panning' for Lymphocytes: A Method for Cell Selection," *Proc. Natl Acad. Sci.* **75**, 2844.
18. Stewart C.C. (1981), "Murine Mononuclear Phagocytes from Bone Marrow," *Methods for Studying Mononuclear Phagocytes* (New York: Academic Press), p. 5.
19. Steinkamp, L.J., and C.C. Stewart (1986), "A Dual-Laser, Differential Fluorescence Correction Method for Eliminating Background Autofluorescence," *Cytometry* **7**, 566.
20. Towbin, H., T. Staehelin, and J. Gordon (1979), "Electrophoresis Transfer of Proteins from Polyacrylamide Gels to Nitrocellulose Sheets: Procedure and Some Applications," *Proc. Natl. Acad. Sci. U.S.A.* **76**, 4350.
21. Laemmli, U.K. (1970), "Cleavage of Structural Proteins during the Assembly of the Head of Bacteriophage T4," *Nature* **227**, 680.
22. Murphy, D.B. (1987), "The I-J Puzzle," *Ann. Review Immunol.* **5**, 405.
23. Feldmann, M., and S. Kontiainen (1976), "Suppressor Cell Induction In Vitro. II. Cellular Requirements of Suppressor Cell Induction," *Eur. J. Immunol.* **6**, 302.
24. Ishizaka, K., and T. Adachi (1976), "Generation of Specific Helper Cells and Suppressor Cells in Vitro for the IgE and IgG Antibody Responses," *J. Immunol.* **117**, 40.
25. Aoki, I., M. Minami, and M.E. Dorf (1984), "A Mechanism Responsible for the Induction of H-2 Restricted Second Order Suppressor T Cells," *J. Exp. Med.* **157**, 1726.
26. Aoki, I., M. Usui, M. Minami, and M.E. Dorf (1984), "A Genetically Restricted Suppressor Factor that Requires Interaction with Distinct Targets," *J. Immunol.* **132**, 1735.
27. Usui, M., I. Aoki,G.H. Sunshine, and M.E. Dorf (1984), "A Role for Macrophages in Suppressor Cell Induction," *J. Immunol.* **132**, 1728.
28. Lowy, A., A. Tominaga, J.A. Debrin, M. Takaoki, B. Benacerraf, and M.I. Greene (1983), "Identification of an I-J+ Antigen-Presenting Cell Required for Third Order Suppressor Cell Activation," *J. Exp. Med.* **157**, 353.
29. Ptak, W., M. Zembala, and R.K. Gershon(1978), "Intermediary Role of Macrophages in the Passage of Suppressor Signals between T Cell Subsets," *J. Exp. Med.* **148**, 424.
30. Nakamura, R.M., H. Tanaka, and T. Takunaga (1982), "In Vitro Induction of Suppressor T Cells in Delayed-Type Hypersensitivity to BCG and an Essential Role of I-J Positive Accessory Cells," *Immunol. Letters* **4**, 295.

31. Nakamura, R.M., Y. Nakamura, A. Nagayama, and T. Tokunga (1986), "I-J Positive Cloned Macrophages as Accessory Cells for the Induction of Suppressor T Cells in Vitro," *Immunol. Res.* **5**, 106.
32. Nakamura, R.M., Y. Nagayama, and T. Tokunga (1985), "Accessory Cell Function of a Lined Macrophage in the Induction of Suppressor T Cells in Vitro," *Surv. Synth. Path. Res.* **4**, 443.
33. Murphy, D.B., K. Yamauchi, S. Habu, D. Eardley, and R.K. Gershon (1981), "T Cells in a Suppressor Circuit and Non T:Non B Cells Bear Different I-J Determinants," *Immunogen.* **13**, 205.
34. Flood, P.M., C. Waltenbaugh, T. Tada, B. Chue, and D.B. Murphy (1986), "Serological Heterogeneity in I-J Determinants Associated with Functionally Distinct T Cell Regulatory Factors," *J. Immunol.* **137**, 2237.
35. Noma, T., M.Usui, and M.E. Dorf (1985), "Characterization of the Accessory Cells Involved in Suppressor T Cell Induction," *J.Immunol.* **134**, 1374.
36. Lee, K.C., and M. Wong (1980), "Functional Heterogenity of Culture Grown Bone-Marrow-Derived Macrophages. I. Antigen Presenting Function," *J. Immunol.* **125**, 86.
37. Walker, W.S., R.B. Hester, D. Gandour, and C.C. Stewart (1981), "Evidence of a Distinct Progenitor for the I-A-bearing (I-A+) Murine Bone-Marrow-Derived Mononuclear Phagocytes," *Heterogenity of Mononuclear Phagocytes*, Eds. O. Forster and M. Landy (New York: Academic Press).
38. Stewart, C.C. (1985), "Constitutive I-A Positive Macrophages," *J. Leukocyte Biol.* **38**, 647.
39. Calamai, E.G. D.I. Beller, and E.R. Unanue (1982), "Regulation of Macrophage Populations. IV. Modulation of I-A Expression in Bone-Marrow-Derived Macrophages," *J. Immunol.* **128**, 1692.
40. Unanue, E.R. (1981), "The Regulatory Role of Macrophages in Antigen Stimulation. Part two: Symbiotic Relationship between Lymphocytes and Macrophages," *Adv. Immunol.* **31**, 1.
41. Lee, S.T., and F. Paraskevas (1979), "Macrophage-T Cell Interactins. II. The Uptake by T Cells of Fragments Bearing Ia Determinants Released from Macrophages," *Cell Immunol.* **48**, 1
42. Zupko, K.C., C. Waltenbaugh, and B. Diamond (1985), "Use of Anti-Idiotypic Antibodies to Identify a Receptor for the T Cell I-J Determinant," *Proc. Nat. Acad. Sci. U.S.A.* **82**, 7379.
43. Yodoi, J., M. Hirashima, and K. Ishizaka (1981), "Lymphocytes Bearing F_c Receptors for IgE. V. Effect of Tunicamycin on the Formation of IgE-Potentiating Factor and IgE-Suppressor Factor by Con A Activated Lymphocytes," *J. Immunol.* **126**, 877.

HANS-GEORG RAMMENSEE
Basel Institute for Immunology

Control of Lethal Autoreactivity without a Network

One mechanism for the induction of self-tolerance is clonal deletion of self-reactive T cells in the thymus.[1] It is not known whether this is sufficient to maintain self-tolerance in the periphery. To be sufficient would require the absence of somatic variation of T cell receptor (TCR) genes in peripheral T cells, of extrathymic differentiation of T cells, and of any leakiness of the thymus negative selection process. Rather than testing directly whether these requirements are fulfilled, we have studied how self-reactive T cells would be handled when introduced artificially into a normal mouse.

In order to make it easier to understand the design of the crucial experiments to be described, I shall briefly explain the veto mechanism, a particular phenomenon of suppression which has been implicated in the induction and/or maintenance of self-tolerance.[2,3] When mature cytotoxic T lymphocytes (CTL) are recognized by CTL-precursors (CTLp), the CTLp are inhibited from being activated (see Figure 1). In a typical assay to study this inhibition, lymphocytes from mouse strain A are stimulated in a mixed leukocyte culture (MLC) with a mixture of irradiated cells of strain B and C (A, B and C are different MHC haplotypes). During the MLC roughly equal numbers of CTL specific for B and C will develop. If, however, some CTL B anti-X (X can be any antigen apart from A, B or C) are added to the MLC, no B-specific CTL develop, whereas the development of C-specific CTL is unaltered.[4,5] The interaction leading to this kind

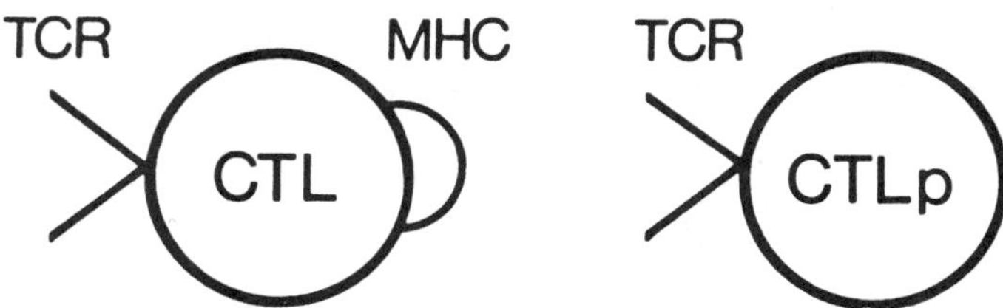

FIGURE 1 Inactivation of CTL precursors (CTLp) by mature CTL. A CTLp recognizes MHC on a mature CTL. Upon this passive recognition, the mature CTL inactivates the CTLp by an unknown mechanism.

of inhibition is well understood. The TCR on a CTLp of strain A recognizes a surface antigen of the mature CTL of strain B. That antigen may be MHC class I, alone or in conjunction with a minor H antigen. Upon this passive recognition, the mature CTL sends an inhibitory signal to the recognizing CTLp. The molecular nature of this signal is not known.[6] The veto function of CTL[7] has two intriguing features: i) the recognition specificity is very simple: a TCR recognizes antigen; ii) such an interaction would seem to be an easy way to get rid of self-reactive CTL in a normal immune system, because any A-specific CTLp arising in a mouse of strain A ought eventually to bind to another CTL, which expresses A antigen, thereby getting inactivated.

Consistent with such a role of CTL in maintaining self-tolerance are some *in vivo* experiments: $CD8^+$ T cells (CD8 is a surface antigen on all CTL), of strain A injected into MHC class I incompatible mice of strain B cause a transient nonresponsiveness to B, which lasts for several weeks and which is followed by a hyporesponsiveness to B lasting several months ([8] and HGR, unpublished). The nonresponsiveness is formally the same as clonal deletion; neither in limiting dilution cultures nor by removal of donor-derived T cells was it possible to recover any donor reactive CTL from nonresponsive recipients.[9,10] Since co-injection of CD4 (a surface antigen on helper T cells) specific antibodies along with donor cells does not affect the induction of nonresponsiveness, helper T cells do not seem to be involved in the induction of nonresponsiveness (HGR, unpublished). More importantly, this experiment also excludes the involvement of $CD4^+$ "T suppressor inducer cells" in this system.

Impressed by the strong and specific effects seen in this system both *in vitro* and *in vivo*, we sought to test its physiological role. For this purpose, we constructed mice which were tolerant by thymic negative selection, but which were devoid of self-antigen-expressing CTL and other bone-marrow-derived cells in the periphery. Athymic nude mice of strain A were grafted with fetal thymuses of strain B (strains A and B are BALB/c and C57BL/6, respectively). Such thymus-grafted nudes develop a normal T cell compartment and are tolerant of A and B antigens.[11] As a control for normal self-tolerance, we used (A x B)F_1 mice expressing all self-MHC antigens on cells of hemopoietic lineage. Both groups of mice were injected intravenously with normal strain A spleen cells (see Figure 2), which contain a substantial number of B-specific CTLp. Six days after injection, cells from the recipient spleen were stimulated against B and third-party antigens, and tested in a killer assay. A striking difference between the two groups was observed: B-reactive CTL were readily detectable in the nudes, but not in the F_1 mice (figure 3). Even when donor-derived CTL were isolated from F_1 recipients, only third-party-reactive CTL could be demonstrated, never recipient-specific ones. Co-injection of CD4-specific antibodies or irradiated B antigen together with A cells into both groups of mice did not reveal donor-derived, recipient-specific CTL from normal F_1. In F_1 mice irradiated (600 rad) before injection, however, recipient-reactive CTL could be readily detected, a result which suggests that trapping of recipient-specific CTL in the recipient body is not the reason for the disappearance of recipient-specific CTL in non-irradiated recipients.

I conclude that CTL reactive against self-antigen in a normal mouse are rigorously eliminated by a mechanism that is a) independent of thymic MHC, b) sensitive to 600 rad irradiation, c) independent of $CD4^+$ T cells (T helper or "T

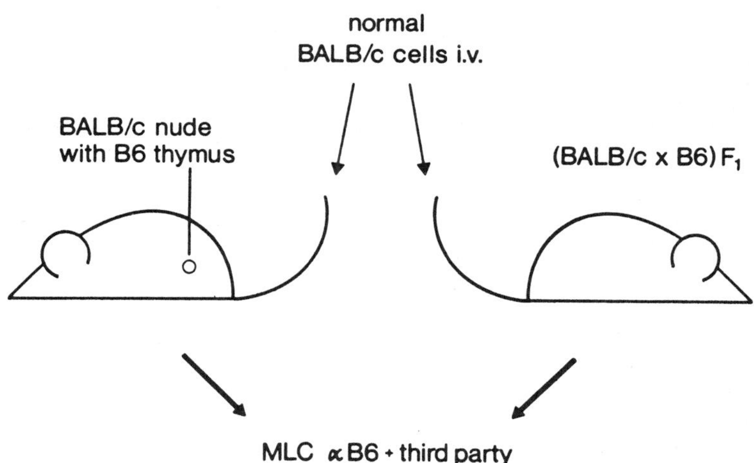

FIGURE 2 Experimental design to test the resistance against self-reactive CTL. Normal mice ($CB6F_1$) or mice tolerant by thymic negative selection only (athymic BALB/c nude mice grafted with B6 fetal thymus) are injected intravenously with normal BALB/c spleen cells which contain B6 reactive CTLp. Six days after injection, recipients are sacrificed and the spleen cells are stimulated against B6 and third-party antigens and tested in a ^{51}Cr release assay.

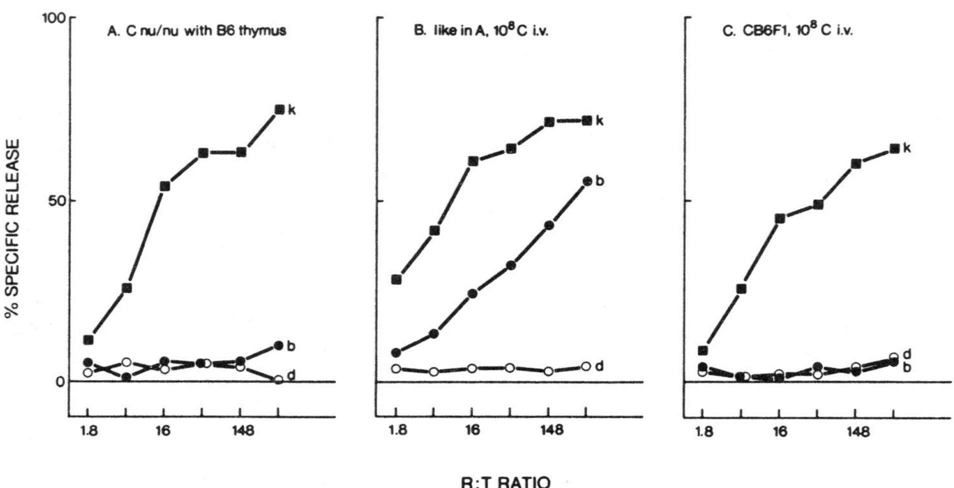

FIGURE 3 Different resistance to thymus MHC-reactive CTL in the two groups of mice from Figure 2. Spleen cells from an uninjected thymus-grafted nude mouse (a), from an injected, thymus-grafted nude and from an injected, $CB6F_1$ (c) were stimulated in MLC with H-2^b and H-2^k. The resulting CTL were tested on H-2^d (○), H-2^k (■), and H-2^b (●) targets in a standard ^{51}Cr release assay.[9]

suppressor inducer" phenotype), and d) dependent on antigen-expressing cells in the periphery. By analogy with the *in vitro* experiments mentioned above, I assume that these antigen expressing cells are of hematopoietic origin.

This mechanism seems not to work for MHC class-II-reactive T helper cells; indeed, class II antigens are not present on mouse CTL. In (A x B)F_1 mice injected with parental A strain cells, lymphoproliferative graft-versus-host disease (GVHD) can be induced by class-II-reactive T cells, but it is usually not lethal.[10,12] When the F_1 is irradiated, however, the GVHD becomes lethal. This lethality in the non-irradiated F_1 hosts may be avoided by the veto mechanism described above. It may be predicted that in a normal mouse, any self-MHC-reactive CTL that arise would be eliminated by this mechanism.

The mechanism I am discussing here is not an antigen-independent network. T cells tolerant by thymic negative selection do not possess the capacity to eliminate other T cells reactive to thymic antigen in the absence of antigen expressed on non-irradiated, presumably hemopoietic cells. Moreover, $CD4^+$ T cells are not required, which is evidence against TCR-TCR interactions involving "T suppressor inducer" cells. Finally, the simple TCR-antigen interaction described in the in *vitro* system above leading to inhibition of CTLp recognizing CTL is also consistent with the *in vivo* data; there is no evidence that more complicated receptor interactions are involved. However, different systems regulate self-class-II-reactive T cells, and it is claimed that these involve TCR-TCR interactions.[13] Thus, it seems that several mechanisms for maintaining self-tolerance exist within the immune system; their usage may depend on the class of immune response.

ACKNOWLEDGMENTS

I thank C. Steinberg and B. Stockinger for improving the manuscript, D. Hügin and K. Hafen for technical and C. Plattner for secretarial assistance. The Basel Institute for Immunology was founded and is supported by F. Hoffmann-La Roche and Company, Basel, Switzerland.

REFERENCES

1. Kappler, J. W., N. Roehm, and P. O. Marrack (1987), "T Cell Tolerance by Clonal Elimination in the Thymus," *Cell* **49**, 273.
2. Miller, R. G. (1980), "An Iimmunological Suppressor Cell Inactivating Cytotoxic T Lymphocyte Precursor Cells Recognizing It," *Nature* **287**, 544.
3. Rammensee, H.-G., M. J. Bevan, and P. J. Fink (1985), "Antigen Specific Suppression of T-Cell Responses—The Veto Concept," *Immunol. Today* **6**, 41.
4. Fink, P. J., H.-G. Rammensee, and M. J. Bevan (1984), "Cloned Cytolytic T Cells Can Suppress Primary Cytotoxic Responses Directed Against Them," *J. Immunol.* **133**, 1775.
5. Claesson, M. H. and R. G. Miller (1984), "Functional Heterogeneity in Allospecific Cytotoxic T Lymphocyte Clones," *J. Exp. Med.* **160**, 1702.
6. Fink, P. J., H.-G. Rammensee, J. D. Benedetto, U. D. Staerz, L. Lefrancis, and M. J. Bevan (1984), "Studies on the Mechanism of Suppression of Primary Cytotoxic Responses by Cloned Cytotoxic T Lymphocytes," *J. Immunol.* **133**, 1769.
7. Miller, R. G. (1980), "How a Specific Anti-Self Deletion Mechanism Can Effect the Generation of the Specificity Repertoire," *Strategies of Immune Regulation*, Eds. E. E. Sercarz and A. J. Cunningham (New York: Academic Press), 507.
8. Rammensee, H.-G., P. J. Fink, and M. J. Bevan (1985), "The Veto Concept: An Economic System for Maintaining Self-Tolerance of Cytotoxic T Lymphocytes," *Transplant. Proc.* **23**, 689.
9. Rammensee, H.-G., P. J. Fink, and M. J. Bevan (1984), "Functional Clonal Deletion of Class I-Specific Cytotoxic T Lymphocytes by Veto Cells that Express Antigen," *J. Immunol.* **133**, 2390.
10. Rammensee, H.-G., and M. J. Bevan (1987), "Mutual Tolerization of Histoincompatible Lymphocytes," *Eur. J. Immunol.* **17**, 893.
11. Von Boehmer, H. and K. Schubiger (1984), "Thymocytes Appear to Ignore Class I Major Histocompatibility Complex Antigen Expressed on Thymus Epithelial Cells," *Eur. J. Immunol.* **14**, 1084.
12. Pals, S. T., H. Gleichmann, and E. Gleichmann (1984), "Allosuppressor and Allohelper T Cells in Acute and Chronic Graft-vs.-Host Disease. V. F_1 Mice with Secondary Chronic GVHD Contain F_1-Reactive Allohelper but no Allosupressor Cells," *J. Exp. Med.* **159**, 508.
13. Kimura, H., and D. B. Wilson (1984), "Anti-Idiotypic Cytotoxic T Cells in Rats with Graft-Versus-Host Disease," *Nature* **308**, 463.

Idiotypic Networks

NEIL S. GREENSPAN† and KENNETH H. ROUX‡
†Institute of Pathology, Case Western Reserve University, Cleveland, Ohio 44106 and ‡Department of Biological Science, Florida State University, Tallahassee, Florida 32306

Categories of Idiotope Overlap and Anti-Idiotypic Mimicry of Antigen

INTRODUCTION

Antibodies (Abs) specific for antigenic determinants (idiotopes or Ids) expressed by the variable domains of other Abs (termed Ab1) are known as anti-Ids (or Ab2). Some anti-Ids, specific for a given Ab1, are able to mimic properties of the antigen also bound by Ab1. Anti-idiotypic mimicry has been observed for a wide variety of molecules, ranging from proteins such as insulin[1] to small molecules such as alprenolol.[2] Anti-Ids that express such properties have been referred to as internal images,[3] homobodies,[4] or Ab2-beta.[5] These Abs have been distinguished from other anti-Ids, sometimes referred to as Ab2-alpha or Ab2-gamma. In this nomenclature, Ab2-alpha refers to anti-Ids able to bind to their corresponding Ids in the presence of hapten, and Ab2-gamma refers to hapten-inhibitable anti-Ids that fail to mimic the relevant antigen.[6] Interest in anti-Ids, and especially those that mimic antigens, stems in part from the fundamental biological and biochemical questions that the mimicry phenomenon raises, and also from the hope that this class of Abs might be useful in research (e.g., in identifying and isolating cell surface receptors) and in clinical medicine (e.g., in manipulating immune responses). For instance, it has been argued that anti-Ids that mimic selected antigens from important pathogens might be useful as surrogate vaccines.[7,8] Experiments designed to evaluate the feasibility of this approach have been carried out over the past few years,[9] and some of the results are encouraging. Thus, understanding the nature and properties of this subclass of anti-Ids could conceivably have practical, as well as theoretical, consequences.

Our main purpose will be to analyze, and hopefully clarify, the concept of anti-idiotypic mimicry. In our view, current classification schemes for anti-Ids fail to adequately define the properties required for an anti-Id to be labeled an internal image of some antigen. We will apply to this analysis a perspective developed through the attempt to map Ids on the surface of an immunoglobulin variable domain. There are three potential relationships between any pair of Ids expressed by an Ab: 1) the sites are identical, 2) the sites are completely distinct and independent, and 3) the sites are overlapping. Consideration of the meaning of the third category, idiotopic overlap, provided the primary stimulus for the argument developed below.

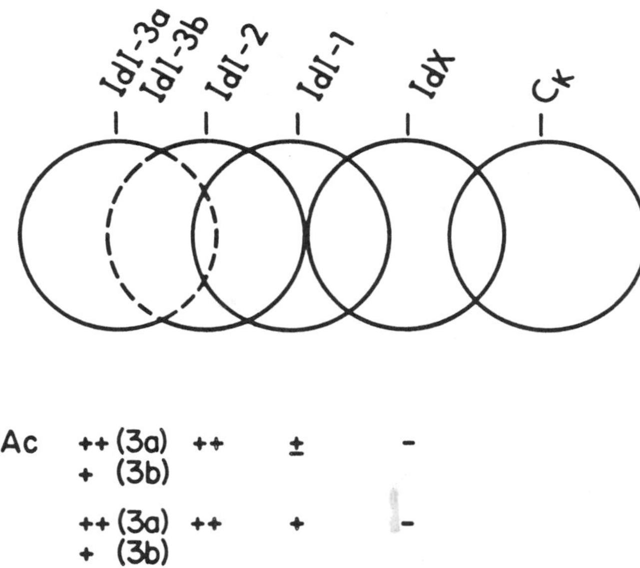

FIGURE 1 Schematic representation of the topographic relationships among HGAC 39 Ids (modified from Figure 7, reference 10). Overlap of circles indicates effective competition (anti-Id versus anti-Id) between corresponding mAbs for binding to HGAC 39 Ids while lack of overlap indicates no competition. Broken lines for IdI-3a (IdI-3b) and IdI-2 indicate uncertainty about the relative placements of these Ids. Relative abilities of GAC and GlcNAc to inhibit binding of each anti-Id to Id (HGAC 39) are indicated by the number of "+" symbols, more +'s meaning that a lower concentration of hapten was required to inhibit binding of 50% of the radiolabeled HGAC 39 (Id) to anti-Id.

IDIOTOPE OVERLAP

In our initial studies on the variable domain of the prototype murine anti-streptococcal group A carbohydrate (GAC) monoclonal antibody (mAb), HGAC 39, we employed a panel of xenogeneic monoclonal anti-Ids (derived from HGAC 39-immune rat spleen cells) in competitive and direct binding assays. The topographic relationships of five Ids (IdX, IdI-1, -2, -3a, and -3b) were operationally determined by evaluating the degrees to which hapten, N-acetyl-D-glucosamine (GlcNAc) or GAC, or other anti-Ids competed with each anti-Id for binding to HGAC 39.[10] Figure 1 provides a schematic summary of the conclusions drawn from these studies. It is apparent from Figure 1 that anti-IdI-1 competes with each of the other anti-Ids in the panel. This fact might be summarized by saying that IdI-1 overlaps with all four of the other HGAC 39 Ids. However, this apparently straightforward statement seems less clear-cut when the same set of Ids is analyzed by another technique.

As stated above, the Id map based on the competitive binding studies is considered operational, in somewhat the same way that a genetic map based on recombination frequency is operational. For example, while a lack of competition between two ligands (simultaneous binding) can be clearly interpreted as binding to separate sites, the presence of competition does not necessarily imply proximity of the two determinants that interact with the two ligands.[11] Therefore, it was important to attempt to map the Ids on HGAC 39 by a technique not subject to the same set of interpretive uncertainties as are associated with competitive binding assays. We chose to construct another Id map based on transmission electron microscopic analysis of negatively stained complexes of idiotopic and anti-idiotopic Abs or their proteolytic fragments.[12] Figure 2 indicates what variables were being measured. Evaluation of complexes such as those in Figure 3 and additional complexes involving three components (data not shown) resulted in the low-resolution three-dimensional model depicted in Figure 4. It is interesting, in light of our previous observations on the overlap (in terms of anti-Id competition) between IdI-1 and the other Ids, that IdI-1 appears quite far away from IdI-2 or IdI-3a. Since an anti-Id Fab fragment must be of equal or greater diameter than the Id being recognized, it may be that reciprocal competition between anti-IdI-1 and anti-IdI-2 and -3a occurs on the basis of steric effects, as opposed to conformational effects between distant sites. Nevertheless, it appears that the degree of overlap at the level of competition for binding is not necessarily an accurate quantitative guide to the magnitude of spatial overlap (overlap of the surfaces on the idiotopic variable domain buried by the respective anti-Ids).

The concept of overlap among Ids (or other epitopes) can also be applied when determinants are defined in terms of the positions in the primary structure of the antibody molecule at which amino acid substitutions affect Id expression. Such substitutions have been defined in a number of systems by comparing variable

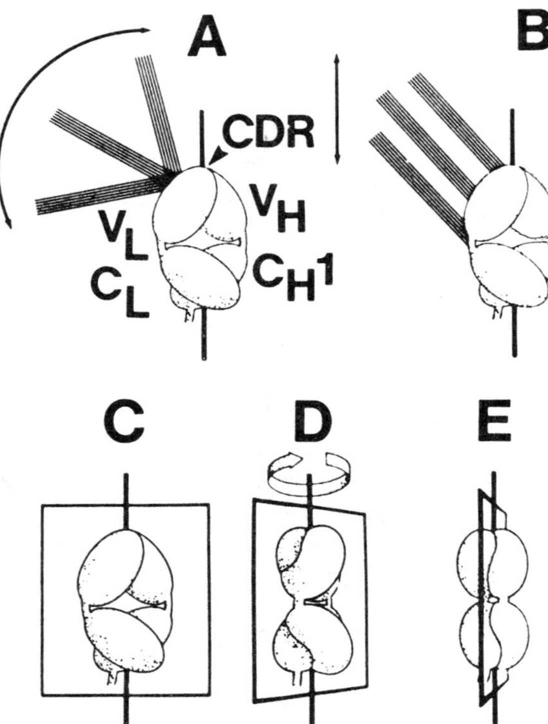

FIGURE 2 Depiction of the three variables evaluated by transmission electron microscopy for each negatively stained Id-anti-Id complex. Heavy chain (H) and light chain (L) domains of the component Fab are labeled in a schematic diagram of an Fab fragment to facilitate understanding of the variables being measured, but are not distinguishable in the electron micrographs. Illustrated are: (A), the angle at which the anti-Id probe (represented by parallel lines) and the target Fab intersect, with reference to the proximal-distal axes; (B), the relative position of the Id-anti-Id intersect on the Id proximal-distal axis; and (C–E), the possible rotational planes of the probe or target Fab fragments. CDR: complementarity determining regions. From Figure 1, reference [12].

domain amino acid sequences and patterns of Id expression either for independently isolated monoclonal Abs of similar antigen-binding specificity[13] (reviewed in [14,15]) or for wild-type and mutant (Id-loss variant) Abs.[16,17] In this context, two Ids can be considered to overlap if the expression of both determinants is influenced by the same amino acid substitution(s). This definition of Id overlap could clearly accommodate a pair of Ids that are non-overlapping in spatial extent or in terms of the competition for binding of the corresponding anti-Ids, but that are each conformationally altered by substitution of an (possibly distant) amino acid.

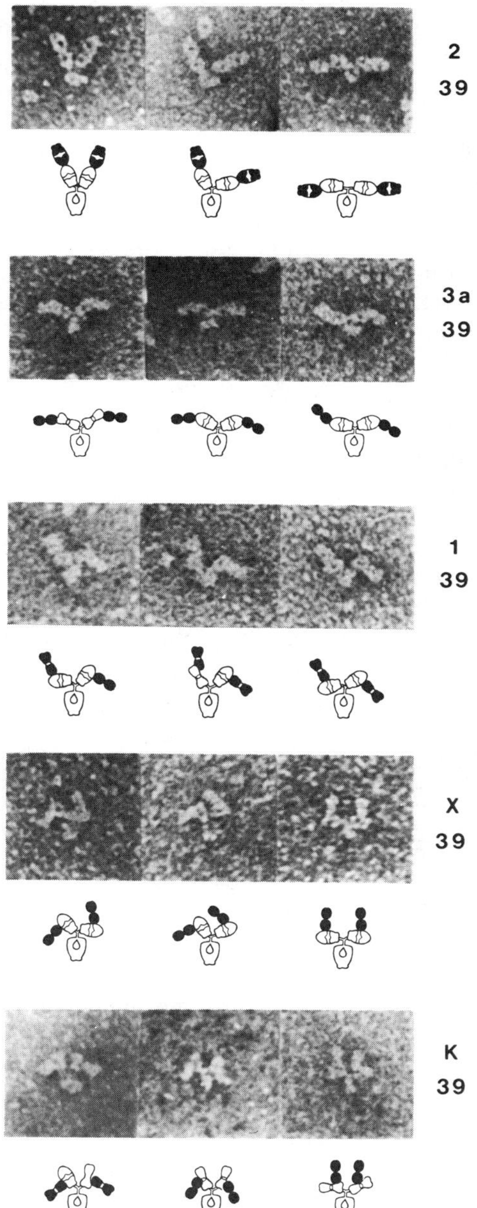

FIGURE 3 Electron micrographs (above) and interpretive diagrams (below) of HGAC 39 in complex with monoclonal anti-Id Fab fragments. HGAC 39 is represented in the diagrams as an open figure and the anti-Id Fab fragments are represented as solid figures. The Fab arms of the Id and anti-Id antibodies are drawn to indicate their rotational orientations as planar (Figure 2C), intermediate (Figure 2D), or perpendicular (Figure 2E). Abbreviations: 39, HGAC 39; 2, anti-IdI-2 Fab; 3a, anti-IdI-3a Fab; 1, anti-IdI-1 Fab; X, anti-IdX Fab; and K, anti-C_K Fab. Magnification: 350,000 X. From Figure 2, reference [12].

The solving of the three-dimensional structure of an Id(Fab)-anti-Id(Fab) crystal by x-ray diffraction represents another potentially useful way to define an Id.

Although no such structure has yet been reported, this approach would potentially allow definition of an Id in terms of the complete set of constituent amino acids that are involved in forming the noncovalent bonds between Id and anti-Id. One can then consider two Ids that share component amino acids as overlapping, but it is certainly conceivable that the magnitude of overlap in this sense would fail to be an absolutely reliable guide to the number of amino acid substitutions affecting two Ids or to the degree of competition of the corresponding anti-Ids. For example, as suggested above, two Ids that are spatially separate, and share no component amino acid residues, might both be substantially altered by the same substitution if it has long-range effects on the conformation of the variable domain.

In summary, these various approaches to the study of Id expression suggest that we can specify at least four ways to define Ids and four levels at which Ids, or other epitopes, can overlap. These four levels of definition or overlap focus on: 1) spatial extent, 2) composition in terms of contact residues, 3) amino acid sequence correlates of Id expression, and 4) competition of the corresponding anti-Ids. For brevity, we will refer to these various kinds of overlap as, respectively, spatial, compositional, sequential (or genetic), and competitive. It is, as argued above, conceivable that a pair of Ids can simultaneously exhibit different degrees of overlap for these four categories of overlap. Furthermore, it is not obvious that any one level is clearly the most important. In some contexts, the amino acid sequence correlates may be most relevant, while in other contexts competitive or compositional definitions (or overlap) may be the most useful. For example, in treating and monitoring human B cell lymphomas with monoclonal anti-Ids, it was observed that variant (Id-negative) tumor cells sometimes arise.[18,19] Such a variant is presumably resistant to the effects of the anti-Id present during its selection. The solution to this problem proposed by one group was the simultaneous use of multiple anti-Ids.[19] This strategy, in theory, would be most likely to succeed if the Ids recognized by the various anti-Ids in the therapeutic mixture were eliminated by distinct sets of amino acid substitutions. Thus, the ideal group of anti-Ids would be specific for Ids that are non-overlapping at the genetic level. However, if the goal was to devise a solid-phase binding assay to detect complexes involving the idiotopic Ab and a particular anti-Id in serum, choosing a second anti-Id as the solid-phase target or the labeled second antibody would be guided most sensibly by a knowledge of which anti-Ids compete for binding to their corresponding Ids. Finally, it should be apparent that no single method for studying Ids will directly provide complete information for all four categories of definition.

ANTI-IDIOTYPIC MIMICRY

How do these arguments apply to an understanding of anti-idiotypic mimicry of antigen? Before answering this question, it is worth asking what exactly is meant by mimicry of an antigen (epitope). Once again, it is important to distinguish

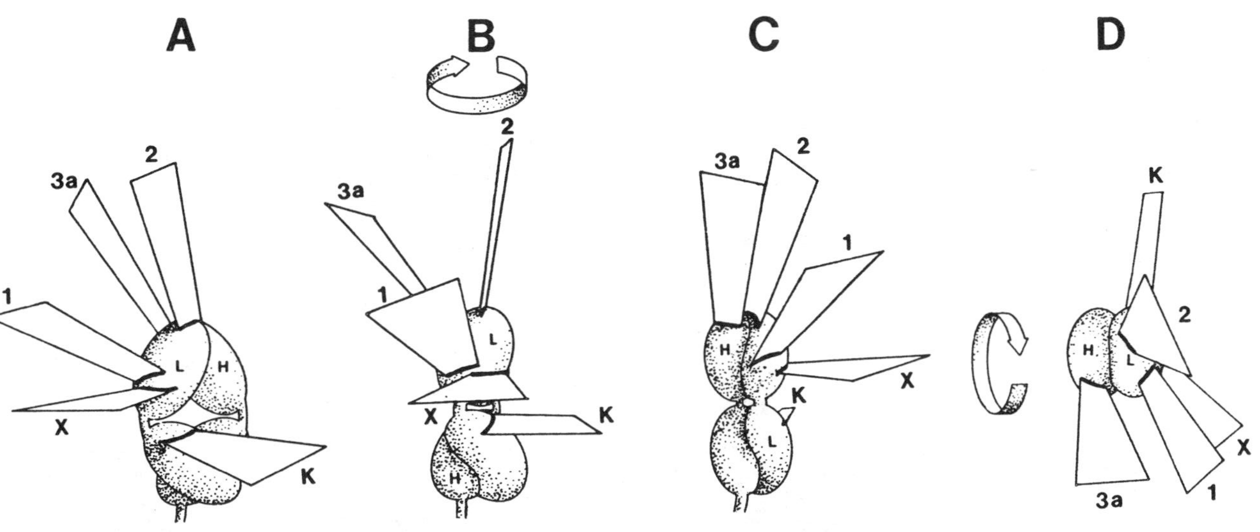

FIGURE 4 Low-resolution three-dimensional model of HGAC 39 Fab (stippled figures) depicting the approximate locations of Ids, the angles of Id-anti-Id intersect, and the rotational orientations of the Id and anti-Id Fab's. Figures A, B, and C represent the rotation of the HGAC 39 Fab through 90° on the proximal-distal axis from planar (A) to intermediate (B) and perpendicular (C). Figure D represents a rotation of the figure in C 90° to allow a view looking directly down on the hypervariable loops. The projections from the HGAC 39 Fab surface represent the planes that pass through the proximal-distal axes of the various anti-Id Fab's at their widest aspects. The thickened bases of the rectangles represent the approximate locations and orientations of the idiotopes (and the C_K isotypic determinant). This model is not intended to indicate the exact sizes of the determinants. From Figure 4, reference [12].

multiple levels of meaning. The purest form of mimicry of one molecule by another would be structural. An ideal structural mimic would express, on some portion of its surface, the identical atoms in the identical three-dimensional distribution as are expressed by the antigen on a portion of its surface. It seems likely that this ideal would be realized infrequently, and that the degree of disparity from the ideal would span a broad range. A second level of mimicry would be immunochemical, at the level of noncovalent binding. The issue under consideration at this level is whether the mimic and the object of mimicry bind to the same receptors and with the same affinities. It has been argued that impressive mimicry at the immunochemical level does not necessarily imply the same degree of structural mimicry.[20–23] A lack of correspondence between these two levels of mimicry could result, for example, from the correlated expression of different sets of contact residues for anti-Id and antigen due to a restricted sample of germ-line, variable-region gene segments or to selective pressures to retain non-paratopic residues (recognized by anti-Id) in association with the paratopic residues. Finally, the degree of mimicry can be assessed on functional grounds, such as induction of an immune response or some other cellular response, as might be elicited by a hormone antigen. Given the complexities of most biological responses, it is easy to conceive of cases where the magnitude of immunochemical and functional mimicry would be imperfectly correlated. In each case, we expect mimicry to behave more like a continuous than a discrete variable.

Since structural mimicry of antigenic molecules at the three-dimensional level has yet to be addressed, and analysis of structural similarities (e.g., primary structure) between antigen and anti-Id is only beginning,[24,25] assessment of anti-idiotypic mimicry of antigen is generally confined to the immunochemical and functional levels. The first criterion generally employed in screening for putative internal image anti-Ids is whether the anti-Ids compete with antigen for binding to the antibody (Ab1) paratope. In other words, it is assumed that an Id recognized by an internal image should overlap competitively with the paratope that binds the relevant epitope. This screening strategy assumes that a useful level of functional mimicry cannot be expressed by an anti-Id that binds to a site that does not overlap competitively with the paratope, a point to which we will return below. It is worth noting, in this context, that the amount of competition between anti-Id and antigen for binding to Ab1 can vary with the methods employed.[26,10,27] Next, the level of functional mimicry is assessed, frequently in terms of the ability of the putative internal image to elicit Ab reactive with the antigen. If the anti-Id can induce Ab to the antigen in multiple species, it is considered unlikely that mere genetic association of the Id (recognized by the putative internal image) and the paratope accounts for the functional mimicry.

Several questions are generally left unaddressed in such investigations. Given that the putative internal image competes with antigen for binding to the Ab1 paratope, how do the relative affinities of the two interactions compare? Thanavala et al. analyzed the binding of Ab1, specific for hepatitis B surface antigen (HBsAg),[28] with Ab2 antibodies and synthetic peptides corresponding in sequence to a part of HBsAg. They found that putative, internal-image, monoclonal anti-Ids had significantly higher affinities for Ab1 than peptide mimics of the native epitope. Interestingly, these Ab2 molecules failed to react with Ab1 molecules that could bind the peptide ligands. These results strongly support the notion that mimicry is a quantitative variable. In the GAC system as well, there appears to be a significant disparity in the affinities for Ab1 of Ab2 versus the affinities for Ab1 of antigen or hapten(see below). Second, if immunization with Ab2 can elicit Ab reactive with the original epitope, one can then ask if all clones stimulated by antigen are stimulated by anti-Id and if anti-Id stimulates clones not stimulated by antigen. It is also important to know if the affinity (for antigen) and isotype distributions of Ab induced by anti-Id are comparable to those elicited by antigen. Since each experimental system may differ substantially from others in terms of immunization protocols and Ab detection procedures, merely demonstrating that anti-Id induces some Ab specific for antigen leaves unclear what magnitude of mimicry is being observed at each of the three levels defined above. In the absence of quantitative information, anti-Ids derived from unrelated systems, and labeled internal images on the basis of the typical criteria outlined above, might exhibit vastly different degrees of mimicry of their respective antigens for any of the three categories of mimicry we have discussed.

Let us now return to the perspective developed from the effort to determine Id topography. Given that anti-Id and antigen (as classes of ligands) both interact with the Ab1 paratope by the same types of noncovalent bonds, it is reasonable to consider overlap between an Id and a paratope as comparable to overlap between two Ids. From this point of view, the degree of mimicry of antigen by anti-Id (particularly at the immunochemical level) can be related to the magnitude of overlap between the paratope and the Id recognized by the putative internal image. The previous arguments about the potential for disparity in the degrees of overlap exhibited for different categories of overlap now take on a new meaning. Competitive overlap between Id and paratope may not be associated with similar magnitudes of compositional, spatial, or sequential overlap, and the Id that most overlaps with the paratope at one level may not be the Id that most overlaps at other levels. A corollary of the former point is that even if anti-Id and antigen mutually compete for binding to the Ab1 paratope, they may not interact with the same sets of contact residues and may not retain the same relative affinities when interacting with other related binding sites. Thus, the complex quantitative comparisons required to assess the relatedness of two Ids also apply to analysis of the relationship between a given Id and a paratope. Furthermore, the difficulty in inferring the relatedness of two anti-Ids by assessing the relationships of their corresponding Ids is equally applicable to inferring the relatedness of anti-Id and antigen by assessing the degrees and types of Id-paratope overlap.

Some of the potential difficulties with the current classification system can be illustrated by considering an example. As mentioned earlier, Anti-IdI-3a is a monoclonal anti-Id specific for a site on the HGAC 39 variable domain. The binding of anti-IdI-3a to HGAC 39 is effectively inhibited by GlcNAc or soluble GAC as well as by three other anti-Ids (anti-IdI-1, -2,and -3b). Analysis of the reactivity of anti-IdI-3a by competitive and direct binding assays revealed that this anti-Id bound more strongly to two thirds of the 38 monoclonal anti-GAC antibodies tested than to isotype-matched control antibodies (not specific for GAC).[10,22] The strength of the binding to the various anti-GAC antibodies varied over a wide range (Figure 5). Thus, anti-IdI-3a is an impressive immunochemical mimic of GAC in that it competes effectively with GAC or GlcNAc for binding to HGAC 39 and also binds, albeit with a wide range of functional affinities, to about two thirds of a large set of paratopes able to bind GAC. It is, however, an imperfect mimic of GAC at the immunochemical level, as it fails to bind a significant number of anti-GAC monoclonal antibodies. Furthermore, although we have not yet formally measured the relevant intrinsic affinities, a variety of binding assays suggest that anti-IdI-3a binds to HGAC 39, and other anti-GAC mAbs with much greater affinity than the hapten, GlcNAc, or the antigen, GAC. This latter result can be interpreted in light of the recently described three-dimensional structures of protein antigen-Ab Fab fragment complexes.[29,30] Both lysozyme and the influenza virus neuraminidase interact with their respective Abs over large surfaces involving many noncovalent bonds. For lysozyme, 16 amino acid residues interact with residues of the Ab. It is very unlikely that epitopes comparable in size to GlcNAc could form as many contacts and, therefore, larger protein epitopes would have the potential to bind more tightly than substantially smaller epitopes. Based on the general properties of Id-anti-Id interactions, it is highly likely that anti-Ids behave like other protein ligands.

Since many carbohydrate and haptenic determinants are small compared to protein epitopes such as those on lysozyme and neuraminidase, it may be common for anti-Id to bind to Ab1 with substantially greater affinity than antigen. The consequences of this relationship for the ability of the anti-Id to mimic antigen may be analyzed in terms of overlap of the relevant Id and the paratope. The major point is that even if the set of contact residues for antigen is contained entirely in a larger set of contact residues (Id) for anti-Id, use of the anti-Id as an immunogen could selectively stimulate receptors on the basis of expression of those Id residues not shared with the paratope. Furthermore, it may frequently be the case that the paratope has constituent residues that are not contact residues for the anti-Id, increasing the probability that antigen and anti-Id will apply divergent selection pressures on the B cell repertoire. A related point is that competitive overlap between Id and paratope does not assure complete genetic overlap. For example, Radbruch et al. have described a single amino-acid substitution in the heavy chain variable domain of a hapten-specific mAb that significantly decreased expression of an Id recognized by a hapten-inhibitable anti-Id, but did not affect the affinity for hapten.[16]

Returning to our example, we have also analyzed the ability of anti-IdI-3a to serve as a functional mimic of GAC in terms of eliciting anti-GAC serum Ab. When anti-IdI-3a and two isotype-matched anti-Ids from the panel were compared as immunogens *in vivo*, anti-IdI-3a was clearly the most effective at eliciting a primary anti-GAC Ab response in C56BL/6J mice even though all three anti-Ids were comparably immunogenic (induced similar levels of serum Abs able to displace radio-labeled HGAC 39 from homologous anti-Id).[31] While it is possible that

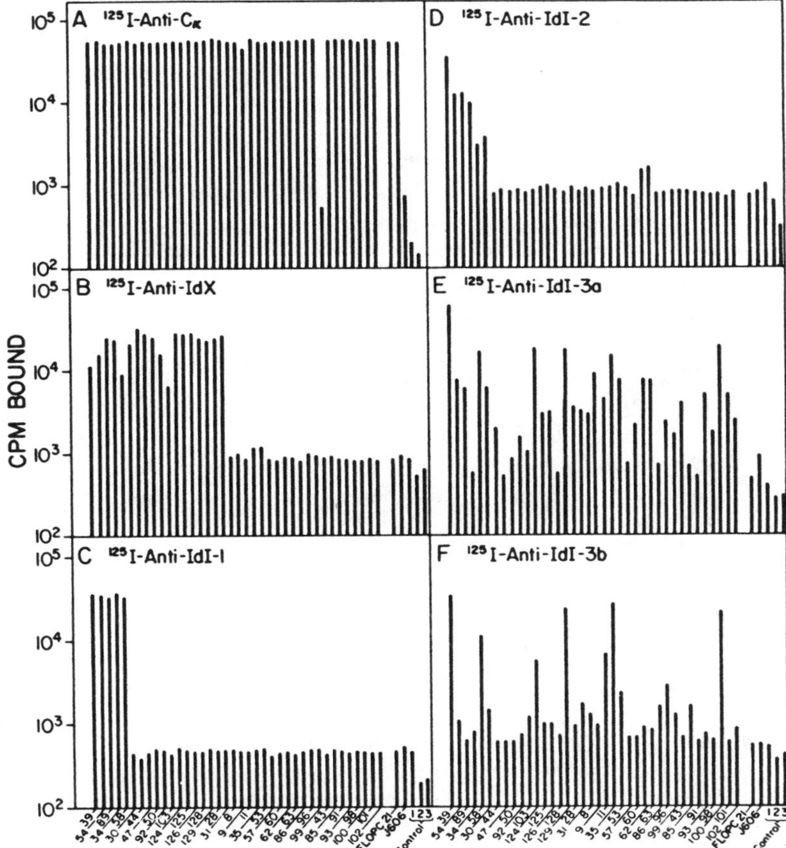

FIGURE 5 Reactivity profiles of radiolabeled anti-Id and anti-C_K determined by direct binding sandwich radioimmunoassay. Vertical bars correspond to arithmetic mean anti-Id or anti-isotype counts bound (determined in duplicate) to each anti-GAC hybridoma tissue culture supernatant (diluted 1:2). All anti-GAC antibodies used were IgG3 [37 kappa, 1 lambda (HGAC 43)]. The antibodies were bound to the wells by previously adsorbed monoclonal anti-mouse gamma 3 antibody. FLOPC 21 and J606 are IgG3K non-GAC binding myeloma proteins. Controls 1, 2, and 3 correspond to: 1) anti-gamma 3 on well but no anti-GAC supernatant, 2) no anti-gamma 3 adsorbed to well plus HGAC 39 supernatant 1:2, and 3) no anti-gamma 3 on well and no anti-GAC supernatant. Modified from Figure 5, reference [22].

other immunization and assay protocols would have resulted in a different relative rankings of these anti-Ids in terms of their abilities to mimic GAC as immunogens, the possibility that the degree of mimicry is protocol-dependent would only add support to the contention that the label internal image carries no uniform implications even at the functional level. Furthermore, for the protocol used in these initial studies, anti-IdX, which does not compete with anti-IdI-3a or with antigen for binding to HGAC 39, appeared to be superior to anti-IdI-3a at priming mice for an Ab response to group A streptococci, even though this antibody was inferior at directly inducing a primary anti-GAC response.[31] These results suggest the possibility that a single monoclonal anti-Id may not be the best mimic of a given epitope in all functional contexts, and that an anti-Id that fails to compete with hapten or antigen for binding to Ab1 (i.e., Ab2-alpha) is not necessarily devoid of the potential for functional mimicry, a point also emphasized by Erlanger.[21]

So, how should we classify anti-IdI-3a? When we informally posed this question to several investigators with appropriate interests, we received all possible answers: Ab2-alpha (Ab2-gamma is not always employed), Ab2-beta, and Ab2-alpha,beta. While the diversity of these responses does not definitively prove the thesis that the current classification scheme is ambiguous, it is certainly consistent with this view. This ambiguity can be discerned at two levels. First, the exact immunochemical and functional criteria for classification as an internal image are unclear, or at least variable from system to system. Second, the the structural and immunochemical implications of a given magnitude of functional mimicry are uncertain.

Another potential source of discrepancy among the various levels of mimicry relates to the molecular contexts of the epitope and the anti-Id binding site (for Id). For example, it has been suggested that the presence of epitopes able to elicit helper T cell responses may have an important influence on the effectiveness of a surrogate immunogen.[32] Consequently, two anti-Ids that comparably mimic an epitope (defined by Ab1) in structural and immunochemical terms might differ substantially as functional mimics for that epitope due to structural differences that determine the induction of unequal magnitudes of T cell help. Differences in physical and chemical properties between anti-Id and antigen may also affect the degrees of correlation among structural, immunochemical, and functional mimicry, particularly for the case where an anti-Id is being compared to a non-protein antigen, such as a polysaccharide. Even if a portion of the surface of an anti-Id variable domain perfectly mimicked (structurally) a carbohydrate epitope, it is extremely unlikely that this site on the anti-Id could be presented at a density (epitopes per unit area or per unit volume) approaching that possible for polysaccharide determinants. Other differences between the rest of the antigen molecule and the rest of the anti-Id, such as net charge or flexibility, might also have consequences for the interactions of these two ligands with the Ab1 paratope. For example, IgG3 anti-GAC monoclonal antibodies bind in a cooperative fashion to the surfaces of heat- killed, pepsin-digested group A streptococci, but do not manifest this same effect, or manifest it to a much smaller extent, in binding to anti-Id.[27] Thus, differences in the general molecular and geometric properties of antigen and anti-Id could decrease the correlations between structural mimicry and mimicry at the immunochemical and functional levels.

It is important to note that recent work on the molecular basis for restriction of T cell specificity by gene products of the major histocompatibility complex has suggested that molecular mimicry may play a fundamental role in the recognition of antigen by T cells.[33] These and related concepts potentially broaden the significance of molecular mimicry for understanding how the immune system functions and also suggest the possibility that anti-Ids may elicit immune responses in molecular forms, peptides being the most likely, other than the native protein. The validity of this paradigm would have important implications for the general understanding of anti-idiotypic mimicry of antigen. It has recently been proposed that self-variable domain peptides may have a role in immunoregulation.[34]

As a final point, we would like to raise a few questions about the general direction of research on the development of anti-idiotypic vaccines. A major goal of such work is to manipulate the immune response in a clonally selective fashion. However, engineering the optimally protective immune response by the host does not necessarily require the maximum level of imitation of the pathogen, or its components. The pathogen has been selected through evolution to elicit an immune response that minimally interferes with its replication. The complex interplay between selective forces acting on the host and on the pathogen will not necessarily yield a host immune response that is optimal from the point of view of the host's survival. Therefore, we believe we should adopt a broader view of the attempt to manipulate immune responsiveness and consider the full potential of the immune repertoire available, instead of limiting our efforts to the subset of the repertoire the pathogen happens to select. For example, anti-idiotypic immunogens should have the potential to activate clones that are normally casualties of the interclonal competition dictated by the antigen (pathogen).[35] These clones, that fail to contribute prominently in the natural immune response, are not necessarily without value in an engineered response. For example, mAbs from the secondary response to the hapten p-azophenylarsonate were examined with respect to idiotype and affinity for antigen.[36] The Abs from the non-dominant idiotypic family were found to have affinities for antigen comparable to those observed for Abs from the dominant idiotypic family. Using separate anti-Ids, at least initially, as immunogens to stimulate two such families of Abs might allow for a more potent overall response, as one set of clones would no longer proliferate at the expense of the other. The point is not that antigen mimicry is necessarily the wrong strategy, but that multiple strategies may be worth pursuing, with different optimal solutions in different cases. It could be that the broader approach we have discussed would fail to achieve a greater net benefit than the pure mimicry approach, but this question can probably only be settled empirically in each system.

CONCLUSIONS

Detailed analysis of the relationships among Ids and among their corresponding anti-Ids suggests that Ids and anti-Ids can be compared for multiple parameters that do not necessarily correlate perfectly. Since there is no physiochemical basis on which to differentiate anti-Ids, as a class, from all other ligands, it seems justified to apply a similar type of analysis to the sites recognized by anti-Ids and antigens and to the corresponding ligands. Therefore, the evaluation of relationships between any two sites of non-covalent interaction, whether both are Ids or one is an Id and the other is a paratope, are seen to be complex and most usefully described in quantitative terms. From this perspective, the concept of anti-idiotypic mimicry of antigen is multi-dimensional. Consequently, the use of qualitative classifications for anti-Ids based on essentially arbitrary and unspecified threshold values for the degree of mimicry of antigen by anti-Id may lead to ambiguity in the meanings of the various anti-Id categories. Furthermore, our current understanding of the relationships among different levels of molecular mimicry do not allow certain prediction of the magnitude of mimicry in one category from that in another category. Understanding the relationships among the structural, immunochemical, and functional levels of molecular mimicry remains an important challenge for future studies.

ACKNOWLEDGMENTS

We thank Bill Monafo and Joe Davie for contributions to studies described in this chapter, and we thank Fred Stevens for discussion.

REFERENCES

1. Sege, K., and P. A. Peterson(1978), "Use of Anti-Idiotypic Antibodies as Cell-Surface Receptor Probes," *Proc. Natl. Acad. Sci. (USA)* **75**, 2443–2447.
2. Schreiber, A. B., P. O. Couraud, C. Andre, B. Vray, and A. D. Strosberg (1980), "Anti-Alprenolol Anti-Idiotypic Antibodies Bind to Beta-Adrenergic Receptors and Modulate Catecholamine-Sensitive Adenylate Cyclase," *Proc. Natl. Acad. Sci. (USA)* **77**, 7385–7389.
3. Jerne, N. K., "Towards a Network Theory of the Immune System," *Ann. Immunol. (Inst. Pasteur)* **125 C**, 373–389.
4. Lindenmann, J. (1979), "Homobodies: Do They Exist?," *Ann. Immunol. (Inst. Pasteur)* **130 C**, 311–318.
5. Jerne, N. K., J. Roland, and P.-A. Cazenave (1982), "Recurrent Idiotopes and Internal Images," *EMBO J.* **1**, 243–247.
6. Kohler, H. (1984), "The Immune Network Revisited," *Idiotypy in Biology and Medicine*, Eds. H. Kohler, J. Urbain, and P.-A. Cazenave (Orlando, Florida: Academic Press, Inc.), 3–14.
7. Nisonoff, A., and E. Lamoyi (1981), "Hypothesis. Implications of the Presence of an Internal Image of the Antigen in Anti-Idiotypic Antibodies: Possible Application to Vaccine Production," *Clin. Immunol. Immunopathol.* **21**, 397–406.
8. Roitt, I. M., A. Cooke, D. K. Male, F. C. Hay, G. Guarnotta, P. M. Lydyard, L. P. de Carvalho, Y. Thanavala, and J. Ivanyi (1981), "Idiotypic Networks and their Possible Exploitation for Manipulation of the Immune Response," *Lancet* **i**, 1041–1045.
9. Kennedy, R. C., G. R. Dreesman, and H. Kohler (1985), "Vaccines Utilizing Internal Image Anti-Idiotypic Antibodies that Mimic Antigens of Infectious Organisms," *BioTechniques* **3**, 404–409.
10. Greenspan, N. S., and J. M. Davie (1985), "Serologic and Topographic Characterization of Idiotopes on Murine Monoclonal Anti-Streptococcal Group A Carbohydrate Antibodies," *J. Immunol.* **134**, 1065–1072.
11. Parham, P. (1984), "Changes in Conformation with Loss of Alloantigenic Determinants of a Histocompatibility Antigen (HLA-B7) Induced by Monoclonal Antibodies," *J. Immunol.* **132**, 2975–2983.
12. Roux, K. H., W. J. Monafo, J. M. Davie, and N. S. Greenspan (1987), "Construction of an Extended Three-Dimensional Idiotope Map by Electron Microscopic Analysis of Idiotope-Anti-Idiotope Complexes," *Proc. Natl. Acad. Sci. (USA)* **84**, 4984–4988.
13. Schilling, J., B. Clevinger, J. M. Davie, and L. Hood (1980), "Amino Acid Sequence of Homogeneous Antibodies to Dextran and DNA Rearrangements in Heavy Chain V-Region Gene Segments," *Nature* **283**, 35–40.
14. Davie, J. M., M. V. Seiden, N. S. Greenspan, C. T. Lutz, T. L. Bartholow, and B. L. Clevinger (1986), "Structural Correlates of Idiotopes," *Ann. Rev. Immunol.* **4**, 147–165.

15. Rudikoff, S. (1983), "Immunoglobulin Structure-Function Correlates: Antigen Binding and Idiotypes," *Contemp. Top. Mol. Immunol.* **9**, 169–209.
16. Radbruch, A., S. Zaiss, C. Kappen, M. Bruggemann, K. Beyreuther, and K. Rajewsky (1985), "Drastic Change in Idiotypic but Not Antigen-Binding Specificity of an Antibody by a Single Amino-Acid Substitution," *Nature* **315**, 506–508.
17. Bruggemann, M., H.-J. Muller, C. Burger, and K. Rajewsky (1986), "Idiotypic Selection of an Antibody Mutant with Changed Hapten Binding Specificity, Resulting from a Point Mutation in Position 50 of the Heavy Chain," *EMBO J.* **5**, 1561–1566.
18. Raffeld, M., L. Neckers, D. L. Longo, and J. Cossman (1985), "Spontaneous Alteration of Idiotype in a Monoclonal B-Cell Lymphoma: Escape from Detection by Anti-Idiotype," *New Engl. J. Med.* **312**, 1653–1658.
19. Meeker, T., J. Lowder, M. L. Cleary, S. Stewart, R. Warnke, J. Sklar, and R. Levy (1985), "Emergence of Idiotype Variants during Treatment of B-Cell Lymphoma with Anti-Idiotype Antibodies," *New Engl. J. Med.* **312**, 1658–1665.
20. Rajewsky, K. and T. Takemori (1983), "Genetics, Expression, and Function of Idiotypes," *Ann. Rev. Immunol.* **1**, 569–607.
21. Erlanger, B. F. (1985), "Anti-Idiotypic Antibodies: What Do They Recognize?," *Immunol. Today* **6**, 10–11.
22. Greenspan, N. S., and J. M. Davie, "Analysis of Idiotype Variability as a Function of Distance from the Binding Site for Anti-Streptococcal Group A Carbohydrate Antibodies," *J. Immunol.* **135**, 1914–1921.
23. Roitt, I. M., Y. M. Thanavala, D. K. Male, and F. C. Hay (1985), "Anti-Idiotypes as Surrogate Antigens: Structural Considerations," *Immunol. Today* **6**, 265–267.
24. Ollier, P., J. Rocca-Serra, G. Somme, J. Theze, and M. Fougereau (1985), "The Idiotypic Network and the Internal Image: Possible Regulation of a Germ-Line Network by Paucigene Encoded Ab2 (Anti-Idiotypic) Antibodies in the GAT System," *EMBO J.* **4**, 3681–3688.
25. Bruck, C., M. S. Co, M. Slaoui, G. N. Gaulton, T. Smith, B. N. Fields, J. I. Mullins, and M. I. Greene (1986), "Nucleic Acid Sequence of an Internal Image-Bearing Monoclonal Anti-Idiotype and its Comparison to the Sequence of the External Antigen," *Proc. Natl. Acad. Sci. (USA)* **83**, 6578–6582.
26. Stevens, F. J., J. Jwo, W. Carperos, H. Kohler, and M. Schiffer (1986), "Relationships between Liquid- and Solid-Phase Antibody Association Characteristics: Implications for the Use of Competitive ELISA Techniques to Map the Spatial Location of Idiotopes," *J. Immunol.* **137**, 1937–1944.
27. Greenspan, N. S., W. J. Monafo, and J. M. Davie (1987), "Interaction of IgG3 Anti-Streptococcal Group A Carbohydrate (GAC) Antibody with Streptococcal Group A Vaccine: Enhancing and Inhibiting Effects of Anti-GAC, Anti-Isotypic, and Anti-Idiotypic Antibodies," *J. Immunol.* **138**, 285–292.

28. Thanavala, Y. M., S. E. Brown, C. R. Howard, I. M. Roitt, and M. W. Steward (1986), "A Surrogate Hepatitis B Virus Antigenic Epitope Represented by a Synthetic Peptide and an Internal Image Anti-Idiotype Antibody," *J. Exp. Med.* **164**, 227–236.
29. Amit, A. G., R. A. Mariuzza, S. E. V. Phillips, and R. J. Poljak (1986), "Three-Dimensional Structure of an Antigen-Antibody Complex at 2.8 A Resolution," *Science* **233**, 747–753.
30. Colman, P. M., W. G. Laver, J. N. Varghese, A. T. Baker, P. A. Tulloch, G. M. Air, and R. G. Webster (1987), "Three-Dimensional Structure of a Complex of Antibody with Influenza Virus Neuraminidase," *Nature* **326**, 358–363.
31. Monafo, W. J., N. S. Greenspan, J. A. Cebra-Thomas, and J. M. Davie (1987), "Modulation of the Murine Imune Response to Streptococcal Group A Carbohydrate by Immunization with Monoclonal Anti-Idiotope," *J. Immunol.* **139**, 2702–2707.
32. Zanetti, M., E. Sercarz, and J. Salk (1987), "The Immunology of New Generation Vaccines," *Immunol. Today* **8**, 18–25.
33. Guillet, J.-G., M.-Z. Lai, T. J. Briner, S. Buus, A. Sette, H. M. Grey, J. A. Smith, and M. L. Gefter, "Immunological Self, Nonself Discrimination," *Science* **235**, 865–870.
34. Kourilsky, P., G. Chaouat, C. Rabourdin-Combe, and J.-M. Claverie (1987), "Working Principles in the Immune System Implied by the 'Peptidic Self' Model," *Proc. Natl. Acad. Sci. (USA)* **84**, 3400–3404.
35. Manser, T., L. J. Wysocki, T. Gridley, R. I. Near, and M. L. Gefter (1985), "The Molecular Evolution of the Immune Response," *Immunol. Today* **6**, 94–101.
36. Rothstein, T. L., and M. L. Gefter (1983), "Affinity Analysis of Idiotype-Positive and Idiotype-Negative Ars-Binding Hybridoma Proteins and Ars-Immune Sera," *Molec. Immunol.* **20**, 161–168.

FRANCO CELADA
Department of Immunology, University of Genoa, Genoa, Italy; present address: Dept. of Rheumatology, The Hospital for Joint Diseases, 301 E. 17th Street, New York, NY 10003

In Search of T-Cell Help for the Internal Image of the Antigen

INTRODUCTION

When immunology was born, more than a century ago, it was not meant to be a theoretical discipline but rather a busy task force with the assignment to apply the knowledge of microbiology to clinical medicine. Immunology has come a long way since, but has carried along, as imprints from the old times, a number of ways of reasoning: typically, the idea that the acid test for any new idea must be its impact or feasibility on the making of *vaccines*. This test may also be applied to the idiotype network and, sure enough, the idiotypic internal image has been proposed for use as vaccine in cases where antigen is too dangerous to handle or too difficult to obtain in reasonable quantities. The internal image is one of the $Ab2 - (Ab2_\beta)$—antibodies directed against the idiotype of $Ab1$—the antibody raised against the external antigen (Ag).[1] The paratope of $Ab2_\beta$ binds to $Ab1$ and crossreacts with Ag—to the limit of stochiometric inhibition of Ag-$Ab1$ binding, as predicted for a virtual identity between $Ab2_\beta$ and the relevant epitope.[2] Internal images have been injected into mice and rabbits and have elicited antibodies capable of binding both $Ab2_\beta$ and Ag; thus, they have become important candidates for third-generation vaccines, along with purified, synthesized or genetically engineered peptides containing the amino-acid sequence of the relevant epitope.

THE PERFECT VACCINE

There are many requirements for a molecule to be considered a suitable vaccine (or a "perfect vaccine," see the article of Greenspan in this volume). Instead of discussing them, I shall note that *vaccine*, this beautiful word inherited from the great victory against smallpox, is still an ambiguous term among immunologists. Some define it as a molecule that—injected—provokes an antibody response capable of binding Ag and, thus, protect the organism. Binding and relative affinity are easy to assess *in vitro*, and both internal image and peptide vaccines can be nurtured by adjuvants and by binding to suitable carriers to enhance the response. Others—myself among them—would consider a vaccine any substance capable of building a *memory* specific for Ag that Ag (the outside invader) is able to recall when it appears, i.e., months, years, or even decades after "vaccination." Both the manufacturing and the testing of a "memory" vaccine may be difficult; on the other hand, only *crossimmunogenicity* will prevent unhappy surprises, such as the failure to mount a secondary response at the time when this may be a matter of life and death. The requirements for crossimmunogenicity is the sharing among Ag and vaccine of relevant epitopes recognized by the B cells (BD) and recognized by the Th cells (HD). Much less is known about HD than about BD. However, the following points can be considered established beyond doubt.

1. BD can be either sequential or discontinuous (conformation-dependent) relative to the amino acid sequence; instead, HD are invariably sequential;[3]
2. BD and HD have a different mode of presentation to their corresponding paratope: direct binding to *Ab* in the case of BD, binding to one of several MHC Class molecules and—in a tertiary complex—to the T-cell receptor (TCR) in a so-called *MHC restricted* presentation;[4]
3. Different determinants act as BD and HD in each immunogen molecule;
4. At least in large multideterminant immunogens, not all HDs are conducive to T help for B cells responding to a given BD; instead, preferential pairings exist between the HD and BD, possibly related to their proximity in the three-dimensional structure of the molecule.[5]

In the present paper, I wish to outline a strategy to understand which are the relevant HD for internal image antigen vaccine and, in particular, whether internal images may present HD crossreaction with the original antigen.

THE E COLI β-GALACTOSIDASE SYSTEM

THE STRUCTURE OF THE MOLECULE

The Z gene product, the product of the gene coding for *E. coli* β-galactosidase, is a polypeptide of 1023 amino acids. Four identical monomers bind isologously to create the wild-type tetramer which has a D-2 symmetry.[6] Two monomers pair with each other by contact sites, some of which are certainly located at the COOH terminus, to form a dimer. Two dimers pair with each other, utilizing crucial N-terminal contact sites to form the tetramer.[7] Both of these interactions are noncovalent, but endowed with high affinity so that no monomer or dimer can be isolated in native form. Monomers and dimers can, however, be obtained from point-mutant or deletion mutants and are usually equipped with substrate binding, but not with catalytic capacity.[8] The tetramer is the only enzymatically active form of it E. coli β-galactosidase and has 4 catalytic sites.[8]

THE EPITOPES RECOGNIZED BY B CELLS (BD)

Although there is no crystallography to link the primary to the tertiary and quaternary structures (as in other proteins, e.g., myoglobin), genetic and immunochemical studies have allowed the distinction of sequential from conformational, and hidden from displayed epitopes, and have provided tools for analyzing single antibody families, by their capacity to induce functional changes in the enzyme's activity.

We list the following three categories of epitopes: sequential epitopes available on the surface of the native molecule (=s); sequential epitopes located inside the native molecule (=i); conformation-dependent determinants. Of these, a subclass that turns out to be essential in Ab-mediated activation and protection of the enzyme is dependent on the *quaternary* conformation of GZ (=c).[3]

THE EPITOPES RECOGNIZED BY T CELLS (HD)

The capacity of different GZ cyanogen bromide peptides to prime for the induction of Th cells has been tested by two types of *in vitro* experiments. In the GZ-hapten system, GZ- or peptide-primed lymph node T cells were cultured for four days with BSA-FITC-primed splenic B cells in the presence of GZ-FITC. Four days later, plaque-forming cells were enumerated by using FITC-RBC as targets.[9] In the "intrinsic epitope" system, GZ- or peptide-primed lymph node T cells were co-cultured with GZ-primed splenic B cells, in the presence of native enzyme. The eight-day supernatants were titrated for three classes of anti-GZ antibodies, *binding*, *protecting* and *activating*: a) binding antibodies represent the total response as measured by ELISA; b) protecting antibodies recognize one set (C1) of "c" conformational determinants and were identified by the supernatants' capacity to prevent the loss of catalytic activity caused by exposing native C12 to 62°C for 10 minutes;[10] and c) activating antibodies, which recognize another set (C2) of conformation epitopes, were measured by their capacity to regenerate enzyme activity in extracts of defective *E. coli* strains, bearing point mutations in the Z gene.[11]

These results have shown that: 1) the T-cell help for B cells recognizing conformational determinants relies on *sequential* T epitopes; 2) helper cells raised against certain peptides of GZ yield marked preferential help for B cells directed against certain, but not other, sites of the native enzyme; 3) a convincing hypothesis to explain the HD-BD pairing phenomenon is that B cells are the principal antigen processing and presenting cells; and 4) the degradation of the antigen which takes place in these cells is influenced by the specific bond that is established between the paratope of the Ig receptors of the B cell and the corresponding epitope of the antigen. This hypothesis is corroborated by studies *in vitro* where macrophages exposed to immune complexes formed by GZ and individual monoclonal antibodies show differential presentation to Th cell clones.[12]

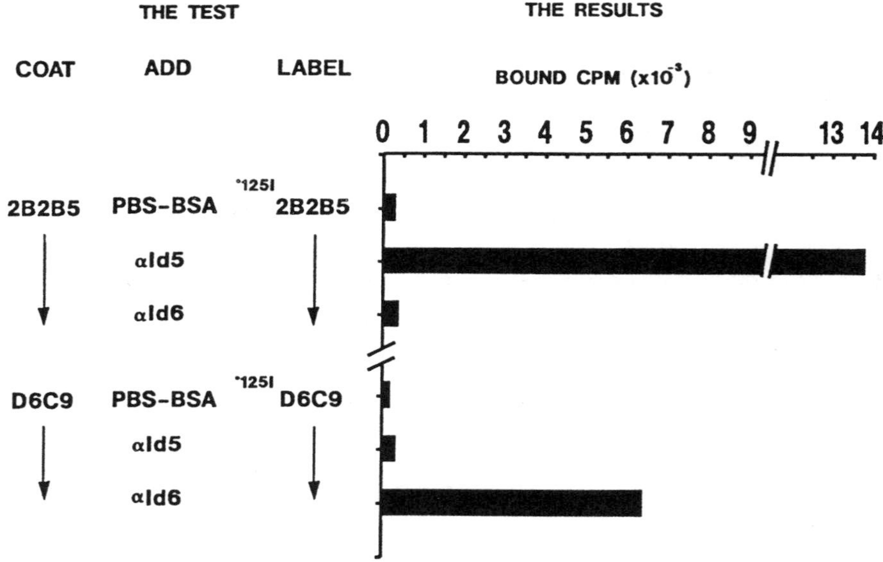

FIGURE 1 Binding properties of anti-Id5 and anti-Id6 polyclonal antibodies: description of the assay and results.

THE IDIOTYPES OF THE GZ SYSTEM

As a first step in the definition of the GZ idiotype, we[12] have recently produced polyclonal and monoclonal syngeneic anti-idiotypes (AB-2) by injecting two IgG Mab anti-β-galactosidase (2B2B5 [a gift from Boris Rotman] and D6C9). 2B2B5 recognizes a sequential determinant of the enzyme and reacts both with the native molecule and the reduced and carboxy-methylated product, RCM-GZ. D6C9, instead, recognizes a conformation-dependent site, does not bind to RCM-GZ and has the remarkable property already described, to "reconform" certain defective products of mutated Z-gene (re-creating an enzymatically active molecule) and to protect wild-type GZ from heat-denaturation. The polyclonal anti-2B2B5 have been named anti-Id5. They bind strongly to the immunogen, not to PBS-BSA and not to D6C9. Polyclonal anti-Id6 bind specifically to D6C9, and not to 2B2B5. A monoclonal anti-Id6 (C12) has the same characteristics as the polyclonal, although its affinity is rather low. The binding results are shown in Figure 1.

The inhibition assay in solid phase has been mounted to better define the distribution of the two anti-Id. The results, in Figure 2, show a remarkable difference in scope of the crossreactivities. Anti-Id5 is inhibited by its corresponding antigen (see curve) and—consistently by a series of nine monoclonal anti-GZ antibodies—endowed with different binding characteristics—but with the notable exception of D6C9. Also, all antisera contain Id5, while normal serum is negative. The characteristics of anti-Id6 are quite different. On the right half, Figure 2 shows the binding inhibition curve by cold D6C9 and the negativity of all Mabs tested, together with the borderline positivity of 2 out of 11 anti-GZ sera. The Mab anti-Id6 has been also tested with respect to inhibition of the antibody-mediated activation of a mutant GZ. As shown in Figure 4, the addition of anti-Id inhibits to about 80% the production of enzymatic activity from the mutant.

The data can be summarized as follows: the anti-β-gal response exhibits at least 2 anti-GZ idiotypes which can be defined as one "public" (Id5) and one (Id6) extremely "private." The Mab anti-Id6 is a good candidate for an internal image of the antigen, thanks to its capacity of competing with the conformation-dependent epitope(s) responsible for the activation phenomenon of mutant GZ: in fact, Figure 3 displays a classic competitive inhibition in a double reciprocal plot.

THE ELICITATION OF MEMORY

With the exception of T-independent (TI) molecules, most protein antigens need to utilize several rounds of "cell cooperation" in the development of a full-fledged

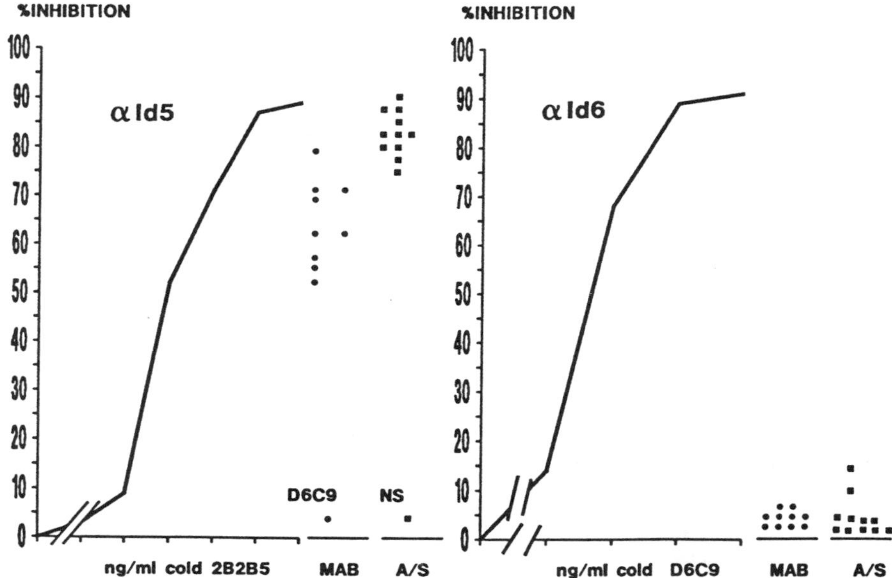

FIGURE 2 Inhibition test of anti-Id5 and anti-Id6. The curves are obtained by using increasing amounts of cold 2B2B5 (2A) and D6C9 (2B) to inhibit the sandwich binding described in Figure 1. Squares (■) and circles (•) represent the inhibition level obtained by admixing a series of anti-GZ Mabs, and a number of polyclonal anti-GZ mouse sera with normal serum, respectively.

immune response. T-cell help is necessary both in the build up and in the elicitation of memory. During the primary response, specific helper T cells are principally stimulated by contact with antigen-presenting macrophages or dendritic cells. When, in turn, the Th-secreted lymphokines reach the B cells, parallel T and B clone proliferation takes place. During the secondary response, the capture and the processing-presentation of antigen are performed with highest efficiency by the expanded B memory cells, which bear Ig receptors on their surface, and the encounter for presentation is made easy by the presence of expanded Th memory clones. Also, it is not necessary that exactly the same molecule that has been injected during the priming be used for the challenge, but the latter should contain at least one HD and one BD in order to address two cooperating clones. If the "preferential pairing" phenomenon[6] turns out to be a general principle, then only a fraction of HD-BD pairs in the immunogen would be effective, thus increasing the difficulties for the vaccine. The problem is—for instance, in the construction of an artificial vaccine—how to predict the good pairings. One obvious criterion proposed is *proximity*, but not sufficient experimental data are available, regardless of the fact that proximity between a conformation-dependent BD and a, say, cryptic HD would be difficult to assess and even to define.

THE PROJECTED EXPERIMENTAL DESIGN

The correspondence of the epitopes among internal image and antigen in the case of using the former as a vaccine is as crucial as for any other "artificial" or "synthetic" vaccine. I am encouraged to begin the search for these relationships using the GZ idiotype system by the following considerations:

a. the fine specificity of internal image epitope will be defined by the capacity to elicit antibodies with the same "conforming" (activating-protecting) properties as the *Ab*1, i.e., by a particularly stringent criterion; and

b. antigen, idiotype-carrying *Ab*1 and internal image (*Ab*2β) are available in amounts sufficient for biochemical and immunochemical purification, degradation and analysis.

The experimental design is illustrated in Figure 4 and should be commented on. The first objective is to build an epitope-specific (activating, protecting) B memory. This will be achieved by priming with native GZ. The full secondary response will be obtained if we challenge with soluble complete GZ. It is anticipated that the challenge with internal image (anti-Id6) will not elicit a secondary response despite the binding of anti-Id6, because of lack of the proper T memory population (and the assumed non-identity of GZ HDs and anti-Id HDs). Should, nevertheless, a response occur, one would have to postulate the partial identity of the two sets of HD.

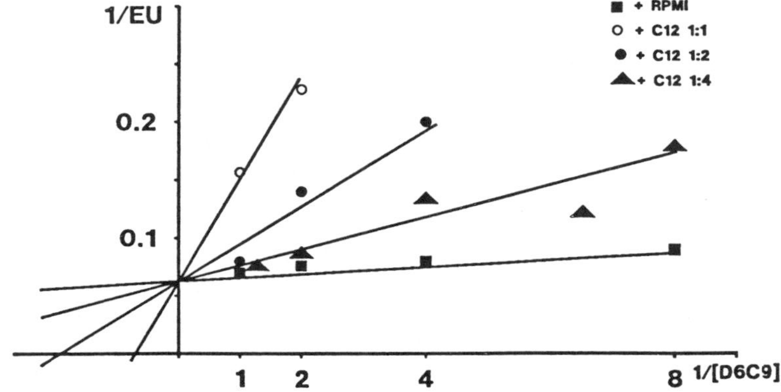

FIGURE 3 Competitive inhibition of antibody mediated activation of mutant GZ 6101 by monoclonal anti-Id6. Double reciprocal plot of enzyme activation vs. D6C9 concentration, in the presence of different concentrations of α Id6.

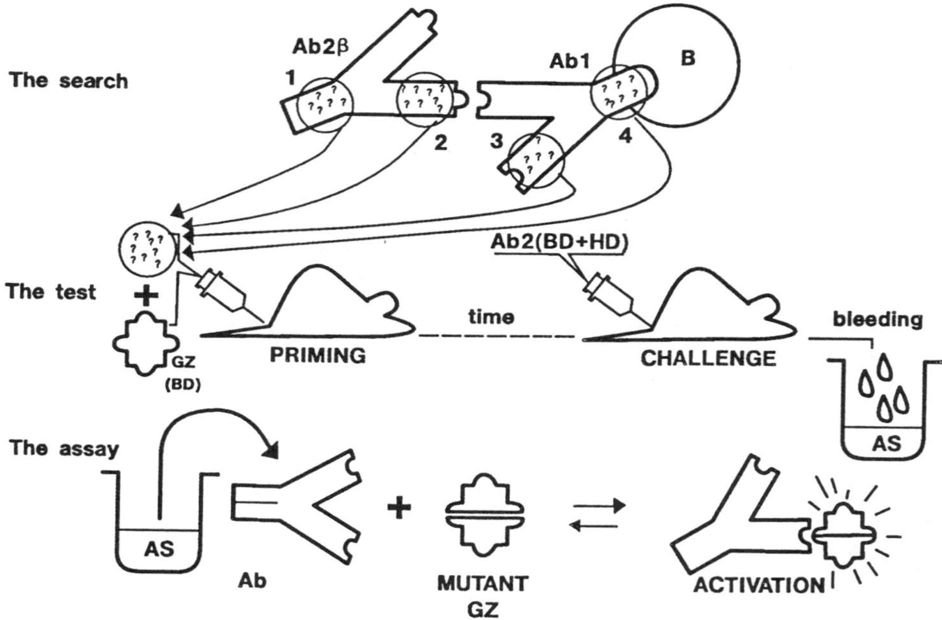

FIGURE 4 Scheme of experiments to search for HD_s of internal image. Explanation in "Summary."

The core of the experiment is designed to answer the question, in which part and in which peptides of Ab2β reside epitopes capable of stimulating those Th clones that will eventually recognize the same anti-idiotype when it is used as a challenge. To this end, a number of fragments will be injected together with the priming dose of GZ, one fragment for each group of mice. Their relative effectiveness as T-primers will be revealed by a high secondary response after the challenge. Initially, the following territories will be tried:

1. the variable part of anti-Id, in its native state;
2. the variable part of anti-Id reduced, carboxymethylated; and
3. the constant part of anti-Id.

Should any of these show positivity, a systematic search to identify the responsible peptides will be started (CNBr fragmentation, enzymatic cleavage, synthesis of single peptides), and each product will be screened by the same T-memory test.

ARE THERE HD IN AB1?

A second hunting ground for HD will be Ab1, and particularly its VH-VL and idiotype. In our system, Ab1 is the Ig receptor of the B cell, i.e., the molecule that recognizes the crossreactivity or identity of Ab2β with Ag and binds to it at the time of challenge. The resulting complex Ab1-Ab2β is then interiorized by the B memory cell and enclosed in endosomal vesicles. In the case of normal Ab-Ag complexes, there seems to be a separation of destinies at this point. The complex comes apart, Ab is returned to the surface to resume its receptor function, Ag follows it path toward enzymatic degradation and finally coupling—as small peptides—to the class II molecules that are lining the inner surface of the vesicles. In the present case, where the molecule bound by the Ig receptor is itself an antibody, the separation of *Ab* as *Ag* from *Ab* as *Ab* would almost need the cunning work of a Cartesian devil. It is feasible that the separation may be imperfect and, therefore, fragments of Ab1 also get to be processed and then coupled to MHC molecules. Will any TCR bind to such an autologous product? Will this be prevented by some tolerance mechanism? From what is known about idiotype network, I would predict that if the Ab1 fragment belongs to the VH-VL part of the molecule, specific binding by T cells may well occur. The search of HD in this hunting ground is therefore justified. Should it be successful, many excruciating doubts about idiotype vaccines would be solved, since the probability of finding a functional T help would definitely exist.

SUMMARY

The proposed experiment to trace the T epitopes (HD) of an idiotype internal image of a well-studied antigen are summarized in Figure 4. Level 1 shows $Ab2_{\beta}$ *ii* making contact with a B cell exhibiting *Ab*1 or *Ab*1-like Ig receptors. The four sections covered with question marks represent the first subdivision of the hunting territory: the two V and C regions of $Ab2_{\beta}$ (2,1), and of *Ab*1 (3,4). Should the test result be positive for one of these, the territory in question will be subdivided into smaller fragments and retested. Level 2 represents the general scheme to study the formation and recall of memory. The mouse will be primed with β-gal (carrying of BD) *plus* the fragment to be tested for HD. After time (months), the challenge will be delivered by *Ab*2 (internal image) that contains both the BD and the HD of internal image. If BDs in priming and in challenge are crossreacting, the eventual response will be positive. Level 3 represents the test for the presence of activating antibodies—as the result of triggering *Ab*1-like (or *Ab*3) B memory cells. The serum antibodies are admixed to mutant GZ, unable to form stable tetramers. If functionally active *Ab* is present in the serum, the mutant is forced in a tetrameric structure and exhibits enzymatic activity.

ACKNOWLEDGMENTS

Part of the work described has been done with Masafumi Ito during a sabbatical year at the Mount Sinai Medical Center, New York, in the lab of Constantine Bona, whom I wish to thank for his hospitality and inspiring discussions. I was supported by a NATO Senior Fellowship and a grant from the American-Italian Foundation for Cancer Research. Work in Genoa is supported by grants from the Italian Research Council (Special Programs "Chimica Fine e Secondaria"—contract n. 85.1792.95—and Biomedical and Clinical Engineering—contract n. 86.01415.57.

The excellent secretarial help of Luisa Di Rosa is gratefully acknowledged.

LIST OF ABBREVIATIONS USED IN THIS ARTICLE

*Ab*1: Antibody elicited by Ag.

*Ab*2: Anti-idiotype, reacting with *Ab*1.

$Ab2_\beta$: Those *Ab*2 bearing internal image, i.e., whose paratope mimicks the determinants of Ag.

BD: B-Determinant; an epitope recognized by specific B cells.

HD: Helper-Determinant; an epitope recognized by specific Th cells in association with a Class II molecule (terminology of E. E. Sercarz).

MHC: Major Histocompatibility Complex.

TCR: T cell receptor.

GZ: *E. coli* β-galactosidase.

s,i,c,C1,C2: Concentrated terms to designate GZ epitopes located on the surface of the molecule (s), inside the molecule (i), and dependent on the quaternary conformation of GZ (c, C1,C2).

FITC: Fluorescein Isothyocyanate.

Z gene: The gene coding for GZ.

BSA: Bovine Serum Albumin.

RBC: Red Blood Cells.

MAB: Monoclonal Antibody.

2B2B5: A monoclonal antibody directed against GZ, recognizing a sequential determinant (a gift from Boris M. Rotman).

D6C9: Another monoclonal antibody directed against GZ, which has the property of activating a set of genetically defective GZs described by Accolla et al.[14]

Id5, Id6: Idiotypes found on anti-GZ antibodies.

C12: A monoclonal anti-idiotype recognizing Id6.

VH: The variable part of the immunoglobulin heavy chain.

VL: The variable part of the immunoglobulin light chain.

REFERENCES

1. Jerne, N. K. (1974), "Towards a Network Theory of the Immune System," *Ann. d'Immunol.* **125c**, 373.
2. Bona, C., and H. Hiernaux (1980), "Immune Response: Idiotype-Anti-idiotypic Network," *Critical Review Immunol.* **2**, 33.
3. Celada, F., A. Kunkl, F. Manca, D. Fenoglio, A. Fowler, U. Krzych, and E. Sercarz (1984), in *Regulation of the Immune System*, Eds. H. Cantor, L. Chess and E. Sercarz (New York).
4. Zinkernagel, R. M., and P. C. Doherty (1979), "MHC-Restricted Cytotoxic T Cells: Studies on the Biological Role of Polymorphic Major Transplantation Antigens Determining T-cell Restriction-Specificity, Function and Responsiveness," *Advances Immunol.* **27**, 51.
5. Manca F., A. Kunkl, D. Fenoglio, A. Fowler, E. Sercarz, and F. Celada (1985), "Constraints in T-B Cooperation Related to Epitope Topology in *E. coli* β-Galactosidase. I. The Fine Specificity of T Cells Dictates the Fine Specificity of Antibodies Directed to Conformation-Dependent Determinants," *Eur. J. Immunol.* **15**, 345.
6. Monod, J., G. Cohen-Bazier, and M. Cohn (1951), "Sur la biosynthèse de la β-Galactosidase (lactase) Chez Escherichia coli: la Spécificité de l'Induction," *Biochem. Biophys. Acta* **7**, 585.
7. Fowler, A. V., and I. Zabin (1966), "Colinearity of β-Galactosidase with its Gene by Immunological Detection of Incomplete Polypeptide Chains," *Science* **154**, 1027.
8. Celada, F. (1979), "Hierarchic Immunogenicity of Protein Determinants," *Lecture Note in Biomathematics. Systems Theory in Immunology. Proc. Working Conference, Rome 1979*, Eds. C. Bruni et al. (Springer-Verlag), 32, 28.
9. Krzych, U., and E. Sercarz (1986), in *IABAS/WHO/UCSF Symposium on Use and Standardization of Chemically Defined Antigens. San Francisco, 1986* **63**, 41.
10. Melchers, F., and W. Messer (1970), "Enhanced Stability against Heat Denaturation of E. coli Wild Type and Mutant β-Galactosidase in the Presence of Specific Antibodies," *Biochem. Biophys. Res. Commun.* **40**, 570.
11. Rotman, B., and F. Celada (1968), "Formation of β-D-Galactosidase Mediated by Specific Antibody in a Soluble Extract of E. coli Containing a Defective Z Gene Product," *Proc. Natl. Acad. Sci.* **60**, 660.
12. Ito, M., D. Fenoglio, A. Kunkl, G. Li Pira, and F. Celada (1987), in preparation.
13. Berzofsky, J. A. (1983), "T-B Reciprocity: an Ia-Restricted Epitope-Specific Circuit-Regulating T Cell-B Cell Interaction and Antibody Specificity," *Surv. Immunol. Res.* **2**, 223.

14. Accolla, R. S., R. Cina', E. Montesoro, and F. Celada (1981), "Antibody-Mediated Activation of Genetically Defective E. Coli β-Galactosidase Enzymes by Means of Monoclonal Antibodies Produced by Somatic Cell Hybrids," *Proc. Natl. Acad. Sci. USA* **78**, 2478–2482.

V. G. NESTERENKO
N. F. Gamaleya Institute for Epidemiology and Microbiology, U.S.S.R. Academy of Medical Sciences, 123098 Moscow, U.S.S.R.

Symmetry and Asymmetry in the Immune Network

INTRODUCTION

Recently the idiotypic–anti-idiotypic (Id–anti-Id) interaction in the immune network has been revealed (Binz and Wigzell, 1976; Nesterenko and Chernyakhovskaya, 1977; Bona et al., 1981; Kazdin and Horng, 1983; Nydegger et al., 1984; Rubakova et al., 1984; Smith et al., 1987). A number of authors believe that the functioning of the immune system is most adequately described by the immune network theory. According to this theory, the different levels of the network are interconnected by the Id–anti-Id reactions (Jerne, 1974; Richter, 1975; Nesterenko, 1982, 1984, 1986; Bona, 1984; Farmer et al., 1986; Guercio and Zanetti, 1987). These interactions result in the inhibition or stimulation of lymphocytes at the corresponding levels. Models of the immune network have been proposed, which can have a limited (Hoffmann, 1975; Hiernaux, 1977; Herzenberg et al., 1980; Cooper-Willis and Hoffmann, 1983) or an unlimited (Jerne, 1974; Richter, 1975; Ivanov et al., 1982; Nesterenko, 1982, 1984) number of levels. The Id-specific effects, transmitted from whatever level, may, in principle, be directed at both the lower and the higher immune network level (Figure 1). However, it is still far from clear what determines the

A_{i-1} LEVEL
↑
A_i
↓
A_{i+1} LEVEL

FIGURE 1 Direction of A_i Stimulus.

symmetry (bidirectional pathway) or asymmetry (unidirectional pathway) in transmitting activation (or suppression) to the neighbouring lower and higher immune network levels. The aim of the present study has been to examine the symmetrical and asymmetrical transmission of a stimulus in the immune network. Our investigations have shown that it is possible to predict the direction of transmission of the signals and the intensity of the non-specific side reactions.

SYMMETRY AND ASYMMETRY OF INTERACTIONS BETWEEN ELEMENTS OF THE IMMUNE NETWORK

It is known that antigen-binding sites (paratopes) and individual markers, the idiotypic antigenic determinants, are localized in the variable domain of antibodies and antigen-recognizing receptors of lymphocytes (Jerne, 1974; Eichmann, 1978; Weigert and Riblet, 1978; Urbain, 1986). The Id can be present both within and outside the domain of the paratopes (Jerne, 1974, 1984; Hoffmann, 1975; Richter, 1975; Eichmann, 1978; Ju et al., 1980; Rajewsky, 1983). Therefore, in describing the antigen-binding lymphocyte receptors in terms of these two characteristics (that is, Id and paratope), the following types of receptors can be distinguished: Id^+-paratope$^+$ (localized together), Id^+-paratope$^+$ (localized separately, but randomly expressed together), Id^+-paratope$^-$ (localized separately), Id^--paratope$^+$ (localized separately), and Id^--paratope$^-$ (Figure 2). The Id^+-paratope$^+$ receptors possess the corresponding Id and paratope. The latter is complementary to the antigen used (epitope). The Id^+-paratope$^-$ molecules express the particular Id, but do not react with the epitope; the Id^--paratope$^+$ molecules, conversely, bind this epitope and do not carry the corresponding Id (but they possess any other Id). The Id^--paratope$^-$ receptors represent all the molecules that possess other paratopes and Id.

First of all, we shall examine a simplified picture (Figure 3) of the immune network in which two conditions are satisfied: (1) the Id and paratope of the same receptor show identical affinity for the complementary sites; and (2) the Id and paratopes of one-type receptors belonging to the same level also have identical affinity for the complementary structures. For the receptors in which the Id

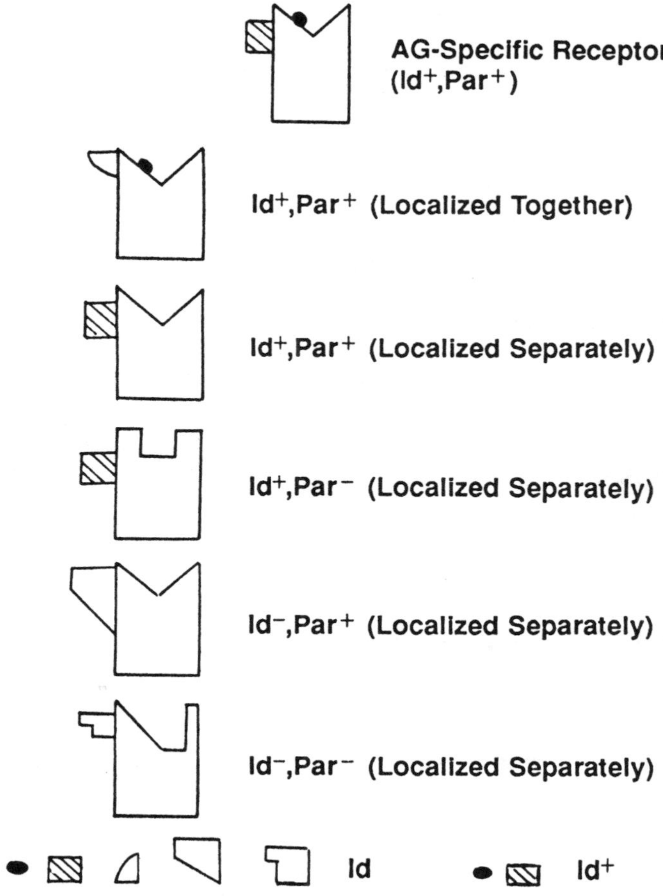

FIGURE 2 Types of receptors.

and paratope are not spatially separated and are localized together, combining Id and paratope with the complementary structures may be regarded as an interaction of the latter with the same region in the variable domain of a receptor, including the paratope and the Id (Figure 4). For such receptors, the immune network develops into pairs of elements complementary to each other (molecules and cells). This variant corresponds to the model of symmetrical Id–anti-Id interactions postulated by Hoffman (1975). The picture is more complex if the Id and the paratope of a receptor are located separately and represent non-identical structures (Jerne, 1974; Richter, 1975; Nesterenko, 1984). Let us accept that the Id and the paratope of interest to us are located at different loci in the variable portion of a receptor. The

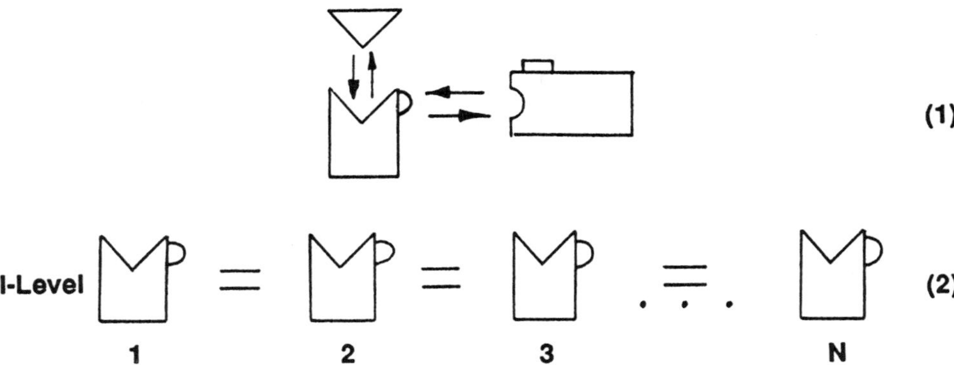

FIGURE 3 Simplified picture of the immune network. (1) Id and paratope of the same receptor show identical affinity for the complementary sites. (2) Id and paratopes of one-type receptors belonging to the same level have identical affinity for the complementary structures.

combining of Id and paratope with complementary structures that belong to the lower and the higher level (under the condition of identical coefficients of binding) will be symmetrical, too (Figure 5). This symmetry of binding may be disturbed only if sites complementary to the Id or the paratope are missing.

Receptors that carry only one of the particular characteristics (Id^+-paratope$^-$ and Id^--paratope$^+$) may also interact with the complementary structures. In this

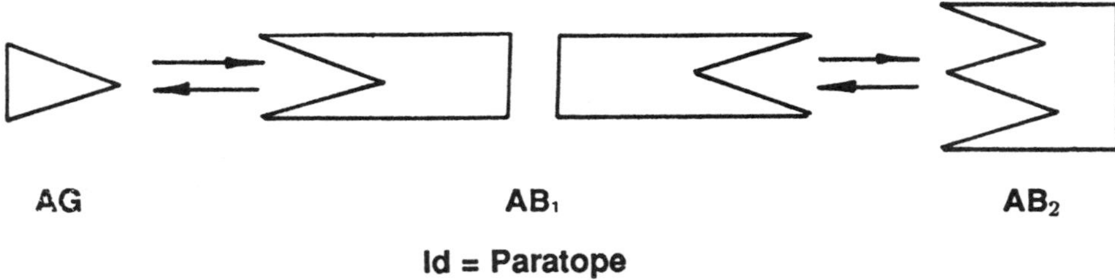

FIGURE 4 Symmetrical immune network with identical spatial localization of idiotopes and paratopes. AG–epitope; AB_1–elements of the first level; and AB_2–elements of the second level, carrying an "internal image" of the epitope.

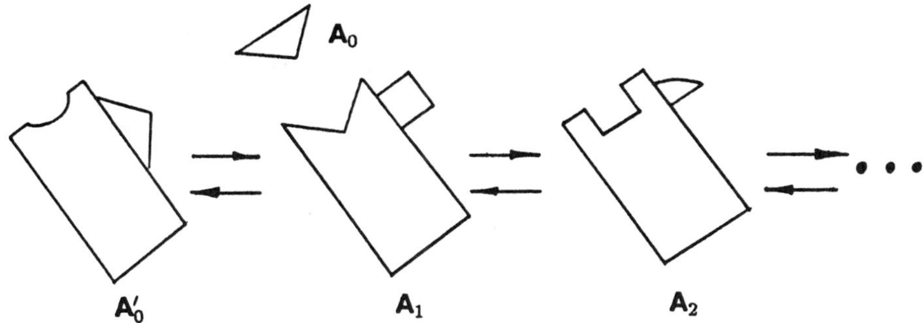

FIGURE 5 Symmetrical network with the different spatial localization of idiotopes and paratopes. A_0–antigen; A'_0–internal image (AB_{2b}); and A_2–classical anti-idiotope (AB_{21}). Receptors are identical within the same level. The probability of the tranmission of signals to the neighbouring lower and higher levels is the same.

case, the interaction of the receptors' segments, designated Id^--paratope$^-$ throughout this article, with the complementary sites will lead to the transmission of activation to other clones of lymphoid cells that are in no way associated with the particular epitope, paratope, and Id. As a result, the non-specific component will be augmented in the course of immune reactions.

We can conclude that two types of immune network may exist, depending on the localization of Id and paratope. If Id and paratope are localized together, only pairs of elements complementary to each other may be stimulated. On the contrary, the spatial separation of Id and paratope may lead to the paratope-induced and Id-induced immune reactions and, consequently, to the activation of a multi-level immune network. With the adopted limitations (see Figure 3), the transmission of activation between levels of these two types of immune network will preferentially be symmetrical in nature. However, even under these conditions, an asymmetrical transmission of stimuli between the lower and the higher levels could take place because of the possible absence of elements of any level.

A different situation may occur in an immune network, if the paratope and Id, localized in the different segments of the variable domain of the same receptor, differ in their affinity for complementary sites (Figure 6). In this case, the transmission of activation through network levels is invariably asymmetrical in nature. Thus, the domination of the paratope's affinity over the Id's affinity will lead to a situation where the activation will be very likely transmitted to the lower level. In contrast, the greater affinity of Id for the complementary structures, compared

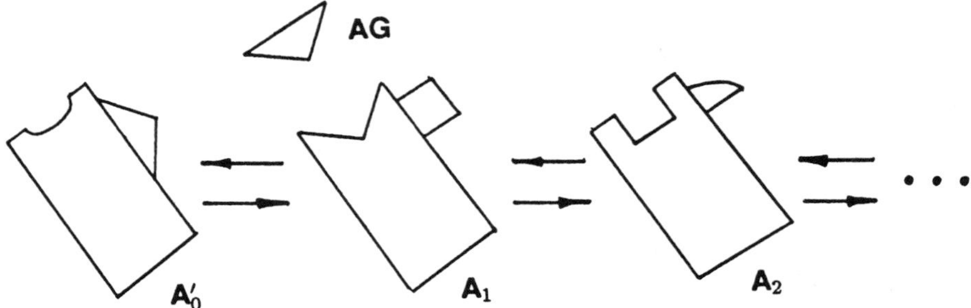

FIGURE 6 Asymmetrical network with the different spatial localization of idiotopes and paratopes. Receptors are identical at the same level. Paratope and Id of the same receptor differ in their affinity for complementary sites. Nonidentical probability of the transmission of signal to the neighbouring lower and higher levels.

to the paratope, is more likely to induce activation of the higher level. But there is the one cause yet, namely, dissimilarity of the population of receptors for the same complementary structure. It may contribute even more substantially to the asymmetry in the transmission of signals.

ASYMMETRY AT THE LEVEL OF CELL POPULATIONS

In the course of an immune response, the antigen selects clones of lymphoid cells at the first network level according to the affinity of their receptors' paratopes. These receptors, selected by great affinity for antigen, may consist of the molecules that possess similar paratopes and different Id. Each paratope and idiotope may react with its complementary structures and may induce a stimulation of the corresponding lymphocytes. So two types of receptors exist at the first immune network level, five types at the second level, eight types at the third level, and so on (Figure 7). Antigen selects clones of lymphoid cells at the first network level according to the affinity of their receptors' paratopes. These receptors are more likely to react with the elements carrying "internal images" of antigen rather than with the elements at the classical second anti-Id level. *Affinity of the receptors' paratopes is therefore a principal, as if produced by Nature, limiting factor in transmission of activation to the higher network levels*. In phylogeny, the affinity of receptors could have been a factor involved in limitation of the activation in the immune network.

From the viewpoints we have presented above, asymmetry in transmitting a stimulus to the lower and higher immune network levels is determined by:

1. The different spatial localization of some paratopes and Id on the same receptor which have non-identical affinity for the complementary sites.

2. Population differences in affinity for the complementary segments in the paratopes and Id that belong to different receptors at the same immune network level.

In the organism, the limitations of activation transmission inside the immune network and localization of activation at the first, lower levels may result from (a) the presence of Id in the region of the paratopes; (b) selection by an immunogenic dose of antigen of lymphocyte populations at the first level with the corresponding high-affinity paratopes and subsequent preferential reactions of the first-level elements with lymphoid cells carrying the internal image of the antigen (it is possible that under certain conditions, elements with an internal image of the epitope may react again with lymphocytes at the first level, by inducing their stimulation or suppression); and (c) the existence of tolerance on anti-Id level to some crossreacting Id (Rajewsky, 1983; Somme et al., 1983).

Again, preferential transmission of a signal inside the immune network (at most to the classical anti-Id level) depends on the stimulation by antigen of elements possessing low-affinity paratopes and high-affinity Id.

$$
\begin{array}{ccccccc}
 & & \text{FIRST LEVEL} & & \text{SECOND LEVEL} & & \text{THIRD LEVEL} \\
 & & & & & & (P_3\text{anti}-X),(Id_x) \\
 & & & & & & P_1\text{anti}-A_0',Id_1 \\
 & & & & (P_2\text{anti}-X),(Id_x) & & P_1\text{anti}-A_0',(Id_x) \\
 & & & & (P_2\text{anti}-X),Id_2 & & P_1\text{anti}-A_0',Id_3 \\
A_0 & \rightleftharpoons & \begin{array}{c} P_1\text{anti}-A_0,(Id_x) \\ P_1\text{anti}-A_0,Id_1 \end{array} & \rightleftharpoons & (P_2\text{anti}-X),A_0 & \rightleftharpoons & (P_3\text{anti}-X),Id_3 \\
 & & & & P_2\text{anti}-Id,Id_2 & & (P_3\text{anti}-X),Id_1 \\
 & & & & P_2\text{anti}-Id),(Id_x) & & P_3\text{anti}-Id_2,(Id_x) \\
 & & & & & & P_3\text{anti}-Id_2,Id_3
\end{array}
$$

FIGURE 7 Asymmetrical immune network with the different spatial localization of idiotopes and paratopes. Receptors aren't identical at the same level. A_0–epitope; P–paratope; Id–idiotope; A_0'–idiotope, carrying "internal image" of the epitope; and X–a foreign antigen or a foreign Id. The family of elements possessing foreign paratopes or foreign Id is in parentheses. Arrows indicate the direction of the reaction.

SOME POSTULATES AND SUGGESTIONS

From the point of view of possible asymmetry in transmitting activation throughout the immune network, it is possible to make the following postulates.

1. ADMINISTRATION OF LOW DOSES OF A_i

Administration of low does of i-level antibodies (A_i) with high affinity for Id of level i–1 receptors will lead to the binding of almost all of these antibodies to level i–1 lymphocytes. If the concentration of A_i antibodies is small, then a fraction of the level i–1 cells will be stimulated and will produce A_{i-1} molecules. On the other hand, the use of low doses of i-level antibodies having low affinity for level i–1 receptors and expressing identical Id induces more easily the stimulation of i+1 level lymphocytes. In order to achieve a stronger, specific stimulation of only level i–1 lymphocytes by A_i antibodies, a mixture of different monoclonal A_i antibodies with high affinity for level i–1 receptors or of affinity-purified A_i serum antibodies rather than one monoclonal high-affinity A_i antibody should be used. In a mixture of A_i antibodies, the concentration of Id of each antibody is smaller than in the same amount of one monoclonal A_i antibody. Therefore, it is possible to construct such a set of A_i antibodies in which the concentration of each Id in the mixture will be smaller than the threshold dose for the stimulation of level i+1 cells reactive to the corresponding Id. Consequently, this mixture of A_i antibodies will stimulate only lymphocytes of the i–1 level (Figure 8).

ADMINISTRATION OF MEDIUM DOSES OF A_i

Administration of medium doses of A_i antibodies with high affinity for Id of level i–1 receptors will also lead to their preferential interaction with level i–1 lymphocytes. Because the concentration of antibodies is higher, their binding will induce two opposite processes in cells of the i–1 level: partial elimination and stimulation of the remaining viable lymphocytes. The use of medium doses of A_i antibodies may also lead to binding small numbers of these molecules to a fraction of the

$$\text{LOW } A_i \text{ DOSES} \longrightarrow \uparrow A_{i-1}$$
$$\text{LOW } A_i \text{ DOSES} \longrightarrow \uparrow A'_0$$

FIGURE 8 Administration of low doses of A_i antibodies. $A'_0 = Ab_{2b}$ = internal image of the epitope.

$$\text{MEDIUM } A_i \text{ DOSES} \longrightarrow \uparrow A_i$$
$$\text{MEDIUM } Ab_1 \text{ DOSES} \longrightarrow \uparrow Ab_1$$

FIGURE 9 Administration of medium doses of A_i antibodies.

level i+1 cells. As a result of these interactions, a small, yet more pronounced, production of antibodies of the i–1 and i+1 levels, compared to the first example, will be observed. These molecules more related to i-level lymphocytes may induce their stimulation. Thus, as a result of the administration of medium doses of A_i antibodies, the synthesis of A_i antibodies is induced (via a number of intermediate stages), many of which have high affinity for level i–1 receptors (Figure 9). The application of medium doses of A_1 antibodies having low affinity for level i–1 receptors and expressing identical Id will induce a preferential binding of A_i antibodies to level i+1 lymphocytes. Consequently, a fraction of the lymphocytes will be killed and the remaining viable lymphocytes will be stimulated and will produce A_{i+1} antibodies. The latter may interact with those lymphocytes that express identical Id, but different paratopes. As a result, molecules that have Id of the i level but do not bind to the receptors at level i–1 lymphocytes will be produced.

ADMINISTRATION OF HIGH DOSES OF A_i

Administration of high doses of i-level antibodies (A_i) with high affinity for Id of level i–1 receptors will result in binding of the receptors of all level i–1 lymphocytes and of a fraction of the level i+1 lymphocytes. One cell of the level i–1 will bind more antibodies. Inasmuch as a large concentration of A_i antibodies is administered, all (or most) of the A_{i-1} lymphocytes and only a small amount of the level i+1 cells will be eliminated. At the same time, a fraction of the A_{i+1} lymphocytes may be stimulated and produce antibodies. Thus, in this case, the synthesis of A_{i+1} antibodies is inducted (Figure 10). On the contrary, the application of high doses of A_i antibodies with low affinity for level i–1 receptors and expressing identical Id will induce the elimination of the level i+1 lymphocytes. It is also possible that a fraction of the low-affinity A_i antibodies that were bound by their paratopes to any lymphocytes may induce their stimulation. Therefore, in this situation, along with elimination of level i+1 lymphocytes, production of non-specific molecules may be observed.

$$\text{HIGH } A_i \text{ DOSES} \longrightarrow \uparrow A_{i+1} + \downarrow A_{i-1}$$
$$\text{HIGH } Ab_1 \text{ DOSES} \longrightarrow \uparrow Ab_{21} + \downarrow Ab_{2b}$$

FIGURE 10 Administration of high doses of A_i antibodies. AB_{21}–classical anti-Id elements and Ab_{2b}–internal image of the epitope.

$$\text{VERY HIGH } A_i \text{ DOSES} \longrightarrow \downarrow A_{i+1} + \downarrow A_{i-1}$$
$$\text{VERY HIGH } Ab_1 \text{ DOSES} \longrightarrow \downarrow Ab_{21} + \downarrow Ab_{2b}$$

FIGURE 11 Administration of very high doses of A_i antibodies.

ADMINISTRATION OF VERY HIGH CONCENTRATIONS OF A_i

Administration of very high concentrations of i-level antibodies (A_i) with high affinity for Id of level i–1 receptors will result in the elimination of level i–1 lymphocytes. The number of A_i antibodies is high enough to induce the killing of level i+1 cells. Analogous results can be obtained by administering very high doses of A_i antibodies that have low affinity for level i–1 receptors and express an identical Id (Figure 11).

THE USE OF XENOGENEIC, ALLOGENEIC AND SYNGENEIC A_i ANTIBODIES FOR STIMULATION

The use of xenogeneic, allogeneic and syngeneic (or autologous) A_i antibodies for stimulation of level i–1 lymphocytes for production of A_{i-1} molecules may, in some cases, evoke different effects. For example, let us assume that the Id family of level i–1 receptors consists of three different antigenic determinants, 1, 2 and 3, being localized each on a separate receptor. Let us admit that the donor of A_i molecules recognizes determinants 1 and 2 of Id family, whereas the recipient recognizes the loci 2 and 3 of Id family. Then the administration of donor A_i molecules (anti-1 and anti-2 antibodies) to a recipient will stimulate the recipient's lymphocytes carrying antigenic determinants 1 and 2 (but not the determinant 3).

We shall examine another example. Let us admit that donor and recipient of A_i molecules recognize determinants 1 and 3 of Id family of the level i–1 receptors, respectively. Then the administration of A_i molecules (anti-1 antibodies) to a recipient will induce the stimulation of those recipient's lymphocytes that express the antigenic determinant 1 (but not determinants 2 and 3). Thus, it can be concluded that the use of xenogeneic and allogeneic A_i antibodies for stimulation of level i–1 lymphocytes will result in the production of those A_{i-1} elements that are "seen" by the donor of A_i antibodies. Moreover, by simultaneously using antibodies against different antigenic determinants of Id that belong to the Id family of the level i–1, it will be possible to induce a more complete picture of this Id family in the immune network.

HARMLESS ANTIGEN-SPECIFIC CORRECTION OF THE IMMUNE RESPONSE

For a harmless antigen-specific correction of the immune response with the aid of the anti-Id effects, it is necessary to use molecules and cells that would interact only with the paratopes on the receptors of the lymphocytes at the first level. These include the cells and their products (for example, antibodies) that (a) carry the internal image of the epitope in the set of their Id, or (b) possess the paratopes for the Id that are parts of a set of the paratopes of the lymphocytes at the first level. Taking this approach, the genetic restrictions of the classical Id–anti-Id interactions that possess individual specificities might be overcome (Nisonoff and Lamoy, 1981; Urbain et al., 1982; Cazenave et al., 1983).

AFFINITY OF THE RECEPTOR'S PARATOPE IN THE IMMUNE NETWORK

Both the degree of involvement of different immune network levels and the expression of the non-specific immune reactions are determined by the affinity of the receptor's paratope in the immune network. In the case of high-affinity paratopes, the immune reaction is limited by the individual immune network levels. On the contrary, domination of low-affinity paratopes induce massive involvement of many immune network levels and numerous non-specific reactions.

LYMPHOCYTE STIMULATION

In this article, emphasis is deliberately placed on the interactions between the paratope and Id of the antigen-recognizing receptors and the complementary sites, but other important aspects of lymphocyte stimulation are not considered. It is known that interactions between lymphocytes stimulated by antigen and the antigen-nonspecific signals for differentiation and for proliferation are also required for a normal immune response to take place (Eardley et al., 1983; Petterson et al., 1983; Schimpl, 1983; Lipkowitz et al., 1984). Consequently, it will be useful to construct a complex molecule of a highly specific stimulator that would be composed of: (a) the antigen's epitopes or of the variable domain of the corresponding anti-Id antibodies; (b) a representative of the family of signals for differentiation; and (c) a representative of the family of signals for proliferation. Obviously the use of such artificial molecules will help effectively regulate the functioning of certain immune network levels and, ultimately, specifically change the immunological reactivity to the corresponding antigens only.

EXPERIMENTAL TESTS OF OUR NETWORK IDEAS

We now have some experimental data concerning our network ideas (Rubakova et al., 1984; Agadyanian et al., 1985; Nesterenko et al., 1986, 1987). (CBA x C57BL/6)F_1 mice were injected with syngeneic affinity-purified serum antibodies against sheep red blood cells (SRBC). To increase the immunogenicity of idiotopes, the antibodies were conjugated either with cellulose or with polyelectrolytes. Earlier (Olovnikov and Gurvich, 1968; Petrov et al., 1985), it was shown that such conjugation of proteins with these nonimmunogeneic carriers led to enhanced immunogenicity of the protein molecules. According to the above-mentioned theory, different doses of Ab_1 should induce nonidentical network effects (Table 2). We have suggested that high concentration of Ab_1 could elicit the strong immunodepressive anti-Id reaction against lymphocytes specific to the homologous antigen (SRBC). On the contrary, administration of low doses of Ab_1 could induce but weak anti-Id perturbation and, thus, contribute to the enhancement of the antigen-specific immune reaction.

To test these ideas mice were inoculated with antibodies conjugated with carriers (Figure 12) and sera of these animals were tested for the presence of the anti-Id antibodies. Also the same mice were immunized with homologous (SRBC) or nonhomologous (rat RBC) antigen and then either number of antibody-forming cells or intensity of delayed-type hypersensitivity were estimated (Table 1).

The experiments revealed that high doses of Ab_1 (200 μkg) conjugated with cellulose-induced strong production of anti-Id inhibitory antibodies, that were shown

TABLE 1 Effect of Ab_1 - Cel and Ab_1 - NA

Characteristics		Ab_1 - cel	Ab_1 - NA
Immune response:	IgM	↓	↑
	IgG	↓	no
	DTH	↓	no
Antigen- and Strain-Specificity		+	+
Inhibitory Anti-Id Antibodies		↑	no
Ts		no	no
Anti-Id Th		↑	no
Enhancement Anti-Id Antibodies		no	?

↑ increase
↓ decrease

TABLE 2 Our Hypothesis

Injection	Anti-Id Induction Strong	Anti-Id Induction Weak	Effect on Antigen-Specific Immune Reaction
High Ab_1 Doses	+		Inhibition
Low Ab_1 Doses		+	Enhancement

to decrease both IgM-IgG-production (10–12 times) and DTH reaction (2–3 times) against SRBC in immunized mice. On the contrary, low doses of Ab_1 (20 μkg) conjugated with polyelectrolyte enhanced the level of IgM-antibody production (3–8 times), but did not induce either IgG-antibody production or DTH reaction. The effects of both high and low concentrations of Ab_1 were antigen and strain specific. They concerned SRBC immune response and were registered in the syngeneic donor-recipient's combination only.

Now we are continuing the network experiments using different antigenic systems which include polysaccharides, proteins, haptens, and cellular surface antigens.

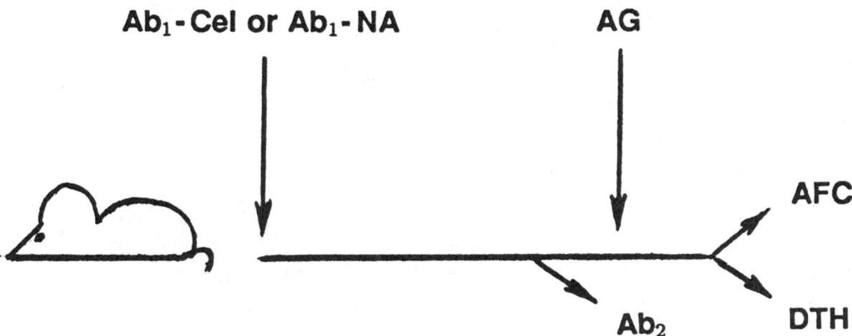

FIGURE 12 Experimental Scheme. AFC–antibody-forming cells; DTH–delayed-type hypersensitivity; Ab_1-cel–anti-SRBC antibodies conjugated with cellulose; Ab_1-NA–anti-SRBC antibodies conjugated with copolymer of acryl acid and PVP; and Ab_2–anti-Id antibodies.

CONCLUSIONS

In this article, we have analyzed the ability to transmit activation signals between immune network levels. We have shown that asymmetry in transmitting a stimulus to lower and higher immune network levels is determined by: (1) the different spatial localization of some paratopes and Id; (2) possible non-identical affinity of such paratopes and Id on the same receptor for the complementary sites; and (3) populational differences in affinity for the complementary segments in the paratopes and Id that belong to different receptors at the same immune network level.

The essential factors limiting transmission of activation are: (a) an identical spatial localization of some paratopes and Id; (b) tolerance at anti-Id level to a number of crossreacting Id; and (c) selection by epitope of the elements possessing high-affinity paratopes.

In light of the above considerations, it can be supposed that in evolution, the affinity of receptors could have been a factor involved in the limitation of activation in the immune network.

The view presented here will help to explain: (1) production of two types of antibody (paratope$^+$-Id$^+$ and paratope$^-$-Id$^+$) which were induced by injection of small doses of anti-Id antibodies to a recipient (Kelsoe et al., 1980; Takemori et al., 1982, 1983; Pierre et al., 1983; Sacks and Sher, 1983); (2) synthesis of antibodies of the first level (A_1) as a result of administration of medium doses of A_1 antibodies to a recipient (Forni et al., 1980; Holmberg et al., 1982, 1983; Ivars et al., 1982); (3) formation of anti-Id antibodies (A_2) immunizing a recipient with large doses of A_1 antibodies (Sakato et al., 1977; Kelsoe et al., 1980; Blaser and Heusser, 1984); (4) detection of different antibodies possessing paratope$^+$-Id$^+$, paratope$^+$Id$^-$, paratope$^-$-Id$^+$, paratope$^-$-Id$^-$ in the immune response to antigen (Avrameas et al., 1976; Bona et al., 1981; Sanchez et al., 1983; Brown, 1984).

On the basis of the findings reported here, I should like to formulate a new principle. Both the degree of involvement of different immune network levels and the expression of the non-specific immune reactions are determined by the affinity of the receptor's paratopes in the immune network. In the case of high-affinity paratopes, the immune reaction is limited by the individual immune network levels. On the contrary, domination of low-affinity paratopes brings about massive involvement of many immune network levels and numerous non-specific immune reactions. Whether or not this conclusion is right must be verified experimentally.

Thus, a number of the theoretical situations we have considered here allow us to give some suggestions concerning stimulation of the lymphocytes of only a certain immune network level (including lymphoid cells of the first level that react to epitopes of foreign antigens).

ACKNOWLEDGMENTS

The author is very grateful to Prof. L. N. Fontalin, Dr. G. P. Ermakov and Dr. E. I. Rubakova for helpful advice and constructive criticism of the manuscript.

REFERENCES

1. Agadyanian, M. G., V. G. Nesterenko, T. B. Megrabian, E. I. Rubakova, and E. V. Sidorova (1985), "Inhibition of Cells Producing Antigen-Dependent Non-Specific Immunoglobulins by Isologous Anti-Erythrocytes Immunoglobulins," *Immunol. Lett.* **9**, 307–309.
2. Avrameas, S., J. C. Antoine, T. Teanynck, and C. Petit (1976), "Development of Immunoglobulins and Antibody-Forming Cells in Different Stages of the Immune Response," *Ann. Immunol. (Inst. Pasteur)* **127C**, 551–558.
3. Binz, H., and H. Wigzell (1976), "Specific Transplantation Tolerance Induced by Autoimmunization against the Individual's Own, Naturally Occurring Idiotypic Antigen-Binding Receptors," *J. Exp. Med.* **144**, 1438–1452.
4. Blaser, K., and C. H. Heusser (1984), "Regulatory Effects of Isologous Anti-Idiotypic Antibodies on the Formation of Different Immunoglobulin Classes in the Immune Response to Phosphorylcholine in BALB/c Mice," *Eur. J. Immunol.* **14**, 93–98.
5. Bona, C. A. (1984), "Parallel Sets and the Internal Image of Antigen within the Idiotypic Network," *Fed. Proc.* **43**, 2558–2559.
6. Bona, C. A., E. Heber-Katz, and W. E. Paul (1981), "Idiotype-Antiidiotype Regulation. I. Immunization with a Levan-Binding Myeloma Protein Leads to the Appearance of Auto-Anti (Antiidiotypic) Antibodies and to the Activation of Silent Clones," *J. Exp. Med.* **153**, 951–960.
7. Brown, A. R. (1984), "Expression and Regulation of Two Idiotype Families and Subsets within an Idiotype Family among BALB/c Antibodies against p-Azophenilarsonate," *J. Immunol.* **132**, 2715–2718.
8. Cazenave, P.-A., J. Roland, and E. Petit-Koskas (1983), "The Idiotypic Network: Internal Images of Rabbit Immunoglobulin Allotypes," *Ann. Immunol. (Inst. Pasteur)* **134D**, 7–14.
9. Cooper-Willis, A., and G. W. Hoffmann (1983), "Symmetry of Effectors Function in the Immune System Network," *Mol. Immunol.* **20**, 865–869.
10. Eardley, D. D., S.-K. Hu, and R. K. Gershon (1983), "Role of Ly-1:Qa1$^-$ and Ly-1:Qa1$^+$ Inducer T Cells in Activation of Ly-23 Effectors of Suppression of Antibody Production in Mice," *J. Immunol.* **131**, 2154–2157.
11. Eichmann, K. (1978), "Expression and Function of Idiotypes on Lymphocytes," *Adv. Immunol.* **26**, 195–221.

12. Estess, P., F. Otani, E. C. B. Milner, J. D. Capra, and P. W. Tucker (1982), "Gene Rearrangements in Monoclonal A/J Anti-Arsonate Antibodies," *J. Immunol.* **129**, 2319–2322.
13. Farmer, J. D., N. H. Packard, and A. S. Perelson (1986), "The Immune System, Adaptation and Machine Learning," *Physica* **22D**, 187–204.
14. Forni, L., A. Coutinho, G. Köhler, and N. K. Jerne (1980), "The IgM Antibodies Induced the Production of Antibodies of the Same Specificity," *Proc. Natl. Acad. Sci. USA* **77**, 1125–1128.
15. Guercio, P., and M. Zanetti (1987), "The CD4 Molecule, the Human Immunodeficiency Virus and Anti-Idiotypic Antibodies," *Immunology Today* **8**, 204–205.
16. Herzenberg, L. A. , S. J. Black, and L. A. Herzenberg (1980), "Regulatory Circuits and Antibody Responses," *Eur. J. Immunol.* **10**, 1–11.
17. Hiernaux, J. (1977), "Some Remarks on the Stability of the Idiotypic Network," *Immunochemistry* **14**, 733–741.
18. Hoffmann, G. W. (1975), "A Theory of Regulation and Self-Nonself Discrimination in an Immune Network," *Eur. J. Immunol.* **5**, 638–643.
19. Holmberg, D., F. Ivars, L. Forni, P.-A. Cazenave, and A. Coutinho (1982), "Idiotypic Characterization of Antibody-Induced Antibody Responses," *Immunology* **162**, 56–65.
20. Holmberg D., F. Ivars, and A. Coutinho (1983), "An Example of Major Histocompatibility Complex-Linked Control of Idiotypic Interactions," *Eur. J. Immunol.* **13**, 82–87.
21. Ivanov, V. V., V. M. Janeko, V. M. Glushkov, L. N. Fontalin, and V. G. Nesterenko (1983), *Mathematical Modelling in Immunology and Medicine* Amsterdam, New York, Oxford: North-Holland Publishing Co.), 141–151.
22. Ivars, F., D. Holmberg, L. Forni, P.-A. Cazenave, and A. Coutinho (1983), "Antigen-Independent, IgM-Induced Antibody Response: Requirement for 'Recurrent' Idiotypes," *Eur. J. Immunol.* **12**, 146–151.
23. Jerne, N. K. (1974), "Towards a Network Theory of the Immune System," *Ann. Immunol. (Inst. Pasteur)* **125C**, 373–387.
24. Jerne, N. K. (1984), "Idiotypic Network and Other Preconceived Ideas," *Immunol. Rev.* **79**, 5–14.
25. Ju, S. T., B. Benacerraf, and M. E. Dorf (1980), "Genetic Control of a Shard Idiotype among Antibodies Directed to Distinct Specificities," *J. Exp. Med.* **152**, 170–182.
26. Ju, S. T., A. Gray, and A Nisonoff (1977), "Frequency of Occurrence of Idiotypes Associated with Anti-p-Asophenylarsonate Antibodies Arising in Mice Immunologically Suppressed with Respect to a Cross-Reactive Idiotype," *J. Exp. Med.* **145**, 540–549.
27. Kazdin, D. S. and W. J. Horng (1983), "Bi-Directional Immune Network Mechanism. Simultaneous Induction of Idiotype and Anti-Anti-Idiotype in Rabbits Immunized with Antibody to a Common Idiotypic Specificity of Anti-VHa2-Allotype Antibodies," *Mol. Immunol.* **20**, 819–824.

28. Kelsoe, G. (1984), "Network Interactions," *Surv. Immunol. Res.* **3**, 169–178.
29. Kelsoe, G., M. Reth, and K. Rajewsky (1980), "Control of Idiotype Expression by Monoclonal Anti-Idiotype Antibodies," *Immunol. Rev.* **52**, 75–96.
30. Lipkowitz, S., W. C. Greene, A. L. Rubin, A. Novogrodsky, and K. H. Stenzel (1984), "Expression of Receptors for Interleukin 2: Role in the Commitment of T Lymphocytes to Proliferate," *J. Immunol.* **132**, 31–35.
31. Nesterenko, V. G. (1982), "Autologous Idiotype-Antiidiotype Interactions and Regulation of the Immune Response," *Immunologiya (Russian)* **2**, 5–15.
32. Nesterenko, V. G. (1984), "Network Interactions and the Regulation of the Immune Response," *Fol. Biol. (Praha)* **30**, 231–252.
33. Nesterenko, V. G. (1986), "Role of Asymmetry in the Immune Network," *Fol. Biol. (Praha)* **32**, 256–272.
34. Nesterenko, V. G., E. I. Rubakova, and L. N. Fontalin (1986), "Inhibition of the Immune Response to Antigen as a Result of Induction of Autologous Anti-Idiotypic Antibodies," *6th International Congress of Immunology, Toronto, Abstracts*, 488.
35. Nesterenko, V. G., L. N. Fontalin, K. M. Khaitov, and E. I. Rubakova (1987), "Antigen-Specific Change of the Immune Response as a Result of the Reaction of Autologous Antibodies to the Antigen," *Periodicum Biologorum. Abstract Book of the 8th European Immunology Meeting, Zagreb*, 183.
36. Nesterenko, V. G., and I. U. Chernyakhovskaya (1977), "Generation of Antibodies against Antigen-Recognizing Receptors of T Lymphocytes in the Syngeneic System," *Bull. Exp. Biol. Med. (Russia)* **12**, 639–643.
37. Nisonoff, A., and E. Lamoy (1981), "Hypothesis, Implications of the Presence of an Internal Image of the Antigen in Anti-Idiotypic Antibodies: Possible Application to Vaccine Production," *Clin. Immunol. Immunopathol.* **21**, 397–408.
38. Nydegger, U. E., K. Blaser, and Hässig (1984), "Antiidiotypic Immunosuppression and its Treatment with Human Immunoglobulin Preparations," *Vox Sang.* **47**, 92–96.
39. Olovnikov, A. M., and A. E. Gurvich (1966), "Study of the Antigenic Properties of the Protein-Cellulose Immunosorvent," *Adv. Med. Chem. (Russia)* **11**, 112–114.
40. Petrov, R. V., R. M. Khaitov, and A. Sh. Norymov (1985), "Phenotypic Correction of the Ir-Gene Control of the Immune Response during an Immunization by the Complex (T,G)-A-L with the Synthetic Polyelectrolytes," *Immunologiya (Russia)* **2**, 21–24.
41. Petterson, S., G. Peber, A. Bandeira, and A. Coutinho (1983), "Distinct Helper Activities Control Growth or Maturation of B Lymphocytes," *Eur. J. Immunol.* **13**, 240–245.
42. Piere, S., D. Juy, and P. A. Cazenave (1983), "Allotypic Restriction of the Expression of MOPC-460 Idiotype after Immunization with Either Anti-2,4-Dinitophenyl (DNP) or Anti-Idiotypic Antibodies," *Eur. J. Immunol.* **13**, 999–1003.

43. Rajewsky, K. (1983), "Symmetry and Asymmetry in Idiotypic Interactions," *Ann. Immunol. (Inst. Pasteur)* **134D**, 133–138.
44. Richter, P. H. (1975), "A Network Theory of the Immune System," *Eur. J. Immunol.* **5**, 350–354.
45. Rubakova, E. I., V. G. Nesterenko, and L. N. Fontalin (1984), "Specific Reduction of the Immune Response to Sheep Red Blood Cells by the Injection of Syngeneic Globulin on a Non-Immunogenic Carrier," *Fol. Biol. (Praha)* **30**, 296–306.
46. Sacks, D. L., and A. Sher (1983), "Evidence that Anti-Idiotype Induced Immunity to Experimental African Tryptanosimiasis is Genetically Restricted and Requires Recognition of Combining Site-Related Idiotypes," *J. Immunol.* **131**, 1511–1516.
47. Sakato, N., C. A. Janeway, and H. N. Eisen (1977), "Immune Responses of BALB/c Mice to the Idiotype of T 15 and Other Myeloma Proteins of BALB/c Origin: Implications for an Immune Network and Antibody Multispecificity," *Cold Spring Harbor Symp. Quant. Biol.* **41**, 719–726.
48. Sanchez, P., C. Le Guern, and P. A. Cazenave (1983), "Incomplete Expression of the MOPC 460 Idiotype in the Sera of BALB/c Mice Immunized Either with DNP Antigen or with Anti-Idiotypic Antibodies," *Mol. Immunol.* **20**, 1405–1410.
49. Schimpl, A. (1983), "Signals in B Lymphocyte Proliferation and Differentiation," *Ann. Immunol. (Inst. Pasteur)* **134D**, 143–161.
50. Smith, L. K., K. L. Bost, and J. E. Blalock (1987), "Generation of Idiotypic and Anti-Idiotypic Antibodies by Immunization with Peptides Encoded by Complementary RNA: a Possible Molecular Basis for the Network Theory," *J. Immunol.* **138**, 7–9.
51. Somme, B. G., C. Roth, M. Jean-Claude, P. Salem, and J. Theze (1983), "Public and Individual Idiotopes in the Antipoly (Glu^{60}, Ala^{30}, Tyr10) Response: Analysis by Monoclonal Antibodies," *Eur. J. Immunol.* **13**, 1023–1031.
52. Takemori, T., H. Tesch, M. Reth, and K. Rajewsky (1982), "The Immune Response against Anti-Idiotope Antibodies. I. Induction of Idiotope-Bearing Antibodies and Analysis of the Idiotope Repertoire," *Eur. J. Immunol.* **12**, 1040–1044.
53. Urbain, R. (1986), "Is the Immune System a Functional Idiotypic Network? Idiotypic Networks: a Noisy Background or a Breakthrough in Immunological Thinking?," *Ann. Immunol. (Inst. Pasteur)* **137C**, 57–64.
54. Urbain, J., M. Slaoui, and O. Leo (1982), "Idiotypes, Recurrent Idiotypes and Internal Images," *Ann. Immunol. (Inst. Pasteur)* **133D**, 179–188.
55. Weigert, M. and R. Riblet (1978), "The Genetic Control of Antibody Variable Regions in the Mouse," *Springer Seminars in Immunopathology* **1**, 13–156.

ROB J. DE BOER
Bioinformatics Group, University of Utrecht, Padualaan 8, 3584 CH Utrecht, The Netherlands

Symmetric Idiotypic Networks: Connectance and Switching, Stability, and Suppression

ABSTRACT

We present a network model that incorporates: 1) symmetric idiotypic interactions, 2) an explicit affinity parameter (matrix), 3) external (i.e., non idiotypic) antigens, 4) idiotypic stimulation at low population densities, and 5) idiotypic suppression at high densities. Such an idiotypic network of two clones has three stable states: a virgin state, i.e., an equilibrium between the normal influx and turnover of cells, and two immune states (one for each clone), which are maintained by idiotypic interactions. In its immune state, a clone suppresses its idiotypic partner and immediately rejects antigen. Introduction of antigen into the virgin state causes a state switch to the corresponding immune state: antigens are thus remembered, i.e., the network displays memory. This symmetric network cannot account for suppression of proliferating clones. Clones that proliferate suppress their anti-idiotypic "suppressors" long before these have grown large enough to become suppressive. This is a consequence of symmetry: asymmetric versions of our model do account for suppression. We here assume that proliferation precedes suppression; if the reverse is assumed (i.e., suppression), the model cannot account for either memory or suppression. We conclude that the model incorporating proliferation before suppression is superior. We next analyse 50-dimensional (50-D) networks of this same model. The network connectance crucially determines the behavior of the network. Only weakly connected networks know a 50-D virgin state in which all clones are in a "resting" state. Switching behavior only occurs in weakly connected systems. The stability of the respective states reached by the systems first decreases, but later increases when connectance increases. Most importantly, highly connected systems are highly unresponsive, i.e., most clones are suppressed; hence, most antigens expand progressively. We conclude that only weakly connected networks have, immunologically, reasonable behavior.

INTRODUCTION

Idiotypic network theory[1] is based on the simple argument that, if the 10^7 (or more) clones of the immune system are capable of recognising any antigen, they should also be able to recognise each other. Thus, the variable regions of the T or B lymphocyte receptors provide unique antigenic determinants (the idiotype) that should, therefore, be capable of stimulating other clones that complementarily match these idiotypes. Such stimulatory (i.e., positive) idiotypic interactions have, indeed, been described.[2,3] Following a period of proliferation, stimulated clones normally acquire effector functions capable of eliminating the stimulatory antigen. This is a minimal definition of negative (suppressive) interactions among lymphocytes. Negative idiotypic interactions have also been described experimentally.[4,5] This argument implies that negative interactions should predominate following the period of clonal expansion, i.e., when populations are enlarged.

This implication, however, contradicts the general opinion in network theory. In the original formulation of the idiotypic network,[1] the basic mode of the network interactions was assumed to be suppression. In the absence of antigens, i.e., in the virgin state, lymphocyte clones suppress each other, thus ensuring a stable virgin state.[1] Stimulation with an external antigen perturbs this suppressed state and enables the clones that complementarily match the antigen to proliferate, i.e., they "escape from suppression". Similar "escape from suppression" systems have been proposed for idiotypic network interactions among T cells.[6] Most theoretical models of the idiotypic network also incorporate some form of "escape from suppression."[7–9] In this paper we will, however, show that "escape from suppression" versions of our model cannot account for any interesting network behavior.

Another important concept in idiotypic network theory is that of symmetry; this was originally put forward by Hoffmann.[9,10] If antigen (or idiotype) recognition is, indeed, based on complementary matching, recognition should be a symmetric event. Moreover, cells are activated by receptor crosslinking[9]: which is a process that cannot discriminate between idiotype and anti-idiotype. Recently, Jerne[11] reviewed some experimental data on idiotypic interactions that fully support the symmetry idea. He suggests that "one of the characteristics of the idiotypic network is the occurrence of pairs of antibodies, of preferred partners." Despite its attractiveness, symmetric network theory[9,10] has been followed by only a few authors.[8,11] Note, however, that symmetric network theory disposes of the distinction between paratope and idiotype; instead, it regards variable regions of receptor molecules as "sticky ends."[10] We here adopt this "sticky end," i.e., symmetry, hypothesis.

In this paper, we investigate the effect that idiotypic interactions can have on the (normal) proliferative immune response to external (i.e., non-idiotypic) antigens. Idiotypic interactions are generally supposed to account for a variety of phenomena described for the immune system. Of these, we will here consider the phenomenon of immunological memory (i.e., immunity),[1] and that of suppression, i.e., the control of proliferation.[6,12] We vary the impact that idiotypic interactions can have by varying 1) the affinity of the idiotypic interactions, and 2) the number of idiotypic interactions (i.e., the size and the connectance of the network). We, thus, require a symmetric idiotypic network model that incorporates external antigens and an explicit affinity parameter. Moreover, in the model, it must be straightforward to vary the dimension (i.e., the number of participating clones) and the connectance (i.e., the number of idiotypic interactions per clone). Because none of the previous models[7–10,13–17] adequately fulfills all these requirements, we will first develop a new model.

THE MODEL

We consider clones of B or T cells (X_i) that have symmetric idiotypic interactions (Eq. 3). Additionally, the model incorporates antigens that grow exponentially (VR_i), e.g., viruses (Eq. 4). Each clone (X_i) has a constant rate of supply (S) from bone marrow or thymus, and a constant rate of death (D). A virgin state for each clone, hence, emerges as the equilibrium between source and death (S/D). All cells are relatively short-lived (about 5 days, $D = 0.2$): we, thus, deliberately ignore the existence of long-lived memory cells. Memory phenomena, if they occur in the model, can therefore only be due to idiotypic network interactions.

Interactions among the clones are defined by the affinity matrix A, and by the respective population densities. Affinity is a parameter ranging between zero and one ($0 \leq A_{ij} \leq 1$), the system is symmetric ($A_{ij} = A_{ji}$), and clones cannot see themselves (all $A_{ii} = 0$). The total interaction terms (Eqs. 1 and 2) of each clone are the sum of all interactions with others (TA = total activation, TS = total suppression).

$$TA_i = VR_i + \sum_{j=1}^{j=n} A_{ij} \cdot X_j \tag{1}$$

$$TS_i = \sum_{j=1}^{j=n} A_{ji} \cdot X_j \tag{2}$$

$$\frac{dX_i}{dt} = S_i - D \cdot X_i + \frac{B \cdot X_i \cdot {TA_i}^M}{KB^M + (F \cdot X_i)^M + {TA_i}^M} - \frac{C \cdot X_i \cdot {TS_i}^M}{KC^M + (F \cdot X_i)^M + {TS_i}^M} \tag{3}$$

$$\frac{dVR_i}{dt} = R \cdot VR_i - \frac{C \cdot VR_i \cdot {X_i}^M}{KC^M + (F \cdot VR_i)^M + {X_i}^M} \tag{4}$$

Idiotypic interactions can both be stimulatory (positive) and inhibitory (negative). We assume that interactions are positive at low population densities and negative at high densities. This is incorporated in the model as the subtraction of two conventional saturation functions (Hill functions). Because KB<<KC proliferation (i.e., the positive term) precedes suppression; because C>B, suppression is dominant; i.e., interactions among large populations are always negative. This precludes the possibility of infinite proliferation due to reciprocal idiotypic stimulation. The model additionally incorporates a buffering term (F), reminiscent of that used by Richter.[7] Buffering ensures that the maximum stimulation or suppression of a very large cell population can never be achieved by a small idiotypic interaction partner (this does occur in conventional saturation terms). Buffering has mainly an effect in the high-D models, most low-D results can be repeated if $F = 0$.

The antigens (viruses) grow exponentially with specific growth rate R, and are eliminated by a process similar to idiotypic suppression. An antigen is numbered according to the (one and only) clone that recognises it (with maximum affinity). Antigens stimulate clones, but never suppress them (Eq. 1 vs. Eq. 2). The parameter setting is: $S \approx 20\,d^{-1}$, $D = 0.2\,d^{-1}$, $B = 0.7\,d^{-1}$, $C = 25\,d^{-1}$, $KB = 10^3$, $KC = 10^6$, $F = 0.01$, $M = 2$, and $R = 0.5\,d^{-1}$. The virgin population density, thus, equals $S/D \approx 100$ cells. The influx is slightly different for each clone (to avoid settlement in unstable equilibria): S has a mean of 20 cells per day with a 10% standard deviation. Maximum proliferation proceeds at a rate $B - D = 0.5$ cells per day (this corresponds to a doubling time of about 16 hours). Viruses grow at the same rate as the lymphocytes ($R = 0.5\,d^{-1}$).

Low-D models are analysed by means of GRIND[18]; GRIND numerically searches for 0-isoclines and performs numerical integration by means of ROW4A.[19] High-D systems are analysed by numerical integration, using Adams-Moulton's algorithm implemented in ACSL.[20]

RESULTS: LOW-D SYSTEMS

First consider a normal proliferative immune response of one clone (X_1) to its virus (VR_1). There are no idiotypic interactions (all A elements equal zero). Because of the absence of idiotypic interactions, suppression and memory phenomena are

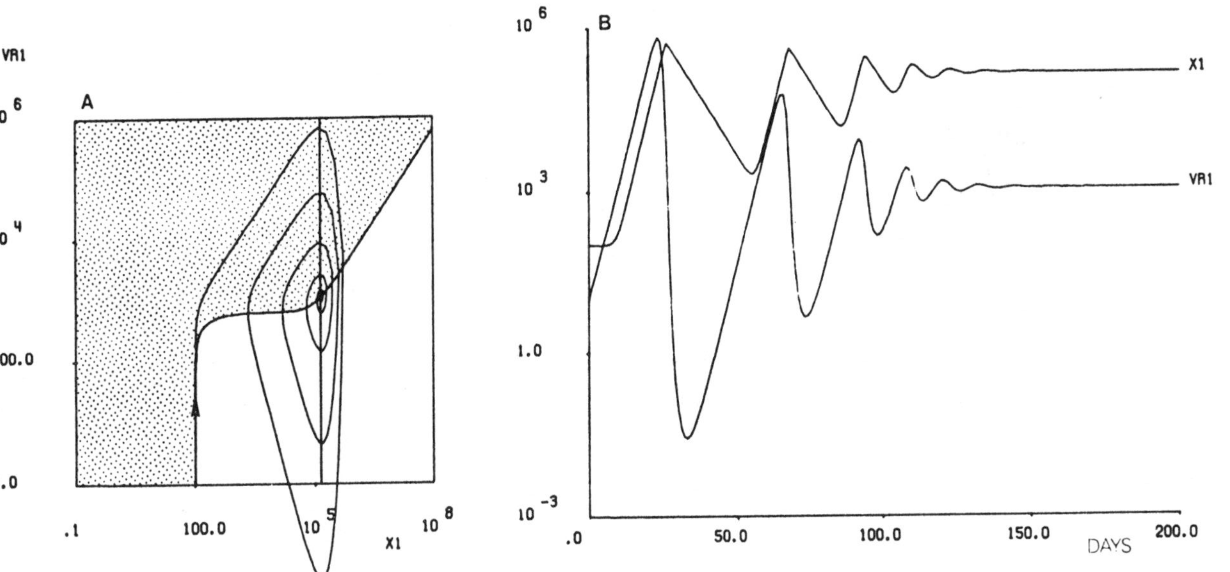

FIGURE 1 The interaction with X_1 and VR_1 without any network interaction. Figure 1a shows the ${X_1}' = 0$ and the ${VR_1}' = 0$ isocline (the ${X_1}' > 0$ region is shaded). The trajectory depicts the immune response following the introduction of $VR_1 = 10$ into the virgin state of X_1 ($X_1 = 100$). The response is highly oscillatory and settles in the stable intersect between the two isoclines. Figure 1b is the time plot of the same reaction.

expected to be absent from this model. We introduce a virus dose of 10 infected cells into the virgin state of the clone.

As shown in Figure 1, the virus expands, and after a while, triggers the proliferation of X_1. This finally results in virus regression, and due to a lack of antigenic restimulation, in a reduction in X_1 numbers. This, in turn, allows for an increase of the virus, and after a number of (predator-prey like) oscillations, the system settles in a stable equilibrium with about 1000 infected cells. We call this virus "dormancy". This (dormancy) equilibrium is a form of immunity: reintroduction of a new dose of this same virus leads to rapid elimination of these extra viruses. Thus, memory can be generated in the absence of memory cells and idiotypic network interactions by continuous stimulation of the virus specific clone (here, X_1), due to the continuous presence of a small virus population. Note, however, that the virus population dropped below the critical level of one infected cell; this would normally correspond to total virus eradication. If we, indeed, remove the virus whenever it consists of less than one cell, the system returns to the virgin state after one oscillation.

A 2-D NETWORK

Now, consider a 2-D network without external antigen; assume that the two clones see each other with maximum affinity ($A_{12} = 1$). The model now incorporates idiotypic suppression and stimulation. Figure 2 shows the $X_1' = 0$ and $X_2' = 0$ isoclines and (2b) a number of different trajectories (all stable equilibria are marked in Figure 2a). This 2-D network knows one virgin state (V) which is situated at the source/decay equilibria of the two clones: this is a state without idiotypic interactions. Additionally, this system knows two immunity equilibria (I1 and I2). In its immune state, a clone is about 1000 times larger than it is in the virgin state. Its idiotypic partner is suppressed (i.e., cannot be stimulated any further), but it is also enlarged (about 10-fold). Note that, although X_1 suppresses X_2 in the $I1$ equilibrium, the net interaction that X_2 experiences from X_1 is still positive; otherwise, X_2 would not be enlarged in this state. It is this enlargement of the (suppressed) idiotypic partner that sustains the large population in the immunity equilibrium.

This figure is highly reminiscent of the phase portraits of the symmetric Hoffmann[9,10,15] models. However, although X_1 and X_2 do suppress each other at high population densities (trajectories move downwards), the present systems lacks Hoffmann's stable suppressed state in this region. Other differences are the absence of interactions in our virgin state (Hoffmann's virgin state is maintained by suppression) and the fact that our immune states are situated above the source/decay (i.e., no interaction) equilibrium whereas those of the Hoffmann models are situated below it.

The existence of these various equilibria of 2-D systems is robust for a large variation in the affinity of the interaction. Figure 3 shows the same 0-isoclines as

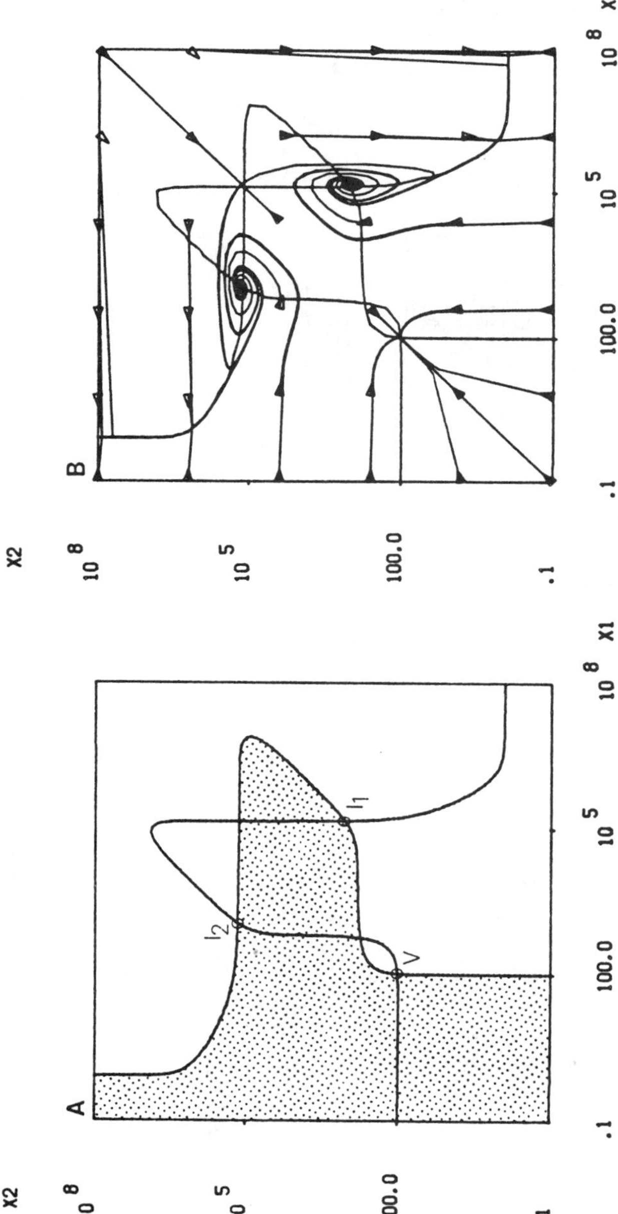

FIGURE 2 The X_1 and $X_2{}' = 0$ isoclines of a 2-D system ($A_{12} = 1$). In Figure 2a, the $X_1{}' > 0$ region is shaded, and the stable equilibria are encircled. Figure 2b shows the same 0-isoclines in combination with trajectories starting at various points in the state space. The picture shows the oscillatory behaviour around the immune states.

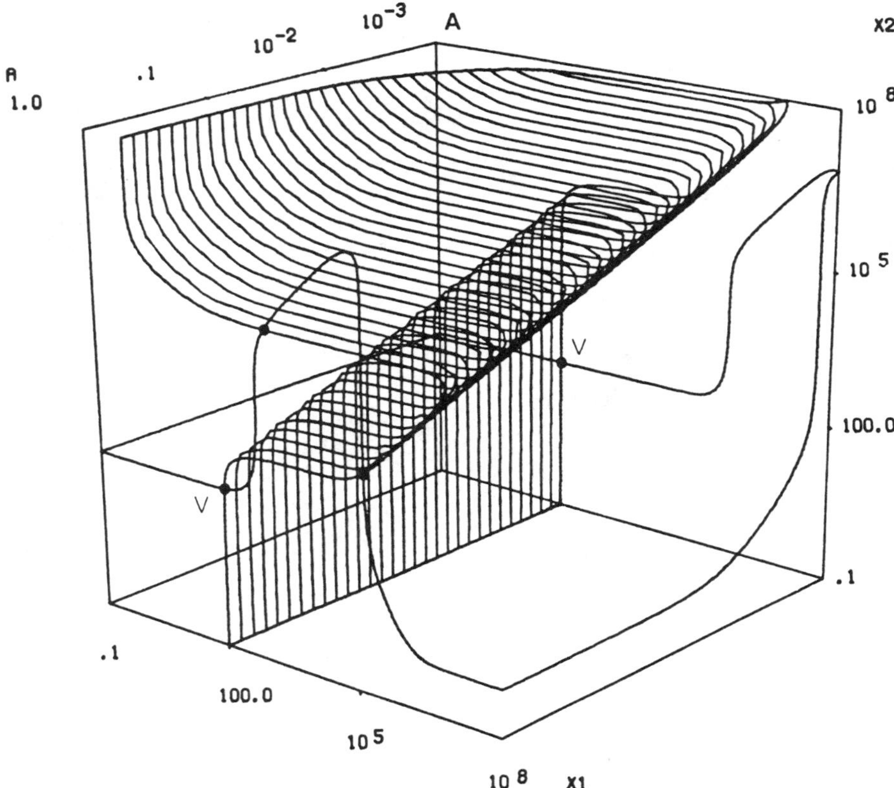

FIGURE 3 The X_1 and $X_2{}' = 0$ isoclines as a function of the affinity (A_{12}) of the idiotypic interaction. The $X_1{}' = 0$ isocline plane is shaded, and all stable equilibria for $A_{12} = 1$ and $A_{12} = 10^{-3}$ are encircled.

Figure 2, but now as a function of affinity (A_{12}). At the front of this 3-D space (at $A_{12} = 1$), one recognises Figure 2a with its 3 equilibria. At the back, by contrast, the 0-isoclines intersect in only one equilibrium: the virgin state. Thus, although we never incorporated an affinity threshold, i.e., some minimum affinity below which clones can never switch states by idiotypic interactions, the model accounts for one. If buffering is removed from the model ($F = 0$), this affinity threshold vanishes.[21] The three stable equilibria exist in this model whenever the affinity is larger than or equal to 0.02. Thus, if affinity values are uniformly distributed between zero and one (this will be the case in the high-D section), 2-D systems are expected to have immunity states in 98% of all cases. Hence, with respect to the existence of multiple stable states, the present model clearly fulfills Hoffmann's[16] Unpredictability Axiom.

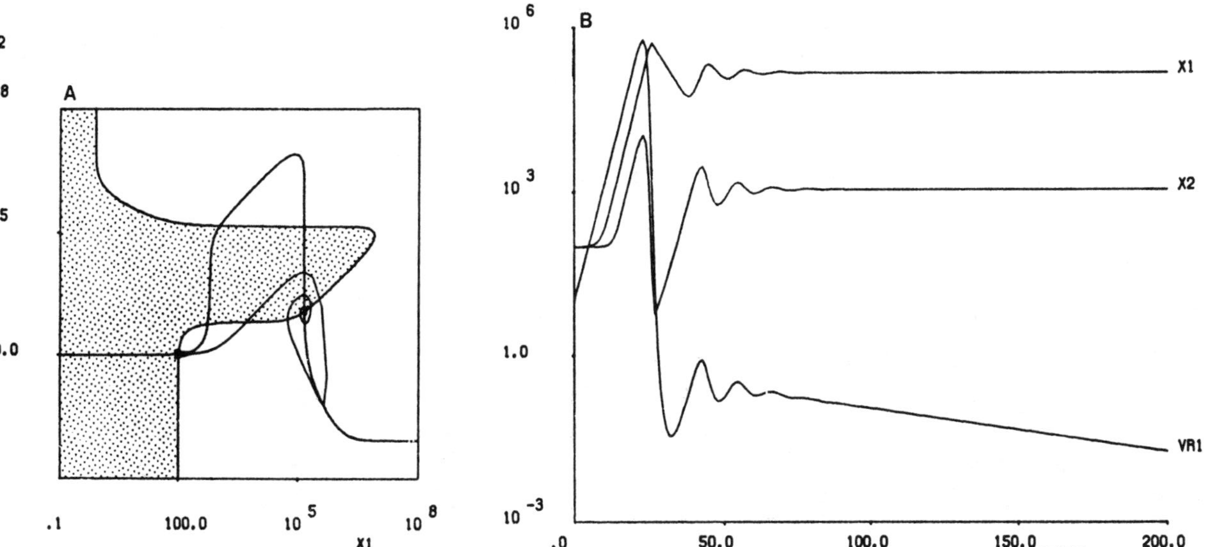

FIGURE 4 The interactions between X_1, X_2 and VR_1, i.e., a 2-D network with antigen. We project the immune responses to VR_1 in the X_1, X_2 state space. This settles in the X_1 immunity equilibrium. The immune response is again shown in a time plot (4b).

2-D NETWORK WITH VIRUS

Now, consider the first situation in which idiotypic interactions are capable of influencing normal proliferative immune responses to antigens: consider the 2-D system of Figure 2 with the virus of Figure 1 (i.e., VR_1). The virus is again introduced in a dose of 10 infected cells into the 2-D virgin state of the network. Figure 4a shows the results projected in the X_1–X_2 state space; Figure 4b shows the same model behavior plotted in time. The first conclusion we draw is that this 2-D system switches correctly after antigenic stimulation: VR_1 activates X_1 which leads to a switch to the X_1 immune state ($I1$). Reintroduction of the same virus into this immune state yields immediate rejection of the virus. Hence, the system truly accounts for immunological memory during a secondary response. This occurs in the absence of memory cells or virus dormancy; immunity is due to a stable state switch of the (2-D) network. The second conclusion is based on the time course of the response. The primary response to the virus, i.e., the proliferative part of the first cycle, is the same in Figures 1 and 4, i.e., is independent of idiotypic interactions. The difference resides in the decline of X_1 following virus regression: the anti-idiotypic clone (X_2) stimulates X_1 and, hence, reduces the fall in X_1 numbers. Hence, X_1 oscillations are reduced. Thus, the first X_1 oscillation is driven by the regression of VR_1; the remainder, however, is driven by the anti-idiotypic partner (X_2). Oscillations of the latter kind were described by Cerny.[23] Secondly, and most importantly, this figure is similar to published experimental data.[3,23] The interpretation of what is going on is, however, different. The normal interpretation of data like that in Figure 4b[3,12] is that the appearance of the anti-idiotypic antibody (here, X_2) suppresses the proliferation of the clone responding to antigen (here, X_1). This would explain the regression of the first clone (X_1). However, in our model, X_2 can never suppress X_1 in this situation because X_2 is necessarily smaller than X_1. If suppression were to occur, it would be X_1 suppressing X_2, simply because X_1 is ahead and because it is the large populations that do the suppression in this model. In the model, X_1 regresses because the antigen concentration (VR_1) reduces. The appearance of X_2 is even responsible for keeping X_1 at relatively high numbers (by idiotypic stimulation); this is what reduces the oscillatory behavior.

NO CONTROL OF PROLIFERATION

We think that most, if not all, symmetric idiotypic network models fail to account for the suppression of proliferating clones (i.e., for proliferation). The proliferating clone is the one to react first (i.e., to antigen), and is, hence, always larger than its anti-idiotypic partners. The symmetry idea implies no direction in the suppression (or activation) interaction. Therefore, the idiotypic clone can also

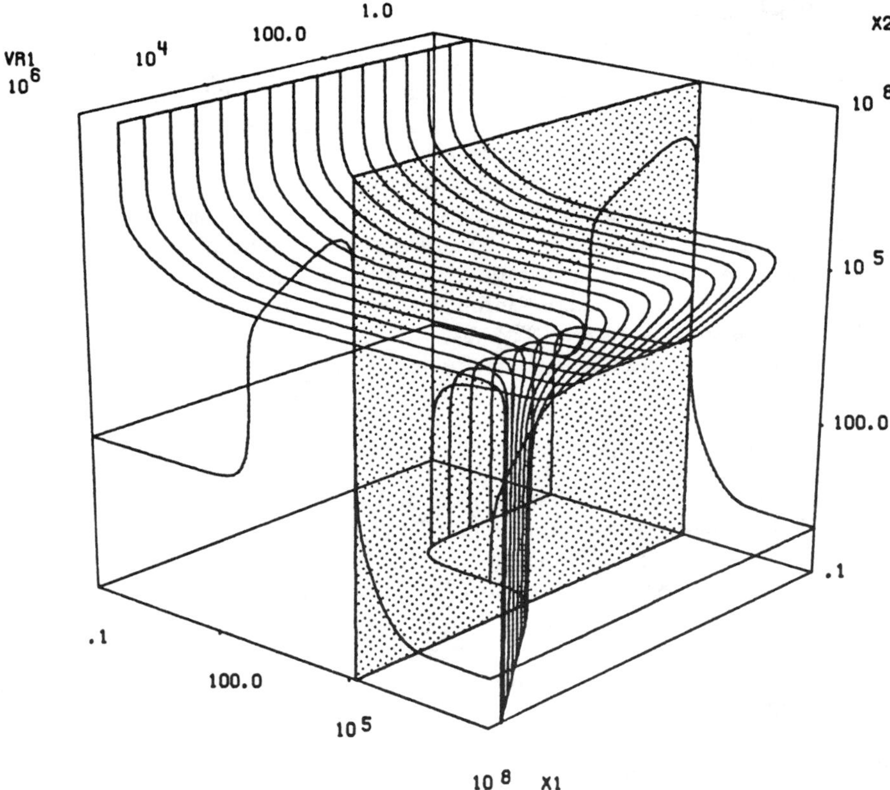

FIGURE 5 A static representation of the interaction between X_1, X_2 and VR_1. The $X_1{}' = 0$, $X_2{}' = 0$ and $VR_1{}' = 0$ isoclines in a X_1, X_2, VR_1 state space. The $X_1{}' = 0$ isocline plane is shaded. The $VR_1{}' = 0$ isocline (dotted) is straight and situated at about $X_1 = 10^5$; the $X_2{}' = 0$ isocline is identical to that of Figure 2.

suppress its anti-idiotypic "suppressors." Just because the idiotypic clone is ahead, it should be the first to become suppressive. Indeed, if we keep on stimulating a clone with antigen, it suppresses its anti-idiotypic suppressors and can subsequently proliferate infinitely (this occurs irrespective of the dimension of the system, not shown). In fact, the same situation occurs in the Hoffmann[9,15] models. If a clone, stimulated somehow by antigen, is approaching its immune state the anti-idiotypic clone is regressing, i.e., is being suppressed. We conclude that an implication of symmetric network theory is the absence of "control of proliferation." This conclusion can be validated further by the observation that asymmetric versions of our model (i.e., $A_{ij} \neq A_{ji}$) easily account for such suppression (not shown).

STATIC ANALYSIS

We analyse this same system (X_1, X_2 and VR_1) statically in Figure 5. At the back of the 3-D state space ($VR_1 = 1$), one recognises Figure 2a. The $X_1' = 0$ isocline plane is shaded by lines: part of it changes as a function of antigen (VR_1). The region in which X_1 proliferates increases if VR_1 increases; the upper part of the $X_1' = 0$ isocline remains constant. This is the suppression part: further stimulation with antigen never changes X_1 dynamics if X_1 is at this isocline. Suppression is dominant ($C >> B$). Suppression in this model, thus, mainly means that a clone cannot be further stimulated. Introduction of an antigen into such a state (e.g., VR_1 in the $I2$ state) thus leads to infinite virus growth because X_2 cannot react. The $VR_1' = 0$ isocline is dotted. The $X_2' = 0$ isocline is independent of VR_1 and, therefore, remains the same as that in Figure 2a.

ESCAPE FROM SUPPRESSION

To test whether the fact that our systems fail to account for the control of proliferation (suppression) is a consequence of our "proliferation before suppression" assumption, we reverse our KB and KC constants. We, thus, assume suppression before proliferation. The maximum proliferation (B) and suppression (C) rates remain identical; if we were to make $B > C$, the model would become capable of infinite proliferation due to reciprocal stimulation of clones. If we simply reverse KB and KC, however, the model loses its ability to reject viruses because proliferation starts too late. We, thus, lowered KB until viruses became rejectable again, now $KB = 10^5$.

This escape from suppression model is analysed statically in Figure 6a and dynamically in Figure 6b. It turns our that such a system is no longer capable of switching between multiple stable points because it only has one steady state: the virgin state (Figure 6a). Moreover, if the system is stimulated with virus (VR_1, Figure 6b), X_1, indeed, escapes from suppression; i.e., it starts to proliferate, but immediately suppresses X_2. Any increase in X_1 should, indeed, increase the suppression of X_2. Thus, X_2 is virtually removed from the system, and X_1 behaves as if it were not connected to any network. Hence, switching (memory phenomena) and suppression (control of proliferation) are impossible.

We conclude, for our low-D systems, that 1) proliferation before suppression is superior to escape from suppression, 2) switching (memory) is easy to obtain, but 3) that control of proliferation (suppression) is impossible due to the symmetry.

RESULTS: HIGH-D MODELS

Although it is common practice to study the 10^7-D immune network with 2-D models, we would here like to approximate the 10^7-D system somewhat better by

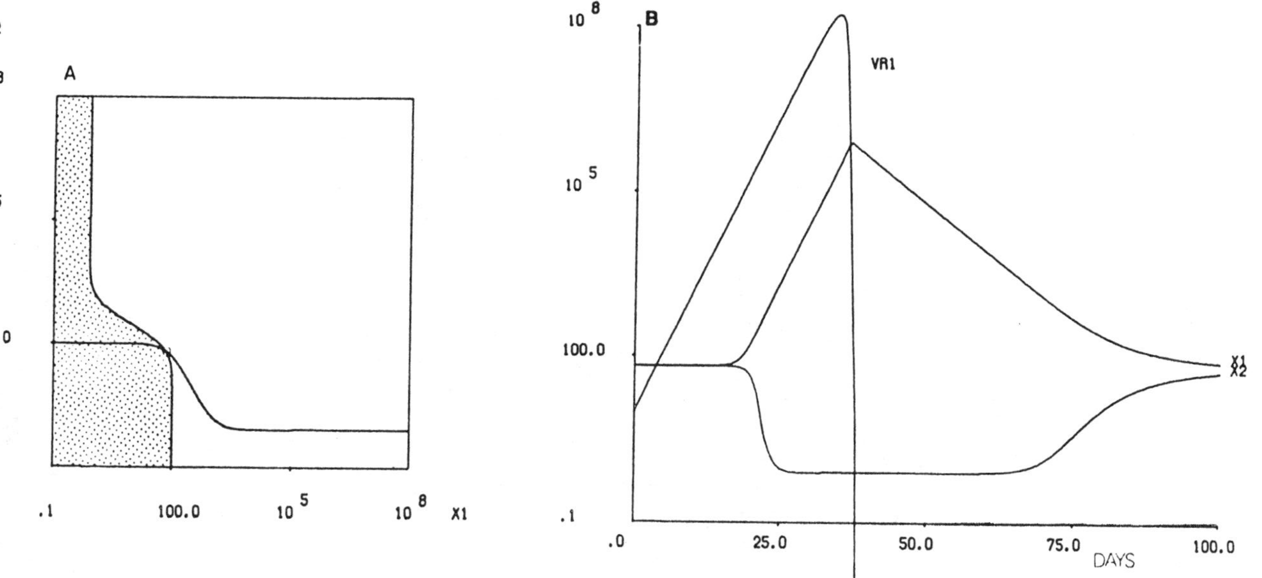

FIGURE 6 A "escape from suppression" model. Figure 6a is the 2-D interaction without virus ($A_{12} = 1$), the $X_1{}' = 0$ and $X_2{}' = 0$ isoclines intersect in only one stable equilibrium (a virgin state). Figure 6b shows the time plot of the immune response to VR_1 introduced into this (suppressed) virgin state.

analysing 50-D models. Elsewhere,[22] we made this step to 50-D more gradually by first analysing 3-D and 5-D models. It turned out that adding extra clones to a 2-D system may both facilitate and diminish switching behavior. Interestingly, the number of qualitative different stable states expands enormously: a 3-D system has 9–12 different reachable stable states (for all its affinity combinations). Here, however, we will step from 2-D to 50-D. Our main questions will be: 1) can the results obtained with the 2-D networks be repeated with 50-D systems, and 2) what is the effect of changing the number of idiotypic interactions (i.e., the connectance) in such a 50-D system?

METHOD OF ANALYSIS

The systems are analysed by simulation. Systems are first allowed to mature in the absence of external antigens; this represents the neonatal maturation of the immune system. Each clone can, therefore, start at a level of zero cells; the influx from thymus or bone marrow ensures the establishment of all clones. Once a stable state is reached (a "virgin" state), we introduce a series of antigens, i.e., VR_1–VR_4. Following its primary introduction, each virus is reintroduced to test whether the system has switched to an immune state. In each state, i.e., after the neonatal period and after each virus, we score 1) whether the system switched, 2) the number of immune and suppressed clones, and 3) the largest eigenvalue of the Jacobian matrix as a measure for the stability of the state reached. Because we study the system by simulation, we can only find "reachable" (hence, stable and feasible) equilibria. Each state is, thus, expected to be stable (i.e., each eigenvalue is expected to be negative). Biotic immune systems do exactly the same: they also ignore unstable and unfeasible equilibria. In this analysis, we (artificially) remove all viruses with a population of less that one cell: this is virus rejection. Conversely, if a virus grows uncontrolled, we "cure" the system by eliminating the virus once it exceeds a population of 10^8 infected cells.

We perform a series of simulations varying the connectance of the system only. Connectance is measured as the average number of idiotypic partners per clone (i.e., the average number of non-zero A elements per row or column). Affinity values are uniformly distributed between zero and one. For each clone connected to the network, we, thus, expect 98% switching (see the 2-D section).

ONE PARTNER PER CLONE

An example of such an experiment is depicted in Figure 7: a system with an average of one partner per clone ($nc = 1$). The data obtained for this particular system can also be found in Table 1: at $nc = 1.00$. At day 1000, i.e., at the end of the neonatal period, each of the 50 clones is in a virgin state. Idiotypic interactions are absent in this stable) state: the maximum stimulation term is below 0.1, the maximum suppression term is below 10^{-6}. All clones are around the influx/decay value; none of them is suppressed or immune. After the introduction of VR_1 (at day 1000), the network switches, but X_1 fails to become immune and returns to its steady state: the immune reaction is scored as VV (from virgin to virgin). X_1 has one connection to the network, i.e., with X_7. X_7 has 3 connections: X_1, X_{38} and X_{46}; the connection with X_1 is weaker than that with the other two. X_{46} is only connected to X_7, but X_{38} is also connected to X_{37}. X_{37} again has 2 new connections. The stimulus provided by the virus thus proceeds through the network leading to a switch after which 4 clones are immune and 4 suppressed (see Table 1). The second virus (VR_2 day 2000) cannot affect the network because X_2 is not connected (VV). The third virus (VR_3, day 3000) induces a switch in its responding clone X_3 (VI: virgin to immune). X_3 is only connected to X_{19} which is only connected to X_3; this is a simple 2-D system as we have analysed above. The fourth virus (day 4000) again induces a switch for X_4 (VI). X_4 was connected to X_{15} and X_{49}; X_{15} had one new connection (to X_{50}) whereas X_{49} could one see X_4. The system switches to 8 immune and 13 suppressed clones. In the virgin state, $\lambda_{\max} = 0.06$; at day 2000 (i.e., after VR_1) $\lambda_{\max} = .08 \pm .31i$. Stability remains at that value until day 5000, see Table 1.

SWITCHING

Table 1 lists the outcome of a series of this type of experiments; we start with weakly connected systems (average connectance: $nc = 0.01$) and connectance increases until $nc = 49$ partners per clone. Network switches are marked by asterisks; double asterisks mark switches that fail to generate immunity for the clone stimulated by antigen. It turns out that switching, and, hence, memory phenomena, can only occur in weakly connected networks, i.e., networks with $nc < 2$ partners per clone. Table 1 also shows the explanation for the restricted switching behavior: in networks with $nc > 2$, the neonatal state no longer corresponds to the 50-D virgin state. For a connectance of 2.16 partners per clone, for instance, the neonatal state comprises 18% immune and 22% suppressed clones. These clones apparently switched although they never saw any antigen except idiotypes. Viruses that are introduced can either be rejected immediately (i.e., if their clones belongs to the 18% immune clones, e.g. VR_3), or grow infinitely in a suppressed system (22% of the cases).

Thus, although we deliberately attempted to keep the virgin state free of idiotypic interactions (i.e., $S/D < KB$), clones appear to react to each other autonomously. This is explained by a positive feedback process: the (positive) idiotypic interactions in the virgin state are initially extremely weak, but the clones do stimulate each other slightly. As a consequence, some clones expand slightly which, in turn, increases the rate of reciprocal stimulation. Thus, clones expand slowly. Once some of the clones approach the stimulation threshold (KB), proliferation commences at full speed. These (few) proliferating clones in turn activate their idiotypic partners and the network will become fully activated in the absence of any initial trigger by antigen. We conclude that 50-D networks lose the virgin state at around $nc = 2$. During a short visit to the Los Alamos Theoretical Biology

TABLE 1 Analysis of a 50-D Network[1]

	nc	ir	%i	%s	lmax				nc	ir	%i	%s	lmax	
	.01	–	0.	0.	-.20	.00			3.36	–	22.	34.	-.05	.01
vr_1:	0	VV	0.	0.	-.20	.00		vr_1:	4	SS	22.	34.	-.05	.01
vr_2:	0	VV	0.	0.	-.20	.00		vr_2:	2	II	22.	34.	-.05	.01
vr_3:	0	VV	0.	0.	-.20	.00		vr_3:	3	SS	22.	34.	-.05	.01
vr_4:	0	VV	0.	0.	-.20	.00		vr_4:	3	SS	22.	34.	-.05	.01
	.12	–	0.	0.	-.18	.00			3.44	–	22.	22.	-.06	.00
vr_1:	0	VV	0.	0.	-.18	.00		vr_1:	2	SS	22.	22.	-.06	.00
vr_2:	1	VI	2.	0.	-.14	.51	*	vr_2:	6	SS	22.	22.	-.06	.00
vr_3:	0	VV	2.	0.	-.14	.51		vr_3:	3	SS	22.	22.	-.06	.00
vr_4:	0	VV	2.	0.	-.14	.51		vr_4:	3	SS	22.	22.	-.06	.00
	.40	–	0.	0.	-.15	.00			5.28	–	24.	44.	-.09	.04
vr_1:	0	VV	0.	0.	-.15	.00		vr_1:	1	VV	24.	44.	-.09	.04
vr_2:	0	VV	0.	0.	-.15	.00		vr_2:	2	VV	24.	44.	-.09	.04
vr_3:	0	VV	0.	0.	-.15	.00		vr_3:	6	SS	24.	44.	-.09	.04
vr_4:	0	VI	0.	0.	-.15	.00	*	vr_4:	6	SS	24.	44.	-.09	.04
	.92	–	0.	0.	-.17	.00			5.44	–	22.	36.	-.02	.17
vr_1:	1	VV	6.	2.	-.11	.52	**	vr_1:	4	SS	22.	36.	-.02	.17
vr_2:	1	VD	12.	2.	-.11	.52	*	vr_2:	5	DD	24.	38.	-.05	.00
vr_3:	0	VV	10.	2.	-.11	.52		vr_3:	6	VV	24.	38.	-.05	.00
vr_4:	0	VI	10.	2.	-.11	.52	*	vr_4:	6	II	24.	38.	-.05	.00
	1.00	–	0.	0.	-.06	.00			10.32	SS	20.	56.	-.08	.31
vr_1:	1	VV	8.	8.	-.08	.31	**							
vr_2:	0	VV	8.	8.	-.08	.31			14.32	SS	12.	66.	-.02	.36

vr3:	1	VI	10.	8.	-.08	.31	*
vr4:	1	VI	10.	8.	-.08	.31	*
	1.68	–	0.	0.	-.09	.00	
vr1:	1	VI	24.	18.	-.05	.28	*
vr2:	2	SS	24.	18.	-.05	.28	
vr3:	2	II	24.	18.	-.05	.28	
vr4:	2	VV	24.	18.	-.05	.28	
	2.16	–	18.	22.	-.01	.00	
vr1:	2	SS	18.	22.	-.01	.00	
vr2:	4	SS	18.	22.	-.01	.00	
vr3:	2	II	18.	22.	-.01	.00	
vr4:	2	SS	18.	22.	-.01	.00	
	2.76	–	26.	24.	-.07	.37	
vr1:	2	DD	28.	26.	-.07	.54	
vr2:	3	II	28.	26.	-.07	.54	
vr3:	2	SS	28.	26.	-.07	.54	
vr4:	2	SS	28.	26.	-.07	.54	

19.68	SS	12.	72.	-.08	.20
25.16	SS	10.	84.	-.14	.51
31.00	SS	10.	88.	-.10	.00
34.60	SS	6.	86.	-.09	.44
39.92	SS	8.	90.	-.13	.39
44.64	SS	4.	94.	-.14	.50
49.00	SS	4.	96.	-.30	.00

[1] The analysis of a 50-D network for various connectance values. We use the following abbreviations: nc, the connectance (average and per clone 1-4); ir, the immune response (either V: virgin, I: immune, D: dormant, or S: suppressed); %i: the percentage of immune clones (n=50); %s: the percentage of suppressed clones; lmax: the largest eigenvalue, its real part followed by its imaginary part. The first line of each box is the neonatal state; the following four represent the equilibria that are reached following the introduction of the four viruses vr1–4. Asterisks denote network switches; double asterisks switches that fail to generate immunity for the responding clone. From a connectance of 10.0 onwards, the first four clones were always suppressed. Thus, the system never switches and behaves in the same way for every virus. We, therefore, omit the data for each individual virus.

and Biophysics group, I was given the opportunity to analyse 100-D and 200-D networks. These preliminary studies suggested that such high-D networks also lose the virgin state around $nc = 2$. The connectance at which the virgin state is lost does, however, depend on the parameters: if the influx of the clones (S) is decreased, the magnitude of the feedback process is reduced. Such systems do remain virgin in absence of antigen, but make an enormous switch after the introduction of the first virus.[22]

Clones that manage to remain virgin may, nevertheless, fail to switch to immunity if the interaction with its partner(s) is disturbed by other (third party) idiotypic interactions. This gives rise to switches without immunity (i.e., the double asterisks in Table 1). Such disturbance already occurs in 3-D and 5-D systems.[22] Disturbance occurs if the interaction between the responding clone and its idiotypic partner is weak and if the idiotypic partners are strongly connected to other clones. We described such a situation in Figure 7 (clone 1).

RESPONSIVENESS

The network responsiveness to external antigens can be measured by the number of suppressed (SS) immune reactions and by the percentage of suppressed clones in Table 1. Suppressed clones are here defined as clones with a negative net interaction coefficient (i.e., $TA_i - TS_i < 0$); note that this does not necessarily include all suppressed clones (see the 2-D system). We depict the percentage of suppressed clones in Figure 8a. If connectance increases, responsiveness decreases (Table 1 and Figure 8a). Figure 8a shows that the first suppressed clones arise during the network switches evoked by the antigens. At higher connectance values, however, suppressed clones arise spontaneously due to internal activation within the network. At a connectance of $nc = 49$, 98% of the clones are suppressed, i.e., 49 clones are suppressed by one immune clone. The percentage of immune clones (Figure 8b) peaks at $nc = 2.76$, 26% of the clones are then in the immune state. These clones also switched in the absence of antigens. If connectance further increases, however, the percentage of immune clones decreases. We conclude that for an idiotypic network to be responsive, the connectance must be sufficiently low.

STABILITY

The largest eigenvalue of the Jacobian matrix in the reached equilibrium is depicted in Figure 8c (its real part) and Figure 8d (its imaginary part). All states are stable, i.e., all real parts are negative; the larger the real part of the eigenvalue, the smaller the (neighbourhood) stability of the equilibrium. Figure 8c, hence, shows that stability first decreases, but later increases. We found such a humped relationship between stability and connectance for various parameter settings of the present model (not shown). Stability increases with increasing connectance in the connectance region where most of the clones are suppressed, i.e., when negative interactions predominate. Conversely the decrease in stability in the first part of the

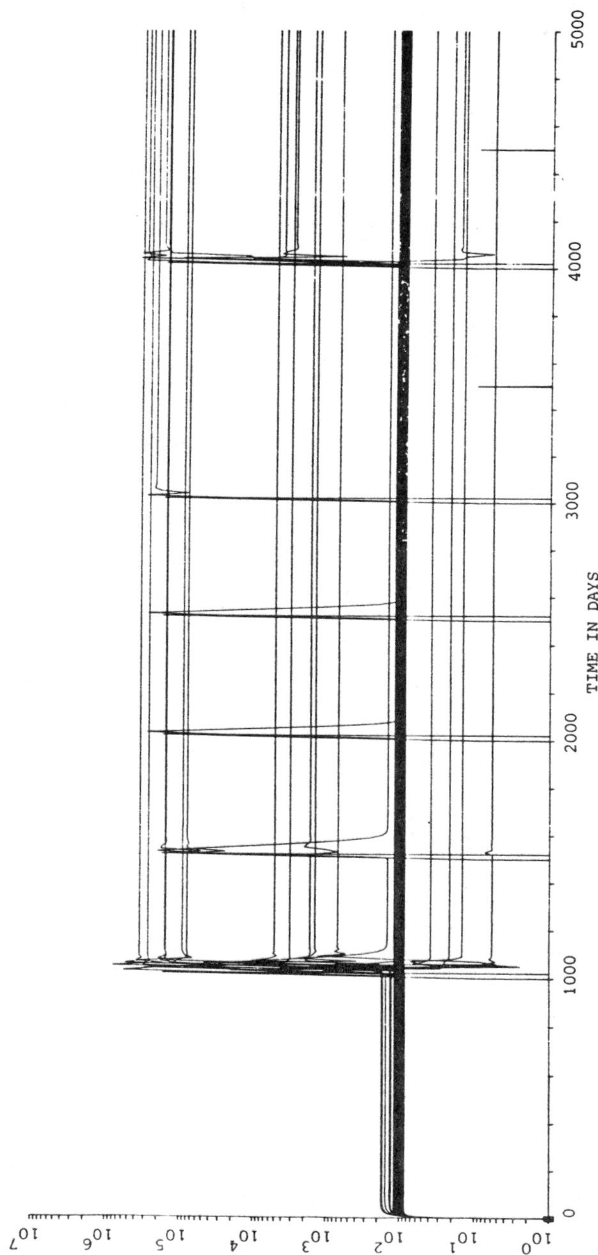

FIGURE 7 The population sizes of 50 clones of lymphocytes and 4 different viruses as a function of time. The average connectance is 1.0 idiotypic interaction per clone. Viruses are introduced at day 1000, 1500, 2000, . . . , 4000 and 4500. See the text for further explanation. The first virus (VR_1) generates a switch but fails to evoke immunity: at day 1500 X_1 again proliferates. VR_2 only evokes a reaction of X_2 at day 2000 and 2500. $VR3$ evokes proliferation and a switch for X_3: the system is immune at day 3500. VR_4 does the same, but this involves more clones.

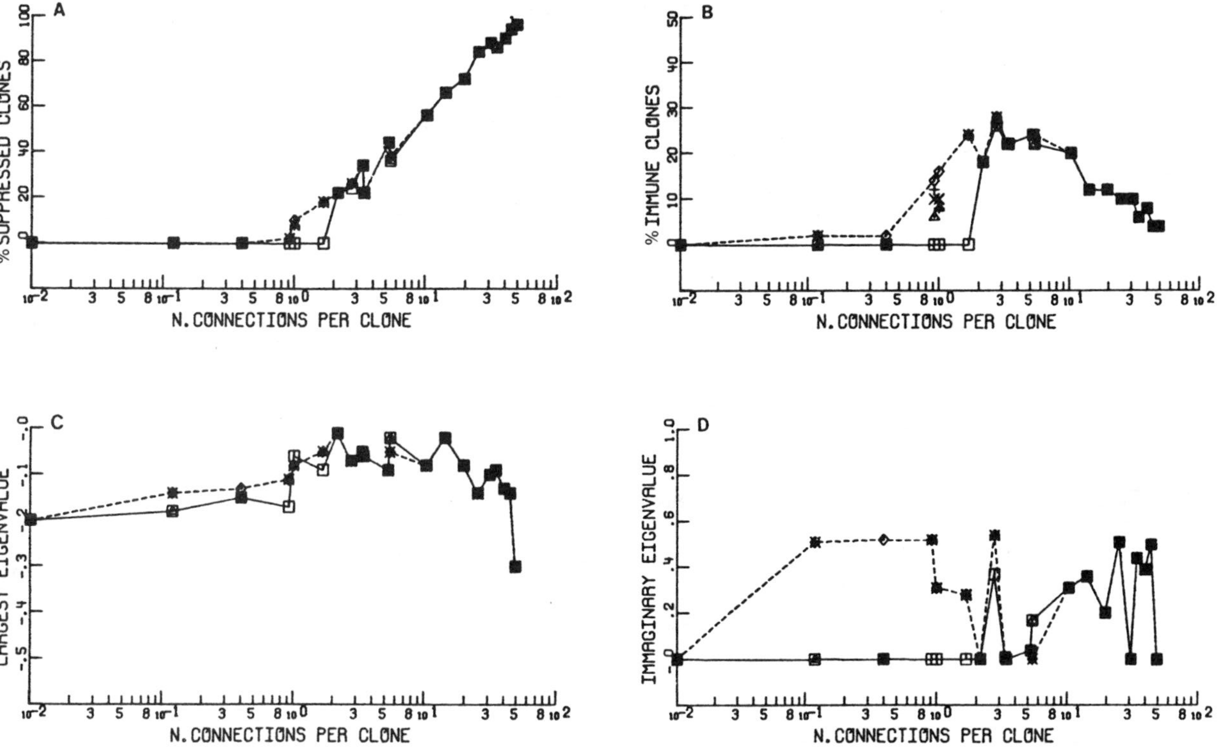

FIGURE 8 The percentage of suppressed (A) and immune (B) clones and the real (C) and imaginary (D) part of the largest eigenvalue are scored in each equilibrium reached by the simulations. The scores are plotted as a function of the connectance of the system. We use the following symbols: □: day 1000 (neonatal); △: day 2000 (after VR_1); +: day 3000 (after VR_2); x: day 4000 (after VR_3); ◇: day 5000 (after VR_4). The neonatal states are connected by solid lines; the day 5000 states by dashed lines. If these points all coincide, the system fails to switch.

curve is a consequence of the destabilisation brought about by the predominance of the positive interactions in this region.

Hoffmann[16] showed for a simple idiotypic network model that only incorporates influx (source) and negative idiotypic interactions (and no decay!), that stability increases with increasing connectance. Moreover, if normal turnover of cells (decay) is incorporated in this model, similar (but not identical!) results are obtained.[22] If we compare these results with the humped curve of Figure 8c, it appears that models based on only negative idiotypic interactions only account for the "suppression" part of the stability versus connectance curve. The first part, the destabilisation part of the humped curve, is ignored by such models. Moreover, because only weakly connected networks are capable of switching and are responsive, we conclude that interesting networks decrease in stability if connectance increases.

For high-D models we conclude that 1) the easy switching behavior of low-D models can only partly be repeated in high-D models, and 2) the system connectance has to be low because, otherwise, 2a) the system never switches and 2b) the system is totally unresponsive. stability of idiotypic networks does not simply increase if connectance increases,[16] but first decreases and then increases.

DISCUSSION

We concluded above that symmetric networks cannot account for suppression (i.e., for proliferation control) Suppression does however occur in the models, but only after "artificial" manipulation, e.g., the introduction of a large anti-idiotypic population. (Note that experimental data on suppression may also hinge upon such "artificial manipulation"). This conclusion is counter-intuitive because the present version of our model explicitly incorporates strong negative interactions among clones (i.e., $C >> B$). The absence of suppression during "physiological" (i.e., non-manipulated) immune reactions, however, suggests that strong negative interactions are unimportant in the model. Note that in the 2-D immune state the suppressed clone still receives a positive signal from its suppressor: the suppressed clone is enlarged. Suppression, therefore, only means that a clone cannot respond any further. An alternative to the strong negative interactions is, therefore, the mere reduction of the stimulation (proliferation) process. This would mean that at high population densities interactions do not become negative but zero. This can easily be incorporated by setting $B = C$.

The analysis of 100-D and 200-D models, which was made possible during my stay in Los Alamos, showed that, after antigenic stimulation, it takes a very long time for such high-D systems to reach a stable state. The stimulatory signal, originally provided by antigen, penetrates very deeply into the network. In 200-D systems, the activation of clones 10 "generations" away from the antigen (i.e., an Ab_{10}) occurs frequently. We conclude that a signal fails to fade out along its path through the network. This did not occur before because paths are not expected to become as long as this in 50-D networks (paths easily form short cycles). In the 10^7-D immune system, however, paths are expected to become extremely long; this suggests that we will have to adjust the present model (or its parameters) to incorporate the fading of signals.

The results showed that the existence of a virgin state depends crucially on the connectance of the system. If the connectance is high, the system activates itself, i.e., displays internal activity, and settles in a state in which most of the clones are suppressed or immune. Interestingly, experimental data show that the network connectance changes during life. Immature immune networks[5,24–27] are highly connected, whereas mature networks are weakly connected.[25,28] We would, thus, predict that mature systems can 1) respond to antigens, and 2) switch to stable immune states; immature systems, on the other hand, are 1) unresponsive and 2) display internal activity. This prediction is in accordance with experimental data.[29] Moreover, it is tempting to speculate that the unresponsiveness of (highly connected) immature systems may account for self/non-self discrimination [work in progress].

Our main questions were about the roles that idiotypic interactions can play in regulating normal proliferative immune responses to antigens. However, one of the generally supposed roles, i.e., control of proliferation (suppression), is seriously questioned by the present results (provided the network is symmetric). Another role, i.e., that of generating memory by stable state switching, is here proved to be possible, but is not at all easy if networks are highly connected. The existence of long-lived memory cells[30], therefore, provides a very simple alternative for explaining the memory phenomenon. The qualitative difference between "network immunity" and "memory cell immunity" is that the former is an active state (the immune clone is maintained by idiotypic stimulation, i.e., proliferation) whereas the latter is a slowly declining population of resting cells. Thus, a memory state maintained by a state switch in an idiotypic network is expected to be stable in time (i.e., to exist for ever), whereas one maintained by memory cells is expected to decline (slowly) in time. The results here demonstrate that understanding the network behavior, and deciding whether it can account for memory, crucially depends on knowing the network connectance. The interesting differences in connectance reported for networks of different developmental stages, therefore, call for further theoretical analysis along the path pursued here.

ACKNOWLEDGMENTS

This work has been done in close collaboration with Pauline Hogeweg, head of our Bioinformatics Group. I would like to thank Byron Goldstein of the Theoretical Biology and Biophysics group in Los Alamos for inviting me to visit the group. Alan Perelson, Doyne Farmer and Irene Stadnyk, kindly incorporated the present model in their network simulation system; moreover, we had many stimulating discussions. I am grateful to the Netherlands Organization for the Advancement of Pure Research (Z.W.O.) and the Los Alamos National Laboratory for financial support. I thank Miss S.M. McNab for linguistic advice.

REFERENCES

1. Jerne, N.K. (1974), "Towards a Network Theory of the Immune System," *Ann. Immunol. (Inst. Pasteur)* **125C**, 373.
2. Eichman, K., and K. Rajewsky (1975), "Induction of T and B Cell Immunity by Anti-Idiotypic Antibody," *Eur. J. Immunol.* **5**, 661.
3. Cosenza, H. (1976), "Detection of Anti-Idiotypic Reactive Cells in the Response to Phosphorylcholine," *Eur. J. Immunol.* **6**, 114.
4. Hardt, D.A., A. L. Wang, L. L. Pawlak, and A. Nisonoff (1972), "Suppression of Idiotypic Specificities in Adult Mice by Administration of Anti-Idiotypic Antibody," *J. Exp. Med.* **135**, 1293.
5. Vakil, M., H. Sauter, C. Paige, and J. F. Kearny (1986), "*In Vivo* Suppression of Perinatal Multispecific B Cells Results in a Distortion of the Adult B Cell Repertoire," *Eur. J. Immunol.* **16**, 1159.
6. Cooper, J., K. Eichmann, K. Fey, I. Melchers, M. M. Simon, and H.U. Weltzien (1984), "Network Regulation among T Cells: Qualitative and Quantitative Studies on Suppression in the Non-immune State," *Immunol. Rev.* **79**, 63.
7. Richter, P.H. (1978), "The Network Idea and the Immune Response," *Theoretical Immunology*, Eds. G. I. Bell, A. S. Perelson, and G. H. Pimbley (New York: Marcel Dekker), 539.
8. Fey, K., M. M. Simon, I. Melchers, and K. Eichmann (1984), "Quantitative Estimates of Diversity, Degeneracy, and Connectivity in an Idiotypic Network among T Cells," *The Biology of Idiotypes*, Eds. M. I. Green, and A. Nisonoff (New York: Plenum Press), 261.
9. Hoffmann, G.W. (1978), "A Mathematical Model of the Stable States of a Network Theory of Self-Regulation," *Systems Theory in Immunology*, Ed. S. Levin; *Lecture Notes in Biomathematics* **32**, 239.

10. Hoffmann, G.W. (1980), "On Network Theory and H-2 Restriction," Ed. N. L. Warner (New York: Plenum Press); *Contemp. Topics Immunobiol.* **11**, 185.
11. Jerne, N.K. (1984)., "Idiotypic Networks and Other Preconceived Ideas," *Immunol. Rev.* **79**, 5.
12. Bona, C.A. (1982), "Inverse Fluctuations of Idiotypes and Anti-Idiotypes during the Immune Response," *Regulation of Immune Response Dynamics*, Eds. C. DeLisi and J. R. J. Hiernaux (Boca Raton, FL: CRC Press), Vol. I, 137.
13. Hiernaux, J. (1977), "Some Remarks on the Stability of the Idiotypic Network," *Immunochem.* **14**, 733.
14. Seghers, M. (1979), "A Qualitative Study of an Idiotypic Cyclic Network," *J. Theor. Biol.* **80**, 553.
15. Gunther, N., and G. W. Hoffmann (1982), "Qualitative Dynamics of a Network Model of Regulation of the Immune System: a Rationale for the IgM to IgG Switch," *J. Theor. Biol.* **94**, 815.
16. Hoffmann, G.W. (1982), "The Application of Stability Criteria in Evaluating Network Regulation Models," *Regulation of Immune Response Dynamics*, Eds. C. DeLisi and J. R. J. Hiernaux (Boca Raton, FL: CRC Press), Vol. I, 137.
17. Farmer, J.D., N. H. Packard, and A. S. Perelson (1986), "The Immune System, Adaptation, and Machine Learning," *Physica* **22D**, 187.
18. De Boer, R.J. (1983), *GRIND: Great Integrator Differential Equations*, Bioinformatics Group, University of Utrecht, The Netherlands.
19. Gottwald, B.A., and G. Wanner (1981), "A Reliable Rosenbrock Integrator for Stiff Differential Equations," *Computing* **26**, 355.
20. Mitchell and Gauthier Associates (1986), *ACSL: Advanced Continuous Simulation Language* (Concord, Mass. 01742, U.S.A.).
21. De Boer, R.J., and P. Hogeweg (submitted), "Symmetric Idiotypic Networks. I. Memory but no Suppression in Low-D Models."
22. De Boer, R.J., and P. Hogeweg (submitted), "Symmetric Idiotypic Networks. II. Stability and Unresponsiveness in High-D Models."
23. Cerny, J. (1982), "The Role of Anti-Idiotypic T Cells in the Cyclical Course of an Antibody Response," *Regulation of Immune Response Dynamics*, Eds. C. DeLisi and J. R. J. Hiernaux (Boca Raton, FL: CRC Press), Vol. I, 137.
24. Holmberg, D., S. Forsgen, F. Ivars, and A. Coutinho (1984), "Reactions among IgM Antibodies Derived from Normal Neonatal Mice," *Eur. J. Immunol.* **14**, 435.
25. Holmberg, D., G. Wennerstrom, L. Andrade, and A. Coutinho (1986), "The High Idiotypic Connectivity of "Natural" Newborn Antibodies is not Found in the Adult Mitogen-Reactive B Cell Repertoires," *Eur. J. Immunol.* **16**, 82.
26. Holmberg, D. (1987), "High Connectivity, Natural Antibodies Preferentially Use 7183 and QUPC 52 V^H Families," *Eur. J. Immunol.* **17**, 399.

27. Vakil, M., and J. F. Kearny (1986), "Functional Characterization of Monoclonal Auto-Anti-Idiotype Antibodies Isolated from the Early B Cell Repertoire of BALB/c Mice," *Eur. J. Immunol.* **16**, 1151.
28. Pollok, B.A., A. S. Bhown, and J. F. Kearny (1982), "Structural and Biological Properties of a Monoclonal Auto-Anti-(Anti-Idiotype) Antibody," *Nature* **299**, 447.
29. Pereira, P., L. Forni, E. L. Larsson, M. Cooper, C. Heusser, and A. Countinho (1986), "Autonomous Activation of B and T Cells in Antigen-Free Mice," *Eur. J. Immunol.* **16**, 685.
30. Levy, M., and A. Coutinho (1987), "Long-Lived B Cells: Mitogen Activity as a Tool for Studying their Life-Spans," *Eur. J. Immunol.* **17**, 295.

GEOFFREY W. HOFFMANN, TRACY A. KION, ROBERT B. FORSYTH, K. GEOFFREY SOGA, and ANWYL COOPER-WILLIS
Departments of Physics and Microbiology, University of British Columbia Vancouver, B. C., Canada

The N-Dimensional Network

ABSTRACT

This paper consists of four theoretical sections and two experimental sections.

We describe a way in which the dynamics of an N-dimensional system can be depicted on a two-dimensional plane. The phase plane axes are the concentration x_i of cells of a clone i, and the connectivity Y_i of clone i, where $Y_i = \sum_j K_{ij} x_j$, and K_{ij} is an interaction strength parameter for clones i and j. The x_i/Y_i phase plane contains several loci of equilibrium; when all the clones in the system have x_i/Y_i values located along these lines, the system is at a steady state.

We describe how IL-1 can be most simply added to the symmetrical network theory, and how the helper and suppressor phenotypes of T cells can then be related to their different cell surface markers and presumed different network connectivities.

We next turn to the I-J paradox, and suggest a solution to that problem in terms of I-J being anti-anti-self and having the role of a network "centre-pole."

The centre-pole model for I-J could help us towards an understanding of acquired immune deficiency syndrome (AIDS). We describe an AIDS network theory that involves a hypothetical human equivalent of I-J, and that is related to the Ziegler-Stites autoimmunity theory. It involves destabilization of the network by both anti-anti-self and anti-anti-anti-self (that is, anti-centre-pole) antibodies. The theory leads to several new experimentally testable predictions and ideas for prevention of the disease. *One of the predictions is that some substances that are being considered for use as vaccines are likely to cause AIDS in high risk groups.* It is, therefore, important that this theory be tested experimentally as quickly as possible.

Motivated by centre-pole and network stability ideas, we looked for and found anti-anti-(self cell surface) antibodies in the autoimmune strains of mice, MRL-(+/+), MRL-*lpr/lpr* and B6-*lpr/lpr*.

Experiments that were designed to detect anti-idiotypic antibodies in ordinary hyper-immune sera yielded the remarkable result that affinity-purified polyclonal chicken antibodies of a particular specificity react with themselves more strongly than with chicken antibodies of different specificities.

INTRODUCTION

Immunology is a science of about 10^6 facts and beliefs. About 8000 papers are published in immunology each year. If we assume that each paper contains an average of, say, five new facts (a typical number of figures and tables), then there are about 40,000 new facts per year, and 25 years of progress at this rate results in the accumulation of a million facts. Many of the facts are not simple. A typical fact might have the form, "If we perform procedure A with antigen B and animals of strain C of the species D, and then apply assay E to cells of type F from organ G after the time period H, we observe result I." Such an experiment is performed with a set of suitable controls, and the conclusion is typically much more terse; for instance, "Suppressor cells express I-J." This is a belief that the investigator has reached, operating, as he or she must, within a particular theoretical framework, which, in turn, he or she has developed from a very large number of experimental facts and accompanying interpretations. The theoretical task we face is to find a minimal set of beliefs that accounts for a maximal number of facts.

The symmetrical network theory is proving to be a fertile framework for encompassing diverse immunoregulatory phenomena. We here describe a refinement of the mathematical model that we have formulated, propose a way of adding the lymphokine IL-1 to the model, discuss the I-J paradox and a related autoimmunity theory of AIDS, and finally, describe some new experimental results concerning autoimmunity and anti-idiotypes, that were obtained while working within the framework of the symmetrical network theory.

DISPLAYING THE DYNAMICS OF AN N-DIMENSIONAL SYSTEM

We previously described a mathematical model of the symmetrical interactions between just two sets of clones, the first set being clones that recognize a particular antigen, and the second set being clones that recognize (and are recognized by) the first set (Gunther and Hoffmann, 1982). The model stimulates network interactions that are considered to be important at and near the stable steady states of the system. We now review that model briefly, and describe an N-dimensional network generalization of it. The dynamics of the N-dimensional system can be displayed in an interesting way on a two-dimensional phase plane.

REVIEW OF THE TWO-DIMENSIONAL MODEL

The minimal two-dimensional model has the form:

$$\begin{aligned} \frac{dx_+}{dt} &= S - k_2 x_+ x_- e_2 - k_3 x_+ (x_-)^2 e_3 - k_4 x_+ \\ \frac{dx_-}{dt} &= S - k_2 x_+ x_- e_2 - k_3 (x_+)^2 x_- e_3 - k_4 x_- \end{aligned} \tag{1a}$$

where

$$e_q = \frac{1}{1 + (x_+ x_- / C_q)^{n_q}} \qquad q = 2, 3. \tag{1b}$$

The first term, S, denotes the influx of cells into the system. There is a killing term that is linear in the concentration of cells of the opposite specificity (with rate constant k_2), a killing term that has a greater-than-linear concentration dependence (rate constant k_3), and a term that is independent of the concentration of cells of the opposite specificity (rate constant k_4). The k_2 term models killing by killer T cells and/or by IgM plus complement. The k_3 term models antibody-dependent cellular cytoxicity and/or killing by IgG plus complement. We are interested in qualitative behaviour of a complex system, and make rather crass simplifications in this model; for instance, we treat the concentration of antibodies as being simply proportional to the concentration of cells that produce the antibodies, and we treat the concentration of specific T-cell factors as being proportional to the product of the concentration of the cells that produce them and the concentration of cells that specifically stimulate those cells. The e_q terms take account of specific inhibition (of killing) by specific T-cell factors. C_2 and C_3 are constants that specify the threshold values of the product x_+x_- at which inhibition by specific T cell factors becomes effective, and n_2 and n_3 are constants that determine the sharpness of thresholds. Interactions between positive and negative cells and antibodies are assumed to be inhibitable by both positive and negative specific T-cell factors.

Implicit in the chosen form of the equations is the suggestion that, at the stable steady states of the system, the specific stimulation of T cells is limiting for the production of specific T cell factors, while specific stimulation of B cells is not so limiting for antibody production by B cells. This might or might not be a reasonable approximation for when the system is at or near its stable steady states; it seems less likely to be accurate during an immune response, when specific stimulation of B cells is certainly very important. The latter aspect does not matter much for our purposes, since at this stage we are primarily interested in modeling the stable states, rather than the more complex events that occur during the switching between stable states. We do not know whether similarly successful mathematical

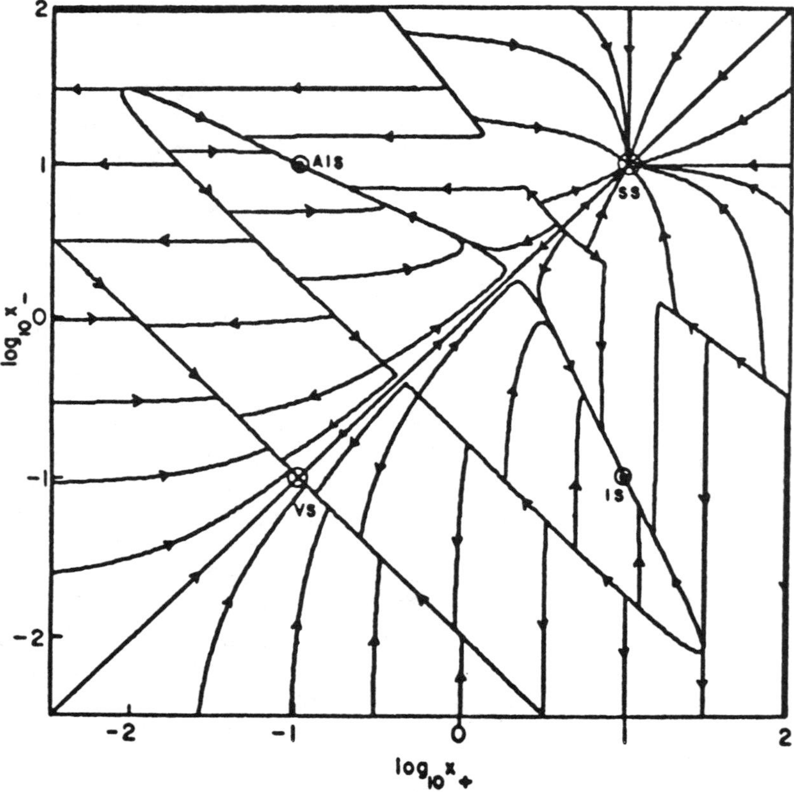

FIGURE 1 The trajectories of a two dimensional model computed using Eq. 1, showing four stable states as attractors: V.S. (virgin state), I.S. (immune state), S.S. (suppressed state) and A.I.S. (anti-immune state). Parameters: $S = 1$, $k_2 = 100$, $k_3 = 0.1$, $k_4 = 0.1$, $C_2 = 0.1$, $C_3 = 10$, $n_2 = 5$, $n_3 = 5$. These values of the parameters are in the region specified by (1d).

models can be devised, in which B cells and T cells both need specific stimulation for the secretion of their specific products.

There can be four stable steady states of this model, as can be seen when trajectories of the differential equations are plotted on the x_+/x_- phase plane; see, for example, Figure 1. They are the virgin state, the immune state, the suppressed state and the anti-immune state. In the virgin state, we have low but not insignificant levels of both positive and negative clones for a given specificity, and this state is characterized by a balance between influx of cells and killing by a mechanism that depends linearly on the concentration of cells with complementary specificities. The immune state is a state in which, due to killing by cells of one specificity (positive cells), we have an elevated level of positive cells and a low level of cells of the complementary specificity (negative cells). In the suppressed state, we have elevated levels of both positive and negative clones, and mutual stimulation between positive and negative T cells (leading to inhibition by specific T-cell factors) as the main network interaction. The anti-immune state is the converse of the immune state: elevated negative and low positive population levels. We showed that there are two regions in parameter space that result in these properties for this system:

$$\frac{k_4}{k_3} < C_3 < \frac{S}{k_2} < C_2 < \left(\frac{S}{k_4}\right)^2 \tag{1c}$$

and

$$\frac{S}{k_2} < C_2 < \frac{k_4}{k_3} < C_3 < \left(\frac{S}{k_4}\right)^2. \tag{1d}$$

AN N-DIMENSIONAL GENERALIZATION OF THE VIRGIN STATE INTERACTIONS

The first step towards generalizing this model was to consider an N-dimensional version of only the interactions that are most important in the virgin state. The first two terms in Eq. (1a) are the terms that are important in the virgin state; C_2 is assigned a value such that e_2 is equal to or close to unity for clones in that state. For the virgin state in the region of parameter space specified by (1c), an N-dimensional version of the differential equation can then be approximated by:

$$\frac{dx_i}{dt} = S - x_i \sum_{j=1}^{N} K_{ij} x_j. \tag{2}$$

The matrix K, with elements K_{ij} for clones i and j, is a symmetrical matrix of inter-clonal interaction strengths ($K_{ij} = K_{ji}$), and x_i is the population size (or concentration) of clone i. Eq. (2) has the unusual and interesting property that it has a single stable steady state solution only if the connectance (fraction of non-zero elements in the matrix K) is above a certain threshold level (Hoffmann, 1982; Spouge, 1986). The threshold level of connectance required for stability decreases with the size of the system (that is, the value of N). This means that unstable sub-systems can be combined to yield a stable system, which is unusual for complex systems (May, 1974; Siljak, 1978). These results are not subject to any restrictions on the magnitudes of the K_{ij}. In the following multi-dimensional network model, the connectivity of clone i, $\sum_{j=1}^{N} K_{ij}x_j$, is an important variable, so for convenience we give it the name Y_i:

$$Y_i = \sum_{j=1}^{N} K_{ij}x_j. \tag{3}$$

The connectivity of a clone defined in this way is something quite different from the connectance of the network, which is the fraction of nonzero elements in the matrix K.

We consider dynamics in the x_i/Y_i phase plane. We have loci of equilibrium that can be drawn in the x_i/Y_i phase plane, and we can observe the path of each of the N clones in this plane relative to the loci of equilibrium. For instance, in the case of the relatively simple system given by Eq. (2), there is just one locus of equilibrium in the x_i/Y_i plane, given by

$$x_iY_i = S. \tag{4}$$

The system, as a whole, is at equilibrium when Eq. (4) is satisfied for all clones, that is, for all i. In Figure 2, we see the dynamics of the system in this phase plane for a case in which $N = 20$, and the connectance C of the matrix K (the fraction of non-zero elements) is 0.5. Here, the clones were given initial values randomly distributed between 0.0 and 1.0, and the equations were integrated numerically until the trajectories reached the locus of equilibrium. If the connectance is below the critical level required for stability, one or more of the clones heads off towards an unbounded value of x_i, and the system does not converge to a stable steady state.

AN N-DIMENSIONAL GENERALIZATION OF THE MODEL WITH MULTIPLE STABLE STATES

One way of generalizing the two-dimensional model of Eq. (1) to N dimensions is to replace x_+ by x_i, and x_- by Y_i throughout. We then have:

$$\frac{dx_i}{dt} = S - k_2x_iY_ie_{2i} - k_3x_iY_i^{\,2}e_{3i} - k_4x_i \tag{5}$$

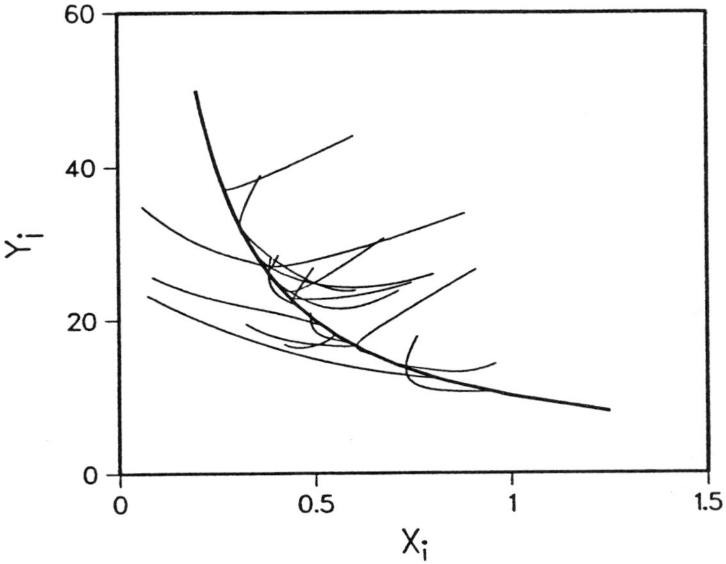

FIGURE 2 Dynamics in the x_i/Y_i phase plane according to Eq. (2), a simple model of the interactions of clones in the virgin state. The x_i/Y_i values converge to the locus of equilibrium $x_iY_i = S$. Parameters: 20 clones with $S = 10$, random initial values between 0.0 and 1.0, a connectance of 0.5, and K_{ij} values randomly distributed between 0.0 and 1.0.

with

$$e_{qi} = \frac{1}{1 + (x_iY_i/C_q)^{n_q}} \qquad q = 2,3;\ i = 1, N. \tag{5a}$$

The e functions are now doubly subscripted, to account of the fact that different specificities are inhibited different amounts. We simplify the model by using step functions for the e_{qi}:

$$e_{qi} = \begin{cases} 1 & \text{if } x_iY_i < C_q \\ 0 & \text{if } x_iY_i > C_q \end{cases} \qquad q = 2,3;\ i = 1, N. \tag{5b}$$

Replacing the smooth functions with these sharp transitions does not change the properties of the equations very much, reduces the number of adjustable parameters, and permits us to solve for the loci of equilibrium explicitly. For the case of the region of parameter space shown in Figure 3a, where $C_3 < C_2$, we have, for $x_iY_i < C_3 < C_2$,

$$\frac{dx_i}{dt} = 0 \quad \text{when} \quad x_i = \frac{S}{k_2Y_i + k_3{Y_i}^2 + k_4}, \tag{6}$$

for $C_3 < x_iY_i < C_2$,

$$\frac{dx_i}{dt} = 0 \quad \text{when} \quad x_i = \frac{S}{k_2Y_i + k_4}, \tag{7}$$

and for $C_3 < C_2 < x_iY_i$,

$$\frac{dx_i}{dt} = 0 \quad \text{when} \quad x_i = \frac{S}{k_4}. \tag{8}$$

In the other region of parameter space of interest, with $C_3 > C_2$, we have the following equilibrium lines:

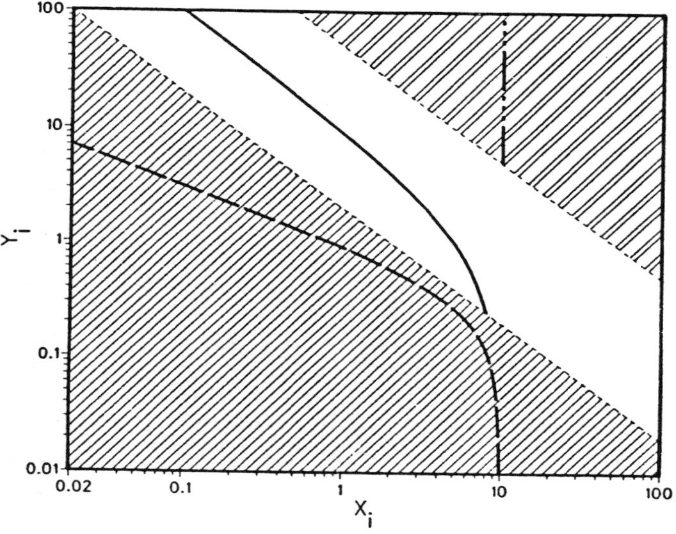

FIGURE 3A The three regions in the x_i/Y_i plane, bounded by the lines $x_iY_i = C_2$ and $x_iY_i = C_3$, each with its own locus of equilibrium, as specified by Eqs. (6), (7) and (8) with the parameters $S = 10$, $k_2 = 1$, $k_3 = 10$, $k_4 = 1$, $C_3 = 2$, $C_2 = 50$. These parameters correspond the condition (1c) of the two-dimensional model. Equilibrium loci: Virgin (——), Immune/Antiimmune (— — —), and Suppressed (—- - - -—).

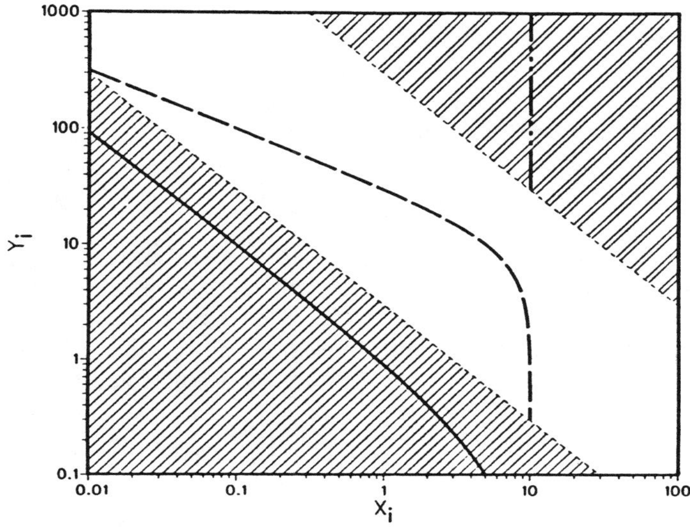

FIGURE 3B The three regions and three equilibrium loci in the x_i/Y_i plane in the N-dimensional model for a point in parameter space corresponding to condition 1d of the two-dimensional model. Parameters: $S = 1$, $k_2 = 1$, $k_3 = .001$, $k_4 = 0.1$, $C_2 = 3$, $C_3 = 300$. Equilibrium loci: Virgin (——), Immune/Antiimmune (— — —), and Suppressed (—----—).

for $x_iY_i < C_2 < C_3$,

$$\frac{dx_i}{dt} = 0 \quad \text{when} \quad x_i = \frac{S}{k_2Y_i + k_3{Y_i}^2 + k_4}, \tag{9}$$

for $C_2 < x_iY_i < C_3$,

$$\frac{dx_i}{dt} = 0 \quad \text{when} \quad x_i = \frac{S}{k_3{Y_i}^2 + k_4}, \tag{10}$$

and for $C_2 < C_3 < x_iY_i$,

$$\frac{dx_i}{dt} = 0 \quad \text{when} \quad x_i = \frac{S}{k_4}. \tag{11}$$

In the x_i/Y_i phase plane, there are, then, three regions, each defined by the value of the product x_iY_i relative to the value of C_3 and C_2, and each with its own locus of equilibrium, as shown for a set of parameters with $C_3 < C_2$ in Figure 3a, and for a set of parameters with $C_2 < C_3$ in Figure 3b.

Figure 4 shows the trajectories in the x_i/Y_i plane for a system in which the initial clone sizes $x_i(0)$ were all set equal to the same value. Here, various clones converge to the virgin state, the immune state or the anti-immune state. It is interesting to note that the clones that are on the virgin state locus of equilibrium cluster at intermediate values of x_i and Y_i, while most of the clones that are on the immune/anti-immune locus of equilibrium cluster at either high x_i/low Y_i (immune) or low x_i/high Y_i (anti-immune). In this simple N-dimensional model, we seem to be able to identify clones with one of the four stable steady states, that were defined in the two-variable model.

These equations are not intended to simulate switching between stable states. (In the theory, switching involves a non-specific accessory cell [macrophage or "A cell"], and mathematically modelling its role would be relatively complicated.) It is, nevertheless, possible to graphically illustrate the stability of the stable states by adding a term to the equation that simulates a perturbing effect of the antigen:

$$\frac{dx_i}{dt} = S - k_2 x_i Y_i e_2 - k_3 x_i (Y_i)^2 e_3 - k_4 x_i + R_i Ag(t) \quad (12)$$

Here $Ag(t)$ is the concentration of antigen as a function of time t, and R_i is the strength of interaction between the antigen and clone i. If we have a set of clones that are all in the virgin state, and perturb the system with a transient pulse of

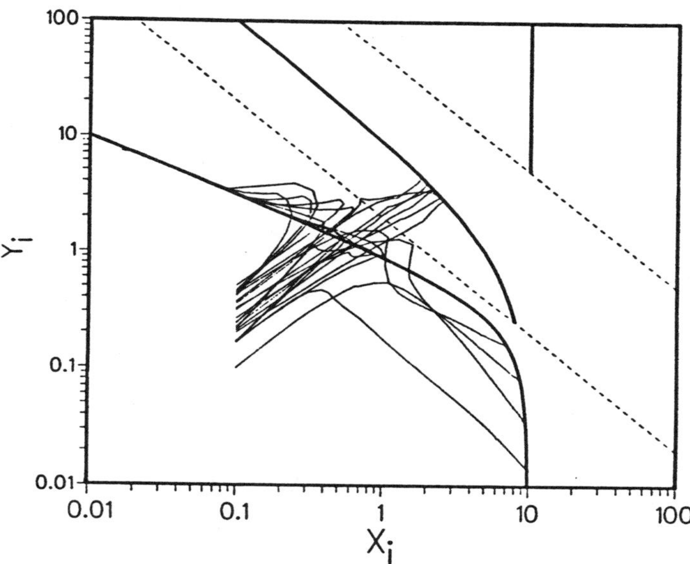

FIGURE 4 Trajectories in the x_i/Y_i phase plane of 25 clones that are randomly connected with a connectance of approximately 0.3, and which are all given initial clone sizes of 0.1. The non-zero K_{ij} are random numbers in the range 0.0 to 1.0; the other parameters are the same as for Figure 3a.

antigen, we obtain trajectories in the x_i/Y_i phase plane as shown in Figure 5. Each of the trajectories returns to its starting point. This,can be seen as a model for the perturbation caused by a T-independent antigen, which typically stimulates each of many clones a little, but does not induce memory. Many different clones responding simultaneously to the antigen cause the elimination of the antigen without any of the clones proliferating enough to leave the region of attraction of the locus of equilibrium for the virgin state. In Figure 6, we see that a larger pulse of antigen results in some of the clones being switched to the suppressed state, and then those in the virgin state return to different equilibrium values. (Since we do not include the A cell in this mathematical model, we do not see switching to the immune state here.) Figure 7 shows a small perturbation by antigen of a system in which we start with clones in all four states. Many of the clones are not directly stimulated by antigen, and most of those adjust to a new position on the locus of equilibrium, without leaving the locus by enough for their trajectories to be seen on this figure.

A possible concern about the above N-dimensional model is that, in one respect, it may not seem to exactly reflect the assumptions of the theory. This has to do with inhibition by specific T-cell factors; it was brought to our attention by Rob

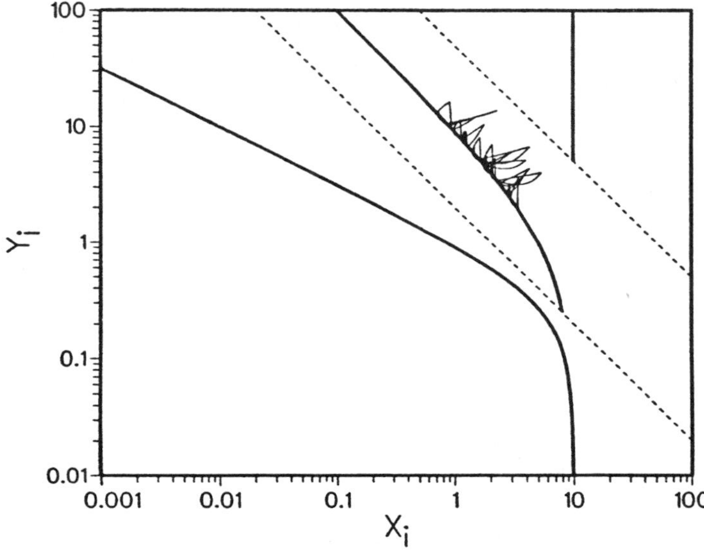

FIGURE 5 A small perturbation of each of many clones that are all in the virgin state results in a transient perturbation of the network. All the clones return to the same virgin state, and there is no memory associated with the response. This is a model of what happens with T-independent antigens; they evoke no memory. Parameters as for Figure 4.

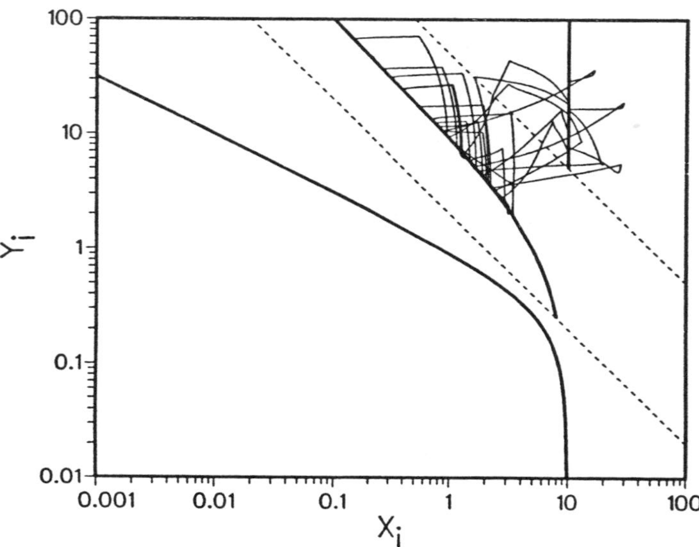

FIGURE 6 A larger pulse of antigen takes some clones into the zone of attraction of the suppressed state. A new equilibrium state results for clones that return to the locus of equilibrium for the virgin state. The mathematical model does not take account of the role played by the A cell, and we see no switching to the immune state here. Parameters again as for Figure 4. Ten of the 25 clones are stimulated directly by the antigen.

De Boer. Cytotoxic interactions between clones i and j are assumed to be inhibited by specific T-cell factors of both i and j types. The amount of the i-clone factor depends on the concentration x_i and the amount of stimulation that clone i receives from the clones to which it is idiotypically connected. This level of stimulation is well reflected by the connectivity of clone i, namely, Y_i. Hence, we can model the amount of factor of specificity i by the product $x_i Y_i$. However, the amount of j factor is similarly given by $x_j Y_j$, which does not appear in our formulation. To include it would obviously complicate the equations. Note that, in the two-dimensional model, $x_i Y_i$ and $x_j Y_j$ are synonymous, so this question does not arise there. An improved version of the above theory that deals with this problem is being developed, and will be published elsewhere. In the improved version, x_i in the differential equation is replaced by X_i, which denotes a weighted sum of i and i-like clones.

Even though it is so far only qualitative rather than quantitative, the above N-dimensional model serves to illustrate several things. Firstly, the idea of the stability of many states in a complex N-dimensional system becomes plausible from the figures. Secondly, the concepts of virgin, suppressed, immune and anti-immune states seem to be defined also in the N-dimensional model. This is perhaps surprising; one might have imagined that the distinction between the various states would become fuzzy in a complex network. Thirdly, we seem to have a very large number of stable states in the N-dimensional model. An interesting and unresolved problem is the question of just how many stable steady states exist in the model. It seems likely that there can be at least of the order of 2^N; in the case of an analogous neural network model, we have almost 2^N attractors (Hoffmann, 1986).

Langman and Cohn (1986) have criticized network theory on the grounds of imagined difficulties when each clone is idiotypically connected to more than two others; they refer to the network being then "tied in a Gordian knot." In the above mathematical model, in which we have used a high connectance, we see that such a network can still self-organize itself quite satisfactorily. There is no evidence of any knots.

IL-1 AND CELL SURFACE MARKERS

An important part of the strategy employed in the development of the symmetric network theory has been to keep it as simple as possible, and to introduce new

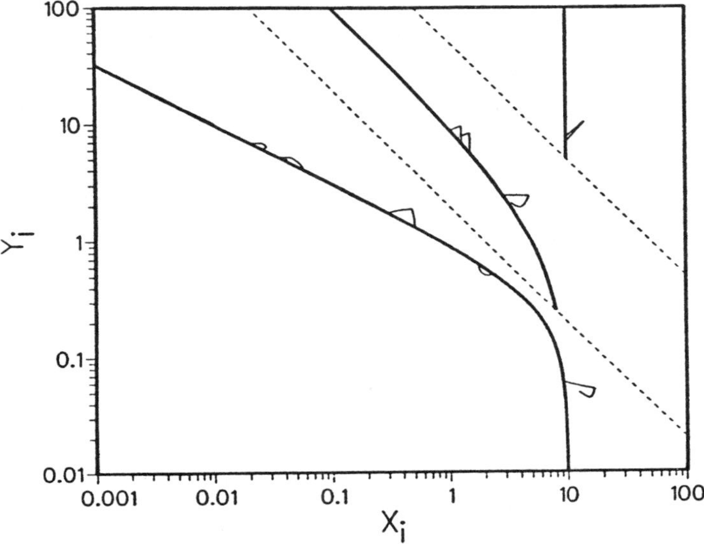

FIGURE 7 Trajectories in the x_i/Y_i phase plane following a small transient perturbation by antigen of a system in which we have clones in all four stable states. Parameters as for Figure 4.

building blocks only with reluctance, and only when they lead to an improvement in the *scope-to-complexity ratio* of the theory. We recently found it expedient to add the well-characterized, non-specific factor IL-1 to the model (Hoffmann, 1987), since doing so leads to simple explanations for the facts that helper cells express CD4 (old nomenclature: L3T4 in mouse and T4 in man), and suppressor cells express CD8 (old nomenclature: Lyt-2 in mouse and T8 in man).

IL-1 is secreted by macrophages ("A cells"), and gives T cells a second signal for proliferation (Gery and Waksman, 1972; Smith, 1984). (The first signal is the crosslinking of the T cells' specific receptors; see, for example, Kaye and Janeway, 1984.) The of T cells is consequently strongly influenced by molecules that are present on the A-cell surface. T cells that have a modest affinity for molecules that are on the A-cell surface have a selective advantage over T cells that do not have this property. (On the other hand, very high affinity for the A cell might be a selective disadvantage for a T cell, as it seems likely to result in phagocytosis of the T cell.) This idea leads to an explanation for the distribution of the above-mentioned cell surface markers in terms of the functions of the corresponding cells. The explanation involves the symmetrical network theory postulate that specific T-cell factors are cytophilic for A cells.

In the theory, the suppressed state is a state of high network connectivity, and the immune state is a state is a state of low connectivity (Hoffmann, 1980). Cells that express CD8 tend to be suppressor T cells (Cantor and Boyse, 1975; Reinherz and Schlossman, 1980). Cells with this phenotype also recognize the ubiquitous class I MHC antigens. Since these antigens are expressed on almost all tissues (Klein, 1975, p. 331), cells that recognise class I MHC receive more stimulation than cells of other specificities. The fact that the CD8 cell surface molecules themselves have affinity for the class I MHC molecules (MacDonald et al., 1982) is a factor that could help to ensure that these cells receive more stimulation than other cells. We would, then, expect that there is more anti-class I MHC-specific T-cell factor on the surface of A cells than specific T cell factors of other specificities. T cells that are, in turn, specific for those specificities are then selected for, which ensures that the anti-class I T cells have high idiotypic connectivity. This is just what we require for suppressor T cells. It is, thus, no accident that CD8 T cells are both anti-class I and suppressor cells.

A similar interpretation can be given to the fact that class II specific cells tend to be helper cells. The CD4 molecule of helper T cells has affinity for the class II MHC molecules (Greenstein et al., 1984; Owens and Fazekas de St. Groth,1987; Biddison et al.,1985), and class II MHC molecules are expressed primarily on B cells and A cells (Sachs, 1984, p. 327). The presence of class II MHC on A cells seems to be particularly important, since T cells that have affinity for molecules on the A-cell surface can be selected (due to the role of IL-1 in T cell proliferation) relatively independently of being connected idiotypically to other V regions in the system. Hence, these T cells can have low network connectivity, which is what we require of helper cells.

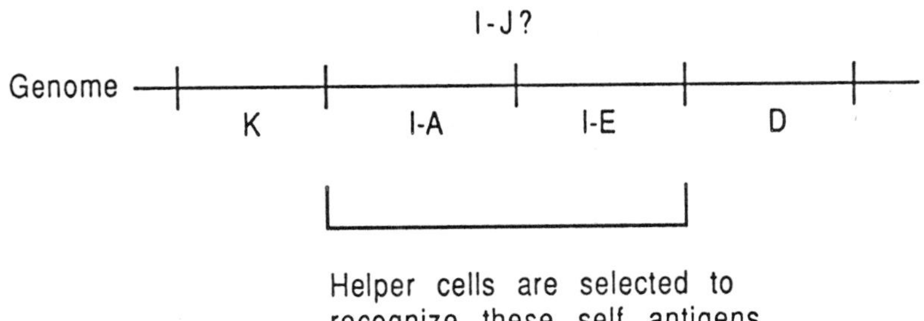

FIGURE 8 The class I (K,D) and class II (I-A, I-E) genes in the major histocompatibility complex of the mouse play an important role in influencing the T-cell repertoire. On the basis of classical genetic mapping studies, genes coding for proteins with I-J antigenic determinants were thought to be located between I-A and I-E, but at the DNA level no genes can be found there.

THE I-J PARADOX

Concentrating our attention on paradoxes is a good way to progress in the formulation and development of theory. Jerne stressed the importance of low zone tolerance when he formulated the network hypothesis (Jerne, 1974), and since low zone tolerance is a particularly simple paradox in the framework of classical clonal selection theory, it attracted much attention in the early days of network modelling. In our view, the prettiest immunological paradox of this decade is the I-J paradox. I-J is a serologically defined marker (or markers) that is present mainly on suppressor T cells, on suppressor T-cell factors and on some accessory cells (Tada, Uracz and Abe, 1986; Waltenbaugh, 1986; Murphy, 1987). Classical genetic mapping studies (Tada, Taniguchi and David, 1976; Murphy et al., 1976) indicate that the gene(s) encoding I-J is (are) located within the I region of the MHC (the region of of the class II antigens, Figure 8), but molecular genetic studies showed that no I-J gene exists in that particular region (Steinmetz et al., 1982; Kronenberg et al., 1983; Kobori et al., 1986). That is the paradox.

A partial resolution of the I-J paradox resulted from the work of Sumida et al. (1985) and Uracz et al. (1985) with chimeras, and the work of Flood et al. (1986) with transgenic mice, which showed that the I-J phenotype of an animal depends on the MHC environment in which the T cells are selected, not on the MHC genotype of the T cells themselves. This suggests that I-J is a V region shape, which is selected due to the presence of class II MHC antigens. Most simply, I-J could be either anti-class II MHC or anti-anti-class II MHC.

Looking at I-J from the perspective of the symmetrical network theory, we suggested that I-J is more likely to be anti-anti-class II than anti-class II (Hoffmann,

1987). The rationale is simple, and is, again, based on the idea that suppressor cells have high network connectivity and helpers have low network connectivity. We already argued above that anti-class II cells should have low network connectivity, so they could not also be suppressors. We therefore assume, by default, that I-J cells (which are primarily suppressors) are anti-anti-class II.

We can then ask how anti-anti-class II clones could have high network connectivity. In Figure 9 and Figure 10, we see two alternative topologies for T cells that are idiotypically connected to class II MHC antigens. In the conventional divergent network topology picture of Figure 9, the anti-anti-class II clones do not have a high network connectivity, and there is no reason why they should all share a serologically detectable shape (I-J). The alternative network focussing model of Figure 10 could result from a natural selection process, and seems to fit the experiments and the theory better. Clones that recognize both class II MHC and I-J would be preferentially selected at the anti-class II level, and clones that recognize as many different anti-class II clones as possible would be preferentially selected at the I-J level. We like to think of this part of the network as being analogous to a circus tent, in which I-J is the centre-pole, and the anti-class II clones are the canvas. The centre pole is stabilized by the presence of the canvas and vice versa.

So far, this is a hand-waving model. The ideas are, however, amenable to mathematical modelling, and an interesting question is whether a simple mathematical model can simulate the development of this centre-pole type of topology.

Another aspect of the I-J paradox is the question of the location in the genome of the difference between two key strains that were important in defining I-J, namely, B10.A(3R) and B10.A(5R). These two strains were thought to be identical, until it

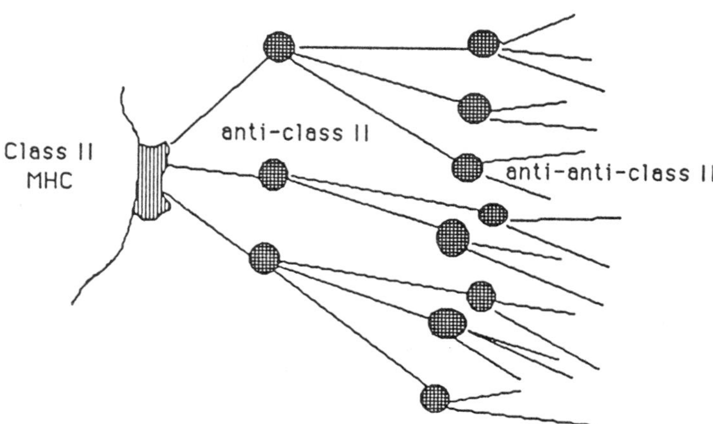

FIGURE 9 A divergent network topology for clones that relate to class II MHC self-antigens.

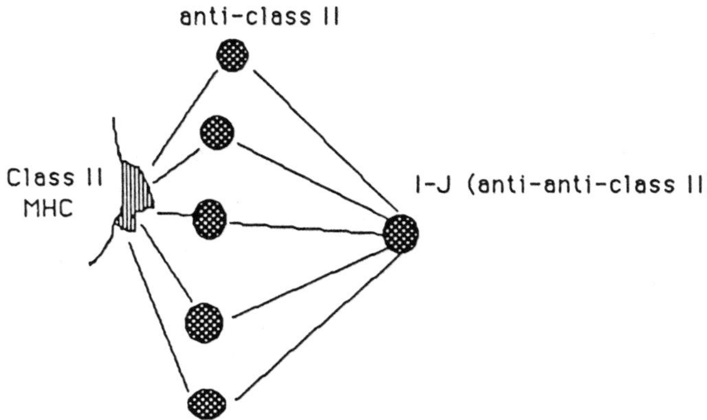

FIGURE 10 A network focussing topology for clones that relate to class II MHC self-antigens. This model fits the postulates of the symmetrical network theory better than the model of Figure 9.

was found that they have different I-J phenotypes; 3R is I-J^b and 5R is I-J^k. Some evidence indicated that genes on a completely different chromosome than that of the MHC, chromosome 4, codetermine I-J phenotype (Hayes et al. 1984), but this model seems to conflict with more recent data (Waltenbaugh et al, 1985), and the question of the nature of the difference between 3R and 5R is generally considered to be unresolved.

The idea of a set of anti-class II clones stabilizing the existence of a set of anti-anti-class II clones, and vice versa, leads to a novel idea about the difference between 3R and 5R. Mutual stabilization of clones with two specificity classes occurs in the symmetrical network theory for T cells in the suppressed state. The continued presence of the antigen is not required for the persistence of this suppressed state. It similarly seems possible that no genetic difference between 3R and 5R self antigens is required for a difference in their sets of T-cell idiotypes to persist from generation to generation. We would have to assume that a maternal effect plays a strong role in deciding which sets of T-cell idiotypes are selected in each generation. The expression of both parental I-J phenotypes in F1 animals (for example, 3Rx5R) could mean that there is also a paternal effect, possibly exerted indirectly via a perturbation of the maternal immune system by lymphocytes in the ejaculate. Evidence supporting this type of paternal effect has been reported by Gorczynski et al. (1983) and by Cooper-Willis et al. (1985). Important results indicating a strong maternal influence on the T-cell idiotypic repertoire has been obtained by Martinez et al. (1986). The maternal effect idea leads to the prediction that 3R embryos transplanted into I-J^k mothers should develop into I-J^k animals, and if these animals are then inbred, their offspring and the offspring of their offspring should be similarly I-J^k (Figure 11). An analogous result is predicted for 5R embryos implanted into I-J^b mothers.

AIDS: A DESTABILIZATION OF THE CENTRE POLE?

Is understanding acquired immune deficiency syndrome (AIDS) primarily a problem in virology or a problem in immunology? The human immunodeficiency virus (HIV) is blamed for destroying helper T cells. There are, however, problems with the simple view that the virus does the damage to the T cells directly. HIV infects less than 0.01% of the cells at any one time *in vivo*, and in some cases less than 0.001% (Harper et al., 1986). HIV is not very cytopathic for CD4 cells *in vitro* (Hoxie et al., 1985; Klatzmann et al., 1984). On the basis of detailed virology and epidemiology considerations, Duesberg (1987) argues that the virus is not the sole causative agent of AIDS, but that HIV is instead the most frequent opportunistic infection associated with the disease, with the disease involving either a cofactor or perhaps even another cause. Several authors have suggested that AIDS is an autoimmune disease; see Andrieu, Even and Venet (1986), Ziegler and Stites (1986), Shearer (1986), and Del Guercio and Zanetti (1987).

Ziegler and Stites (1986) proposed an interesting specific mechanism whereby AIDS could be an autoimmune disease triggered by the virus. They suggested that the virus mimics class II antigens by binding to CD4 (Figure 12). The gp120 glycoprotein of the HIV envelope has complementarity to CD4 (McDougal et al. 1986), and class II MHC also has complementarity to CD4, so gp120 could have a shape

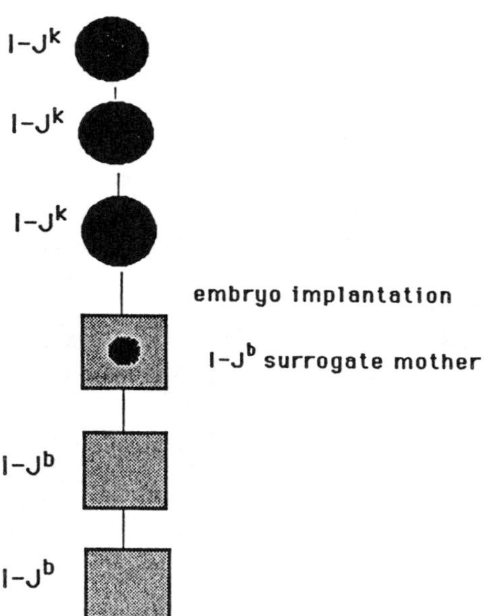

FIGURE 11 Our hypothesis about the difference between the I-J phenotypes of 3R and 5R mice predicts that a 3R (I-J^b) embryo implanted in a 5R (I-J^k) surrogate mother should develop into an I-J^k mouse, and that the I-J^k phenotype should then be maintained in subsequent inbred generations.

similar to that of class II MHC. This idea is supported by the fact that antibody to HLA class II specificities can give rise to false positives in the HIV-ELISA antibody screening system (Kuhnl et al., 1985; Weiss et al., 1985; Hunter et al., 1985). Sattentau et al. (1986) have shown that some monoclonal anti-class II MHC antibodies react with HIV. Similar results have been obtained by Montagnier et al. (1986). Ziegler and Stites accordingly suggested that the immune response to HIV could crossreact with class II antigens, and that an anti-idiotypic response to the anti-HIV response could crossreact with CD4 molecules, leading to the immune destruction of CD4 helper cells.

Another important theoretical contribution has been made by Shearer, who suggested several years ago that allogeneic leukocytes could be a factor in the induction of AIDS in homosexual men (Shearer, 1983; Shearer and Levy, 1984). Shearer and his collaborators have shown that homosexual men without evidence of AIDS differ from heterosexual controls in their immune responses to allogeneic cells (Tung et al., 1985).

Recent findings by Kohler and his collaborators support the idea of an immunological cofactor for AIDS, that could be related to allo-immunity (Chang, Huang and Kohler, 1985; Chang, Muller and Kohler,1987). They demonstrated that multiple immunizations with allo-antigens cause a decrease in immune responsiveness, and a decrease in the helper to suppressor T-cell ratio for a specific antigen. They suggested that allo-antigenic stimulation could predispose one to the emergence of AIDS.

We have suggested a network model for AIDS that combines the Ziegler-Stites autoimmunity hypothesis with Shearer's and Kohler's ideas about allo-immunity, and with our own ideas about the of I-J (Hoffmann, 1987). It invokes network interactions to a greater degree than the Ziegler-Stites theory; we suggest that AIDS involves both an anti-(centre pole) response and an anti-anti-self response of the A anti-(B anti-A) variety. The idea is that two complementary responses synergize to destabilize the network. But, first, we need to say a little about anti-anti-self antibodies.

In an extension of work by Ramseier and Lindenmann (1972), Binz and Wigzell (1975) and Bellgrau and Wilson (1979), we found recently that anti-anti-self antibodies are routinely produced during immune responses to allogeneic lymphocytes (Hoffmann, Cooper-Willis and Chow, 1986). The foreign cells have receptors that recognize self, including self-class II MHC. An immune response occurs against those receptors, and it is anti-anti-self, or A anti-(B anti-A), where A is self and B is foreign. People in high-risk groups for AIDS typically receive foreign lymphocytes together with the virus. Ejaculates contain lymphocytes (especially pathological ejaculates), and blood contains lymphocytes, so we might expect the recipients of these cells to make such anti-anti-self responses.

If we denote the class II MHC antigens as "self," the CD4 molecule is anti-self, since it has affinity for class II MHC, as shown in Figure 12. The gp120 glycoprotein of the HIV envelope binds to CD4, so it is anti-anti-self, and anti-gp120 is anti-anti-anti-self or Ab3. The centre pole of the network is the human equivalent

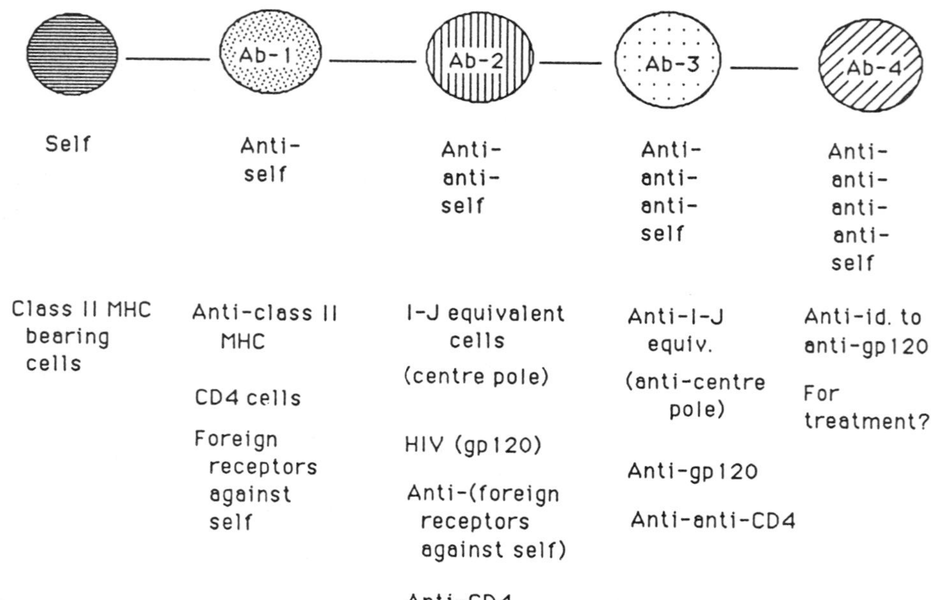

FIGURE 12 The sequence of cells and specificities from self class II MHC to anti-anti-HIV.

of I-J, and Ab3 is not only anti-gp120, but also anti-centre pole. The immune response to a combination of HIV and allogeneic cells can thus include a combination of Ab3 and the A anti-(B anti-A) variety of Ab2, and could result in a simultaneous attack on both the centre pole and on the cells that stabilize it.

These considerations lead to several predictions and ideas for prevention of the disease.

A combination of anti-(murine CD4) antibodies and allogeneic lymphocytes could produce AIDS in mice, which would be a valuable animal model. The anti-CD4 antibodies would play the role of gp120, inducing an Ab3 (anti-centre pole) response, and the allogeneic cells would induce an anti-anti-self response. It might be possible to induce AIDS in monkeys, that do not contract the disease from the virus alone, by inducing also an anti-anti-self response with allogeneic cells.

The theory similarly leads us to the conclusion that immunization of people, who also receive allogeneic cells from some source or other, with any one of three leading candidates for a vaccine, namely gp120, gp160 (from which gp120 is derived) or anti-(human CD4), could easily cause AIDS. (Anti-CD4 is being considered because anti-anti-CD4 is anti-HIV, see Figure 12). Such immunizations could have the same effect on people as HIV infection itself; the antibodies so raised could also be anti-centre pole. People in high-risk groups sometimes receive allogeneic cells, so they are the ones most at risk of being adversely affected by the vaccines. Some investigators are already doing small-scale testing of a vaccine containing gp160 (Zagury et al., 1987), and gp120 and anti-CD4 (see Joyce and Kingman, 1987) are being considered for use as vaccines. It is thus a matter of urgency that research be directed towards determining whether or not the above theory is correct.

It is known that immune complex formation is prevalent early in the disease (Boyko et al., 1987). The theory suggests that these complexes are mixtures of Ab2 and Ab3, where Ab3 is anti-HIV and anti-(I-J equivalent).

The theory predicts that anti-gp120 antibodies include anti-(I-J equivalent), or anti-centre pole antibodies. Since the antibodies rather than the virus are postulated to do the actual damage, we might want to make people tolerant to gp120 rather than immune to it. Alternatively, we need to devise forms of immunization that evoke immune responses that do not endanger the centre pole. If the anti-anti-self response to allogeneic cells is an important component in the development of the disease as postulated, then it is possible that most anti-HIV responses that occur in the absence of allogeneic cells are harmless.

Anti-(murine CD4) antibodies should be tested for reactivity to anti-I-J, and anti-(human CD4) antibodies should be tested for reactivity to anti-gp120.

Antibodies that are anti-idiotypic to anti-gp120 antibodies might be invaluable for prognosis (not all anti-gp120 antibodies necessarily being anti-centre pole), and also for treatment. Anti-gp120 responses that are also anti-centre pole are assumed to play a central role in causing the disease. Hence, by analysing the idiotypes of the anti-gp120 antibodies, we might be able to tell whether someone who is sero-positive is at high risk for contracting the disease. The anti-gp120 idiotypic profile could well be narrow in diseased patients, since the selection process that gives rise to the disease is slow and presumably stringent. An appropriate treatment for people who are sero-positive and are at high risk for the disease could be with anti-idiotypic antibodies against the anti-gp120 antibodies (Ab4, Figure 12). The same treatment could be applicable for patients that already have the disease, although it might be very difficult or impossible to restabilize a completely destabilized network.

Some of the experiments suggested above are simple, and can be used to fairly quickly determine whether the theory is or is not likely to be correct. It is important that trials of vaccines that are designed to induce immunity to gp120, gp160 or anti-CD4, and that involve people in high risk groups, should be approved only if the predictions of the theory cannot be substantiated.

ANTI-ANTI-(SELF CELL SURFACE) ANTIBODIES IN AUTOIMMUNE MICE

The idea that anti-anti-self antibodies could play a role in destabilizing the network centre pole caused us to look in the sera of autoimmune mice, to see whether they make anti-anti-self antibodies. MRL-(+/+) and MRL-*lpr/lpr* are strains of mice that are models for the human autoimmune disease, systemic lupus erythematosus. When they are 5 to 6 months old, these mice develop autoimmune disease, and produce antibodies to nuclear antigens such as double-stranded DNA, ribonucleoprotein and a nuclear protein known as Sm. The assay we used to detect anti-anti-self antibodies is the inhibition of cytotoxicity assay that we developed for detecting anti-anti-self in allo-immune sera, in the "second symmetry" experiments (Hoffmann, Cooper-Willis and Chow, 1986).

We found that sera of 5 to 6 month old, autoimmune MRL mice (H-2^k) inhibit the cytotoxic activity of hyperimmune Balb/c anti-CBA serum against CBA cells (H-2^k), while the sera of young (3 to 5 week) MRL mice do not have this property (Kion and Hoffmann, submitted). While this result is consistent with the idea of anti-anti-self antibodies in MRL autoimmune sera, it might also be explained on the basis of rheumatoid factor (anti-Ig antibody, or RF) in the sera; indeed, MRL mice have been reported to have multispecific RF in their sera (Rubin et al., 1984). We showed that this is not the case, by absorbing the sera with insolubilized Fc ragments or normal mouse Ig, prior to using them in the assay; in both cases the sera remained inhibitory. We verified that the inhibitory factor is, indeed, immunoglobulin; the activity is absorbed out by rabbit anti-mouse immunoglobulin. We have obtained similar results with the related autoimmune strains MRL-(+/+) and B6-*lpr/lpr*.

ANTI-IDIOTYPES IN ORDINARY ANTISERA

One of the reasons why the network paradigm is not yet universally accepted is that immunologists do not routinely see anti-idiotypic antibodies being produced in the course of ordinary immune responses to garden-variety antigens like bovine serum albumin (BSA) and keyhole limpet haemocyanin (KLH). Network theory needs to be able to account for this. According to first symmetry ideas, if a particular antibody is made, it will tend to kill off cells with specificities that are complementary to itself, so it is perhaps not surprising that anti-idiotypes are not readily observed during most immune responses. According to the theory, it is in the suppressed state that one expects to observe anti-idiotypic T cells, and these have, in fact, been observed (for example, Hirai and Nisonoff, 1980). Idiotypic studies are typically done using an antigen and a strain of mice that is special, in the sense that a particular idiotype is reproducibly produced with that antigen in that strain. If a particular mouse strain always produces antibodies with a particular idiotype, this presumably means that it is predisposed genetically to do so; perhaps, there is a germ-line gene that codes for antibodies with that specificity. For example, one much studied system is the response of Balb/c mice to phosphorylcholine (PC) with antibodies bearing the T15 idiotype; in this case, we know that there are two germ-line anti-PC V_H genes coding for that idiotype (Gearhart et al., 1981). In such cases, the anti-idiotypes would be expected to be kept at a low level, due to killing by these idiotypes, which have a head start. With most antigens in most strains of mice, we do not have such dominant idiotypes, and then the technical problem is that we do not seem to have anything reproducible in the way of idiotypes to study.

We did some exploratory studies using two mutually specific idiotypes (the T15 bearing hybridoma HPC-M2 and the anti-T15 hybridoma B36-82, both of them IgM antibodies) as immunogens in the "wrong strain," namely, CBA instead of Balb/c. We discovered that, in addition to an anti-T15 response to the T15 idiotype of HPC-M2 as immunogen, sometimes we could also detect an anti-anti-T15 response, and similarly, when we used B36-82 as immunogen, we could sometimes detect a response to both the immunogen and to HPC-M2 (A. Cooper-Willis, unpublished). We reasoned that these results might be understood along the following lines. The perturbation of the immune system by a particular foreign shape (antigen) causes the stimulation of both a variety of clones that are specific for that antigen, and a variety of anti-idiotypic clones, with the latter being most rapidly stimulated via antigen-specific T-cell factors adsorbed onto A cells. We envisaged that there could then be a "battle" between the various clones on the two sides (the idiotypes and the anti-idiotypes), with some clones on each side eventually proving "victorious," by killing off their complementary clones. We would accordingly expect both idiotypes and anti-idiotypes to be present in normal immune sera, but the idiotypes that survive would interact only weakly, if at all, with the surviving anti-idiotypes.

We decided to investigate this idea by looking at the responses to a given antigen in two very different animal species. Then the set of idiotypes that survive on the Ab1 (idiotype) side in the first species is likely to be quite different from those that survive on that side in the second species. Which anti-idiotypes (Ab2) survive in the second species would not be constrained by which idiotypes survive in the first species. We thought we might then be able to see a stronger reaction between Ab1(species P) and Ab2(species Q) than we would see between Ab1(species P) and Ab2(species P). We chose chickens and mice as species P and Q for these experiments, because chickens are phylogenetically very distant from mice. As antigens, we chose several proteins, including BSA, KLH and diptheria toxoid (DT).

One of our first results was that anti-BSA antisera from mice do, indeed, react specifically with anti-BSA antibodies from chickens, as we had hoped. We then obtained an initially surprising result: affinity-purified anti-BSA antibodies from hyper-immune chickens reacted specifically with anti-BSA antibodies from the same serum (Forsyth and Hoffmann, submitted). These experiments were done using the avidin-biotin ELISA method (Bayer and Wilchek, 1980). A first sample of affinity-purified anti-BSA antibodies was used to coat the ELISA plate, and a second sample of the same affinity-purified anti-BSA antibodies was biotinylated and allowed to react with the first sample. We saw a significantly stronger reaction of the anti-BSA antibodies with each other than between anti-BSA and anti-KLH or between anti-BSA and anti-DT. We obtained similar results for KLH and DT; anti-KLH reacts preferentially with anti-KLH, and anti-DT reacts preferentially with anti-DT.

Kang and Kohler (1986) have previously reported a phosphorylcholine-specific monoclonal antibody with binding specificity for itself. That result might have been regarded as a freak, but our results suggest that self-specificity could be a general characteristic of antibodies in hyperimmune sera.

The idea that antibodies are selected during the immune response to be simultaneously anti-antigen and anti-anti-antigen is an intriguing new twist to the concept of multispecificity. In the context of the symmetrical network theory, it is not hard to imagine how such a selection process might occur. During an immune response, both antigen and specific T-cell factors (initially "plus", then both "plus" and "minus") would be bound to accessory cell (A cell) surfaces, where they would form immunogenic arrays for both B and T cells. B cells that are specific for both antigen and plus factors, (or for both plus and minus factors) would have a selective advantage over B cells that are specific for only antigen.

ACKNOWLEDGMENTS

We thank J. Douglas Waterfield for introducing us to the MRL and B6-*lpr/lpr* systems, and providing us with sera. We thank Pat Gearhardt and Günter Hämmerling for providing us with the monoclonal antibodies HPC-M2 and B36-82, and Alan Perelson for helpful comments on the manuscript.

This work is supported by grants from the Medical Research Council of Canada (Grant A6770) and the Natural Sciences and Engineering Research Council of Canada (Grant 6729).

REFERENCES

1. Andieu, J. M., P. Even, and A. Venet (1986), "AIDS and Related Syndromes as a Viral-Induced Autoimmune Disease of the Immune System: An Anti-MHC II Disorder. Therapeutic Implications," *AIDS Research* **2**, 163–174.
2. Bayer, E. A. and M. Wilchek (1980), "The Use of the Avidin-biotin Complex as a Tool in Molecular Biology," *Methods of Biochemical Analysis*, Ed. David Glick (New York: Wiley), Vol. 26, 1–45.
3. Bellgrau, D., and D. B. Wilson (1979), "Immunological Studies of T-cell Receptors. 2. Limited Polymorphism of Idiotypic Determinants on T-cell Receptors Specific for Major Histocompatibility Complex Alloantigens," *J. Exp. Med.* **149**, 234–243.
4. Biddison, W. E., P. E. Rao, M. A. Talle, G. Goldstein, and S. Shaw (1984), "Possible Involvement of the T4 Molecule in T-cell Recognition of Class II HLA Antigens. Evidence from Studies of CTL-target Cell Binding," *J. Exp. Med.* **159**, 783–797.
5. Binz, H., and H. Wigzell (1975), "Shared Idiotypic Determinants on B Lymphocytes and T Lymphocytes Reactive against the Same Antigenic Determinants. 1. Demonstration of Similar or Identical Idiotypes on IgG Molecules and T-cell Receptors for Alloantigens," *J. Exp. Med.* **142**, 197–211.
6. Boyko, W. J., M. T. Schechter, J. P. Craib, B. Willoughby, B. Douglas, P. Sestak, W. A. McLeod, and M. O'Shaughnessy (1987), "The Vancouver Lymphadenopathy-AIDS Study: 7. Clinical and Laboratory Features of 87 Cases of Primary HIV Infection," *Can. Med. Assoc. J.* **137**, 109–113.
7. Cantor, H., and E. A. Boyse (1975), "Functional Subclasses of T Lymphocytes bearing Different Ly Antigens. I. The Generation of Functionally Different T-cell Subclasses is a Differentiative Process Independent of Antigen," *J. Exp. Med.* **141**, 1376–1389.
8. Chang, H. C., J. H. Huang, and H. Kohler (1985), "Accelerated Decline of the T15 Idiotypic Dominance in Balb/c Mice Injected with Allogeneic Autoblasts and Primed Spleen Cells," *Immunobiol.* **170**, 68–69.
9. Chang, H. C., S. Muller, and H. Kohler (1987), "Induction of the Network Perturbation by Multiple Immunizations with Alloantigens: An Animal Model of Immunological Predisposition for AIDS.," *Fed. Proc.* **46(4)**, 1352.
10. Cooper-Willis, C. A., J. C. Olson, M. E. Brewer, and G. A. Leslie (1985), "Influence of Paternal Immunity on Idiotype Expression in Off-spring," *Immunogenetics* **21**, 1–10.
11. Del Guercio, P., and M. Zanetti (1987), "The CD4 Molecule, the Human Immunodeficiency Virus and Anti-Idiotypic Antibodies," *Immunology Today* **8**, nos. 7 and 8.
12. Duesberg, P. H. (1987), "Retroviruses as Carcinogens and Pathogens," *Cancer Research* **47**, 1199–1220.

13. Flood, P. M., C. Benoist, D. Mathis, and D. B. Murphy (1986), "Altered I-J Phenotype in E_α Transgenic Mice," *Proc. Nat. Acad. Sci. USA* **83**, 8308–8312.
14. Gearhart, P. J., N. D. Johnson, R. Douglas, and L. Hood (1981), "IgG Antibodies to Phosphoryline Choline Exhibit More Diversity than their IgM Counterparts," *Nature* **291**, 29–34.
15. Gery, I., and B. H. Waksman (1972), "Potentiation of the T Lymphocyte Response to Mitogens. II. The Cellular Source of Potentiating Mediators," *J. Exp. Med.* **136**, 143–155.
16. Gorczynski, R. M., M. Kennedy, S. Macrae, and A. Ciampi (1983), "A Possible Maternal Effect in the Abnormal Hyporesponsiveness to Specific Alloantigens in the Offspring Born to Neonatally Tolerant Fathers," *J. Immunol.* **131**, 1115–1120.
17. Greenstein, J. L., J. Kappler, P. Marrack, and S. J. Burakoff (1984), "The Role of L3T4 in Recognition of Ia by a Cytotoxic, H-2D^d-specific T Cell Hybridoma," *J. Exp. Med.* **159**, 1213–1224.
18. Gunther, N., and G. W. Hoffmann (1982), "Qualitative Dynamics of a Network Model of Regulation of the Immune System: A Rationale for the IgM to IgG Switch," *J. Theoret. Biol.* **94**, 815–855.
19. Harper, M. E., L. M. Marselle, R. C. Gallo, and F. Wong-Staal (1986), "Detection of Lymphocytes Expressing Human T-Lymphotropic Virus Type III in Lymph Nodes and Peripheral Blood from Infected Individuals by *In Situ* Hybridization," *Proc. Natl. Acad. Sci.* **83 (3)**, 772–776.
20. Hayes, C. E., K. K. Klyczek, D. P. Krum, R. M. Whitcomb, D. A. Hullett and H. Cantor (1984), "Chromosome 4 *Jt* Gene Controls Murine T Cell Surface I-J Expression," *Science* **223**, 559–563.
21. Hirai, Y., and A. Nisonoff (1980), "Selective Suppression of the Major Idiotypic Component of an Anti-hapten Response by Soluble T-cell Derived Factors with Idiotypic or Anti-idiotypic Receptors," *J. Exp. Med.* **151**, 1213–1231.
22. Hoffmann, G. W. (1980), "On Network Theory and H-2 Restriction," *Contemp. Topics in Immunobiol.* **11**, 185–226. This paper contains an unfortunate typographical error; half a sentence was omitted. On page 207, the statement, "T cells [with] low connectance would preferentially suppress," should read "T cells with low connectance would preferentially help, and those with high connectance would preferentially suppress." The term "connectivity" is now used in this context instead of "connectance."
23. Hoffmann, G. W. (1982), "The Application of Stability Criteria in Evaluating Network Regulation Models," *Regulation of Immune Response Dynamics*, Eds. C. DeLisi and J. Hiernaux (Boca Raton, FL: CRC Press), pp. 137–162.
24. Hoffmann, G. W., A. Cooper-Willis and M. Chow (1986), "A New Symmetry: A Anti-B is Anti-(B anti-A), and Reverse Enhancement," *J. Immunol.* **137**, 61–68.
25. Hoffmann, G. W. (1986), "A Neural Network Model Based on the Analogy with the Immune System," *J. Theoret. Biol.* **122**, 33–67

26. Hoffmann, G. W. (1987), "On I-J, a Network Centre Pole and AIDS," *The Semiotics of Cellular Communication in the Immune System*, Eds. E. Sercarz, F. Celada, N. A. Mitchison and T. Tada, (New York: Springer-Verlag), in press.
27. Hunter, J. B., and J. E. Menitove (1985), "HLA Antibodies Detected by ELISA HTLV-III Antibody Kits," *The Lancet*, August 17, 1985, 397.
28. Jerne, N. K. (1974), "Towards a Network Theory of the Immune System," *Ann. Immunol. Inst. Pasteur* **125C**, 373–389.
29. Joyce, C., and S. Kingman (1987), "Researchers Want Trials with Anti-antibodies, *New Scientist*, 11 June 1987, 26.
30. Kang, C.-Y., and H. Kohler (1986), "Immunoglobulin with Complementary Paratope and Idiotope," *J. Exp. Med.* **163**, 787–796.
31. Kaye, J., and C. A. Janeway (1984), "The Fab Fragment of a Directly Activating Monoclonal Antibody that Precipitates a Disulphide-linked Heterodimer from a Helper T-cell Clone Blocks Activation by Either Allogeneic Ia or Antigen and Self-Ia," *J. Exp. Med.* **159**, 1397–1412.
32. Klatzmann, D., F. Barre-Sinoussi, M. T. Nugeyre, et al. (1984), "Selective Tropism of Lymphadenopathy Associated Virus (LAV) for Helper-Inducer T Lymphocytes," *Science* **225**, 59–63.
33. Klein, J. (1975), *Biology of the Mouse Major Histocompatibility-2 Complex* (Berlin: Springer-Verlag).
34. Kobori, J. A., E. Strauss, K. Minard, and L. Hood (1986), "Molecular Analysis of the Hotspot of Recombination in the Murine Major Histocompatibility Complex," *Science* **234**, 173–179.
35. Kronenberg, M., M. Steinmetz, J. Kobori, E. Kraig, J. A. Kapp, C. W. Pierce, C. M. Sorensen, G. Suzuki, T. Tada, and L. Hood (1983), "RNA Transcripts for I-J Polypeptides are Apparently not Encoded between the I-A and I-E Subregions of the Murine Major Histocompatibility Complex," *Proc. Nat. Acad. Sci. USA* **80**, 5704–5708.
36. Kuhnl, P., S. Seidl, and G. Holzberger (1985), "HLA-DR4 Antibodies Cause Positive HTLV-III Antibody ELISA Results," *The Lancet*, May 25, 1222–1223.
37. Langman, R. E., and M. Cohn (1986), "The 'Complete' Immune System is an Absurd Immune System," *Immunology Today* **7**, 100–101.
38. Levy, J. A., and J. L. Ziegler (1983), "Acquired Immunodeficiency Syndrome is an Opportunistic Infection and Kaposi's Sarcoma Results from Secondary Immune Stimulation," *Lancet*, 1983, ii, 78-81.
39. MacDonald, H. R., A. L. Glasebrook, C. Bron, A. Kelso, and J.-C. Cerottini (1982), "Clonal Heterogeneity in the Functional Requirement for Lyt-2/3 Molecules on Cytolytic T Lymphocytes (CTL): Possible Implications for the Affinity of CTL Antigen Receptors," *Immunol. Rev.* **68**, 89–115.

40. McDougal, J. S., J. K. A. Nicholson, G. D. Cross, S. P. Cort, M. S. Kennedy, and A. C. Mawle (1986), "Binding of the Human Retrovirus HTLV-III/LAV/ARV/HIV to the CD-4(T4) Molecule: Conformation Dependence, Epitope Mapping, Antibody Inhibition and Potential for Idiotypic Mimicry," *J. Immunol.* **137**, 2927–2944.
41. Martinez-A. C., M. L. Toribio, A. De La Hera, P. A. Cazenave, and A. Coutinho, (1986), "Maternal Transmission of Idiotypic Network Interactions Selecting Available T Cell Repertoires," *Eur. J. Immunol.* **16**, 1445–1447.
42. Montagnier, L., J. Gruest, D. Klatzmann, and J. C. Gluckman (1986), *Anti-class II Monoclonal Antibodies Inhibit LAV Infection in CEM Cells* (Abstract), Poster 261, 2nd International AIDS Conference, Paris.
43. Murphy, D. B., L. A. Herzenberg, K. Okumura, L. A. Herzenberg, and H. O. McDevitt (1976), "A New I Subregion (I-J) Marked by a Locus (Ia-4) Controlling Surface Determinants on Suppressor T Lymphocytes," *J. Exp. Med.* **144**, 699–712.
44. Murphy, D. B. (1987), "The I-J Puzzle," *Ann. Rev. Immunol.* **5**, 405–427.
45. Owens, T., and B. Fazekas de St. Groth (1987), "Participation of L3T4 in T Cell Activation in the Absence of Class II Major Histocompatibility Complex Antigens. Inhibition by Anti-L3T4 Antibodies is a Function both of Epitope Density and Mode of Presentation of Antireceptor Antibody," *J. Immunol.* **138**, 2402–2409.
46. Ramseier, H., and J. Lindenmann (1972), "Aliotypic Antibodies," *Transplant. Rev.* **10**, 57.
47. Reinherz, E. L., and S. F. Schlossman (1980), "Differentiation and Function of Human T Lymphocytes," *Cell* **19**, 821–827.
48. Rubin, R. L., R. S. Balderas, E. M. Tan, F. J. Dixon, and A. N. Theofilopoulos, (1984), "Multiple Autoantigen Binding Capabilities of Mouse Monoclonal Antibodies Selected for Rheumatoid Factor Activity," *J. Exp. Med.* **159**, 1429–1440.
49. Sachs, D. H. (1984), "The Major Histocompatibility Complex," *Fundamental Immunology*, Ed. W. E. Paul (New York: Raven Press), 303–346.
50. Sattentau, Q. J., A. Dagleish, P. Clapham, M. Exley, R. Weiss, and P. C. L. Beverley (1986), "Cross-reactivity between HTLV-III/LAV and MHC Class II Antigen," *Communication* **159**, S21d (abstract) at the 2nd International AIDS Conference, Paris.
51. Shearer, G. M. (1983), "Allogeneic Leukocytes as a Possible Factor in Induction of AIDS in Homosexual Men," *N. Engl. J. Med.* **308**, 223–224.
52. Shearer, G. M. (1986), "AIDS: An Autoimmune Pathologic Model for the Destruction of a Subset of Helper T Lymphocytes," *Mount Sinai J. Med.* **53**, 609–615.
53. Spouge, J. L. (1986), "Increasing Stability with Complexity in a System Composed of Unstable Subsystems," *J. Math. Analysis and Applications* **118**, 502–518.

54. Smith, K. A. (1984), "Lymphokine Regulation of T Cell and B Cell Function," *Fundamental Immunology*, Ed. W. E. Paul (New York: Raven Press), 559–576.
55. Steinmetz, M., K. Minard, S. Horvath, J. McNichol, C. Wake, E. Long, B. Mach, and L. Hood (1982), "A Molecular Map of the the Immune Response Region from the Histocompatability Complex of the Mouse," *Nature* **300**, 25.
56. Sumida, T, T. Sado, M. Kojima, K. Ono, H. Kamisaku, and M. Taniguchi (1985), "I-J as an Idiotype of the Recognition Component of Antigen-specific Suppressor T-cell Factor," *Nature* **316**, 738–741.
57. Tada, T., M. Taniguchi, and C. S. David (1976), "Properties of the Antigen-specific T Cell Factor in the Regulation of Antibody Response of the Mouse. IV. Special Subregion Assignment of the Gene(s) that Codes for the Suppressive T Cell Factor in the H-2 Histocompatibility Complex," *J. Exp. Med.* **144**, 713–725.
58. Tada, T., W. Uracz, and R. Abe (1986), "Are There Unique I Region Controlled Determinants on T Cells?," *Paradoxes in Immunology*, Ed. G. W. Hoffmann, J. G. Levy and G. T. Nepom (Boca Raton, FL: CRC Press), 253–265.
59. Tung, K. S. K., F. Koster, D. C. Bernstein, P. W. Kriebel, S. M. Payne, and G. M. Shearer (1985), "Elevated Allogeneic Cytotoxic T Lymphocyte Activity in Peripheral Blood Leukocytes of Homosexual Men," *J. Immunol.* **135**, 3163–3171.
60. Uracz, W., Y. Asano, R. Abe, and T. Tada (1985), "I-J Epitopes are Adaptively Acquired by T cells Differentiated in the Chimaeric Condition," *Nature* **316**, 741–745.
61. Weiss, S., D. L. Mann, C. Murray, and M. Popovic (1985), "HLA-DR Antibodies and HTLV-III Antibody ELISA Testing," *The Lancet*, July 20, 1985, 157.
62. Waltenbaugh, C., L. Sun, and H.-Y. Lei (1985), "I-J Expression is not Associated with Murine Chromosome 4," *Eur. J. Immunol.* **15**, 922–926.
63. Waltenbaugh, C. (1986), "I-J - Immunoregulatory Molecules in Search of a Gene," *Paradoxes in Immunology*, Eds. G. W. Hoffmann, J. G. Levy and G. T. Nepom (Boca Raton, FL: CRC Press), 271–281.
64. Zagury, D., Z. Lurhuma, K. Mbayo, J. J. Salaun, R. Leonard, J. Bernard, M. Fouchard, B. Reveil, B. Goussard, and J. Wane (1987), *Abstract T.16.5*, from III International Conference on Acquired Immunodeficiency Syndrome (AIDS).
65. Ziegler, J. L., and D. P. Stites (1986), "Hypothesis: AIDS is an Autoimmune Disease Directed at the Immune System and Triggered by a Lymphotropic Retrovirus," *Clin. Immunol. Immunopath.* **41**, 305–313.

LEE A. SEGEL† and ALAN S. PERELSON‡
†Department of Applied Mathematics, Weizmann Institute of Science, Rehovot, Israel and
‡ Theoretical Division, Los Alamos National Laboratory, Los Alamos, NM 87545, USA

Computations in Shape Space: A New Approach to Immune Network Theory

ABSTRACT

This study is based on the use of generalized shape as the fundamental variable to characterize the binding properties of antigens, antibodies, receptors, etc. Within this framework the investigation concentrates on a network view of the immune system with particular stress on the effects of cross-reactivity. The schematic model considered postulates cells whose probability of entering a proliferative state depends on the interplay of activating and suppressing factors, which in turn are functions of the degree of binding to corresponding sets of receptors. A one-dimensional shape space is considered. Emphasis is placed on the likelihood that the unstimulated immune system is resistant to perturbation, but not too resistant (control vs. stability tradeoff). Such a "nearly unstable" immune system is constructed with the aid of activation that is relatively short range (i.e., relatively specific in binding) compared to suppression. Surprisingly, it is found that instability, or near-instability, can also occur when activation is relatively long range. Various computer simulations illustrate the properties of the model, for example, focussing or defocussing of the antigenic input.

1. INTRODUCTION

The spirit of this work may be characterized as an approach to *computational immunology*. The term is an adaptation of the computational approach to vision first stressed by Marr (1982). Marr pointed out that there are advantages to separately considering three matters: (i) the computations made by the visual system (e.g., edge enhancement, formation of the "primal sketch"), (ii) the algorithms used to make these computations (e.g., serial or parallel), and (iii) the "hardware" (or "wetware") used by neurons to carry out the computations. In this spirit, the approach here is to focus on certain aspects of the possible computational strategy and tactics of the immune system. Not too much regard will be paid to the details of the cell machinery for implementing the tactics, but of course, in formulating conjectures as to possible tactical procedures, one must at the very least keep in mind known general features of identified cells and factors that constitute the implementation machinery.

Our main object here is to explore a network view of the immune system (Jerne, 1974) with an emphasis on the effects of cross-reactivity. The "smeared out" character of an accurate view of complementarity, taking account of the fact that a variety of molecules can bind one another with varying affinities, results in a formulation in terms of integral equations. We begin with the essential novelty of our formulation, the use of generalized "shape" as the fundamental descriptive variable.

2. SHAPE SPACE

The immune system, and arguably all biology, is driven by myriad encounters between pairs of molecules whose degree of binding governs the outcome of the encounter (e.g., fleeting ineffective binding, formation of a complex that is sufficiently long lived to be "useful", and binding that induces a conformational change.) The degree of binding, in turn, depends on the *generalized shapes* (Perelson and Oster, 1979) of the molecules involved. By generalized shape, we mean not only the average geometric shape of the active site of the molecule, but also such factors as the electric charge distribution, the position of dipole moments, and the ability of groups of atoms to undergo geometric fluctuations in time. One is, thus, led to conceive of the immune system as a vast collection of shifting molecular populations each characterized by a generalized shape x.

In spite of the fact that "true" shape space may be infinite dimensional, it is quite conceivable that an adequate characterization of essential "shape" variables can be accomplished by a relatively small number of measurements (Perelson and Oster, 1979). (The recent findings of Geysen et al (1987) are encouraging in this regard.) This would give a shape space of some small finite dimension N. In our first model, we shall take $N = 1$. This drastic simplification may be permitted in exploratory calculations provided that one tries to check whether, at least in principle, one's major conclusions can be extended to more realistic values of N.

We shall thus assume that a single (positive or negative) number x characterizes the generalized shapes of antigens, antibodies and receptors, the key molecules in our hypersimple immune system. We shall further assume that $y = -x$ describes the shape that is exactly complementary to x. More generally, we assume that the smaller is $x - (-y) \equiv x + y$, the tighter the fit there is of shape x with shape y.

Figure 1 provides two possible realizations of our shape assumption. In one realization [Figure 1(a)], portions of molecules that are relevant to the immune system are imagined to be wedges of equal width. Positive x indicates a protuberance of height x units, while negative x characterizes an indentation. Shapes x and $-x$ fit together perfectly, other shapes less perfectly. Caricatured here is the fit of one piece of molecule into a cleft on another molecule, as seems to be characteristic of hapten binding (Marrack, 1987) and molecular recognition generally (Rebek, 1987).

A second realization imagines that patches of charge, in stereotyped positions, fully characterize the various molecules. For definiteness, one can imagine three patches with each patch containing agglomerations of atoms whose net charge ranges up to one unit. "Perfect fit" would occur if 3 negative patches were opposed to 3 positive patches, a 2–2 confrontation would be less perfect, etc. [See Figure 1(b)]. This scenario is somewhat reminiscent of recent work from the laboratory of R. Poljak (Amit et al., 1986) on the antibody binding of hen egg lysozyme wherein the interacting surfaces appear virtually flat with mutual contact via

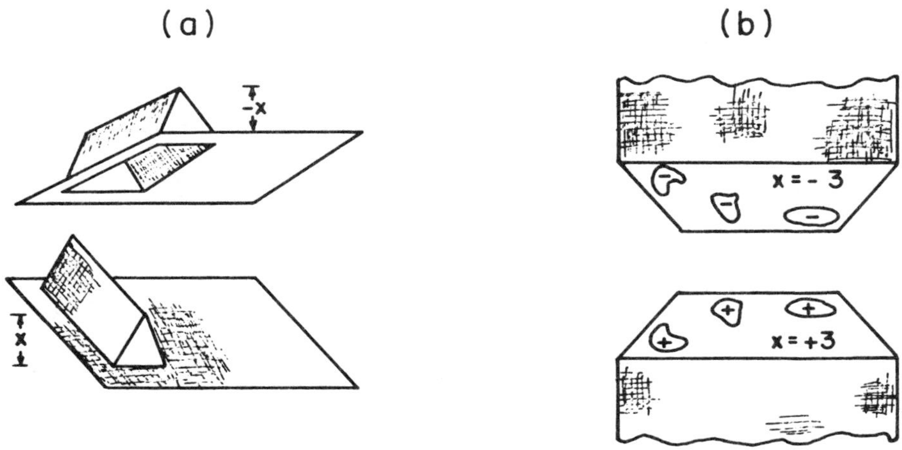

FIGURE 1 Possible idealized one-dimensional shape spaces. Alternative (a): Shape is defined by the height of wedge-shaped epitopes, positive for protuberances and negative for indentations. Alternative (b): The "shape" variable gives the charge, which is distributed in homologously located patches.

many amino acids. Note that if we restrict the permissible values of x to a domain of 6 to 10 units, then we have a range of values which is similar to the range of the logarithms of affinities in the binding of a selected antigenic determinant to a random collection of antibodies.

3. FORMULATION

In the model considered here, we shall not distinguish between cells and molecules. Thus, our fundamental unknown function will be $b(x,t)$, which can be regarded for definiteness as the number of lymphocytes with receptors of shape x that constitute the immune system at time t. The lymphocytes could be T cells or B cells. Here we are only keeping track of receptor shape. Antibody can be imagined as present, if one assumes that there always is a constant number of solution-phase antibodies per lymphocyte. This would be the case if a steady state were reached in which the production of antibodies by stimulated B cells were balanced by the natural elimination of antibody from the body.

At this point, readers who are not theorists may be tempted to give up in disgust, fed up with the oversimplifications (and there will be more!). We beg indulgence. We know about the various classes of T and B cells, growth factors, MHC, etc., etc. We hope before too long to make our models more biologically realistic. But we believe that we can convince you that our somewhat novel point of view provides some interesting conclusions, and the promise of more, even though our models contain only a minimum of biological reality.

We imagine the cell population to be divided into two classes, stimulated and unstimulated cells. The fraction of stimulated cells will be denoted by $\alpha(b)$, $0 \leq \alpha \leq 1$, so that $1-\alpha$ gives the fraction of unstimulated cells. The presence of the argument b in α indicates that the fraction of stimulated cells may differ at different cell population levels, because of mechanisms that will be discussed shortly.

The structure of our model can be represented as

$$\begin{array}{c} \text{rate of} \\ \text{population} \\ \text{increase} \end{array} = \begin{array}{c} \text{influx} \\ \text{from} \\ \text{bone marrow} \end{array} - \begin{array}{c} \text{death of} \\ \text{unstimulated} \\ \text{cells} \end{array} + \begin{array}{c} \text{reproduction} \\ \text{of stimulated} \\ \text{cells} \end{array} + \begin{array}{c} \text{source} \\ \text{of} \\ \text{antigen.} \end{array} \tag{1}$$

The bone marrow source term and the death rate will be taken as constants m and d. The reproduction rate r of x-cells will be a function not only of the population level $b(x,t)$ of such cells but also of the average lymphocyte population size $B(t)$. (Dependence on B can account for effects of nonspecific factors that can either inhibit or enhance a given response.) Since the only characteristic of cells and molecules that our model follows is shape, we necessarily represent antigen as a shape. Assuming that antigen is something we inject from outside the system, we represent antigens of shape x at time t by a source function, $\gamma(x,t)$. With this, the mathematical counterpart of (1) is

$$\partial b/\partial t = m - db[1-\alpha(b)] + r(B,b)\alpha(b)b + \gamma(t). \tag{2}$$

If shape ranges between $-L$ and L, then the expression for the average population size B is (Appendix 1)

$$B(t) = \frac{1}{2L}\int_{-L}^{L} b(x,t)\,dx. \tag{3}$$

Our next task is to specify with more precision the nature of the function α which determines the fraction of stimulated cells of shape x. In a large population of cells, α is the probability that a cell of shape x is stimulated. We assume that whether or not a cell is switched into an active state depends on a competition between activating and suppressing influences, A and S. The influences may be generated by the same receptor in different states of aggregation as in the immunon theory (Dintzis, this volume) or by different receptors. For definiteness, imagine that a cell contains two sets of receptors, one set contributing to an activating influence and one set to a suppressing influence. (On B cells, these may be the immunoglobulin and Fc receptors, respectively). For any fixed shape x, let $a(x,y)b(y,t)dy$ be the fraction of an x-cell's activating receptors that are bound by entities (antibodies or cells, we do not distinguish) whose shape variable ranges between y and $y+dy$, where dy is a small number. Then

$$A(b) = \begin{array}{l}\text{fraction of } x\text{-cells'}\\ \text{activating receptors}\\ \text{bound when the}\\ \text{distribution in shape space}\\ \text{of cells is given by } b(y,t)\end{array} = \int_{-L}^{L} a(x,y)b(y,t)\,dy. \tag{4}$$

In essence $a(x,y)$ is the association constant for y binding to activating receptors on x-cells. It has been assumed that such a small fraction of receptors is bound that the amount of binding to x receptors of one shape, say, y_1, can be calculated independently of the amount of binding of any other shape, y_2.

In a manner completely analogous to Eq. (4), we calculate the fraction of suppressing receptors bound

$$S(b) = \int_{-L}^{L} s(x,y)b(y,t)\,dy. \tag{5}$$

Here $s(x,y)$ is the association constant for y binding to suppressor receptors on x cells. With Eqs. (4) and (5), we can specify the interaction between activation and suppression by the assumption that the probability of stimulation α is some function F of the amount of binding to the two classes of receptors:

$$\alpha = F(A, S). \tag{6a}$$

We assume that α increases if more activator receptors are bound and decreases if more suppressor receptors are bound. Mathematically this is expressed by

$$\partial F/\partial A > 0, \qquad \partial F/\partial S < 0. \tag{6b, c}$$

A relatively simple specific example that can be used for F is

$$F(A, S) = \frac{A}{(p + qS) + A}, \tag{7}$$

where p and q are positive constants. The right side of Eq. (7) is of the familiar Michaelean form. The maximum value of F is unity, and the half-maximal concentration is $p+qS$. Thus, an increase of the suppression factor S shifts the graph of F to the right (as in ordinary enzyme inhibition) and, thus, decreases the probability of stimulation α. DeBoer (this volume) also incoroporates a term in his activation function which can shift the graph of the function to the right.

To embody our assumption that the fit of shape y to shape x decreases from a maximum when $y = -x$ (perfectly complementary shape), we assume Gaussian functions of $x-(-y)$ for the activation and suppression association constants a and s. Thus,

$$a(x, y) = a_M (2\pi\sigma_a^2)^{-1/2} \exp[-(x+y)^2/2\sigma_a^2], \tag{8a}$$
$$s(x, y) = s_M (2\pi\sigma_s^2)^{-1/2} \exp[-(x+y)^2/2\sigma_s^2], \tag{8b}$$

where a_M, s_M, σ_a and σ_s are constants. If each shape is complementary to only a small fraction of all possible shape, then

$$\sigma_a, \ \sigma_s \ll L. \tag{8c}$$

In the sequel, we use the fact that if a_M and s_M are chosen to be one, then with Eq. (8c)

$$\int a(x, y)dy = 1 = \int s(x, y)dy. \tag{8d}$$

Complete specification of our model requires assignment of a functional form to the dependence of the proliferation rate r of activated cells. We take

$$r(b, B) = r_0 e^{-\lambda b} e^{-\eta B^n} \qquad (r_0 = \text{constant}). \tag{9}$$

The constant η will be assumed positive, corresponding to a suppressive effect of an increase in the overall population size. If the positive parameter n is not equal to unity, then suppression can be regarded as cooperative in nature. If λ is negative, then there is a self-reinforcement effect induced by the growth of a clone, or a family of clones bearing similar shapes. If λ is positive, this effect is suppressive.

4. THE STABILITY-CONTROLLABILITY TRADE-OFF

The equations of our model are formulated. What can we learn from them? Here we shall focus on only one matter, the trade-off between good stability properties and good controllability properties.

Aircraft (for example) and the immune system both should remain largely unaffected by relatively small random disturbances, but should modify their trajectories in response to a "purposeful" command—by the pilot or by virtue of a significant appearance of antigen. A system whose state more or less remains unaffected by small disturbances is called *stable* in the language of engineering and applied mathematics. If a system is too stable, it will be insensitive to commands. Thus, an airplane that remains virtually unbuffeted by even quite strong gusts of wind will also respond very sluggishly to a deflection of its rudder.

Let us then consider the stability of our model system given by Eqs. (2)–(9). We wish to examine the behavior of the system without continual outside stimulation, so we equate to zero the antigen input term γ of Eq. (2). We first note that Eqs. (2)–(9) [with $\gamma = 0$] have a *uniform* steady state solution (Appendix 2)

$$b(x,t) = \bar{b}, \qquad \bar{b} \text{ a certain positive constant.} \tag{10}$$

This solution corresponds to a "virgin" immune system, wherein all shapes (e.g., populations of antibodies and lymphocytes) are equally represented. (Theorists will wish to know about boundary conditions; these are discussed in Appendix 2.)

We now inquire whether this solution is stable to small perturbations. By this we mean the following. Suppose that at some instant (denoted by time $t = 0$), the distribution of antibody and lymphocytes were such that some shapes were slightly overrepresented and some slightly underrepresented. In mathematical terms

$$\text{at } t = 0, \quad b = \bar{b} + b_0(x), \quad |b_0(x)| \ll \bar{b} \text{ for every } x. \tag{11}$$

If no further outside stimulation is given to the system, how will it evolve in time? If the evolution is such that the disturbances fade away, i.e.,

$$b(x,t) \to \bar{b} \qquad \text{as } t \to \infty, \tag{12}$$

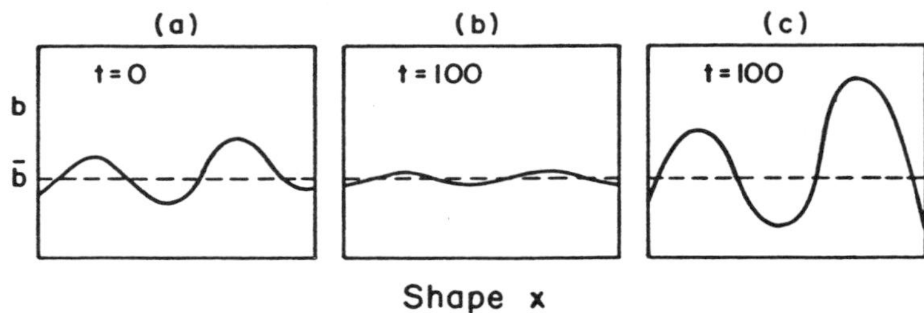

FIGURE 2 Stability vs instability for a population $b(x,t)$. In the case of stability, an initial deviation (a) from a uniform distribution of shapes (possible steady solution $b(x,t) = \bar{b}$, dashed horizontal line) evolves in time to a more nearly uniform distribution (b). In the case of instability, (a) evolves to a less uniform distribution (c).

then we say that the system is *stable to small perturbations*, or just *stable* for short. (See Figure 2.) If disturbances fade away rapidly, i.e., if $b(x,t)$ becomes very close to $\bar{b}$ after a short time, then the system can be regarded as "very stable"—an undesirable situation according to the second paragraph of this section. If $|b(x,t)-\bar{b}|$ *increases*, for some ranges of x, the system is termed *unstable*.

We wish to find conditions for which our system is stable but not very stable. Our strategy is to utilize known techniques that allow us to find parameter ranges corresponding to which the system is unstable. We determine these, and then change conditions slightly to make the system slightly stable.

It is debatable whether an arbitrary choice of a model immune system will provide the possibility of instability. Fortunately, general circumstances that promote instability are well known in theoretical biology. These are *short-range activation* and *long-range inhibition*, which promote nonuniformity in various contexts ranging from developmental biology to neurobiology (see the review of Levin and Segel, 1985; Oster, 1988). In the present context, these general conditions correspond to the assumption that the width (i.e., the standard deviation) of the activation Gaussian is smaller than the corresponding quantity for the suppression Gaussian:

$$\sigma_a < \sigma_s. \tag{13}$$

Indeed, calculations show (Appendix 2) that under the assumption (13), there is a range of parameters for which the steady state is stable and a complementary range for which it is unstable. To illustrate our results, let us consider an especially simple model for which calculations can be easily and explicitly carried out. We shall take $\lambda = \eta = 0$ in Eq. (9) so that the proliferation rate satisfies

$$r = r_0 = \text{constant}. \tag{14}$$

For the fraction of cells activated, we take

$$\alpha = F(A,S) = A/(qS), \qquad q \text{ a constant.} \tag{15}$$

This can be regarded as an independent assumption, or as an approximation to Eq. (7) that will be valid if the suppression factor S is sufficiently large. With Eqs. (14), (15) and (8d), the steady state of the virgin immune system is given by

$$b(x,t) = \overline{b} = m/[d - q^{-1}(d + r_0)], \tag{16}$$

where we must have

$$q > 1 \quad \text{and} \quad d > r_0/(q-1), \tag{17}$$

in order to insure that $\overline{b} > 0$.

The results of the calculations specialized to this case [see Eq. (A30b) of Appendix 2] are summarized in Figure 3. Five parameters influence the stability of the solution $b = \overline{b}$, but they only appear in two parameter groups. Instability is predicted when

$$\frac{dq}{d+r_0} - 1 < \phi(\theta) \quad \text{where} \quad \phi(\theta) \equiv (\theta - 1)\theta^{-\theta/(\theta-1)}, \qquad \theta \equiv \sigma_s^2/\sigma_a^2. \tag{18a, b, c}$$

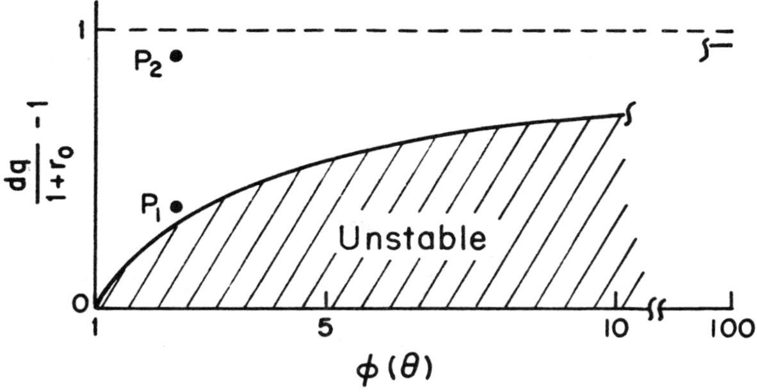

FIGURE 3 Parameter domains for stability and instability of the uniform state $b(x,t) = \overline{b}$. [From (A30b).] One of several ways to interpret the diagram is that at a fixed value of $\theta \equiv \sigma_s^2/\sigma_a^2$, the uniform state is unstable if the death rate d is too small. (Also see text.) It is suggested that the immune system will be such that parameters correspond to a slightly stable point such as P_1 rather than a very stable point such as P_2.

If parameters correspond to a point in the shaded region of Figure 3, then the uniform state is unstable and any small disturbance to uniformity will become magnified as time increases. After some time, the system may settle down to a new uniform state, but one that is inhomogeneous as a function of the shape descriptor x.

What is most important for our purposes is the existence of the heavy line in Figure 3 that divides stable from unstable parameter groups. One way to view the existence of this line is mentioned in the caption to Figure 3. Another is to rearrange Eq. (18a) to give as an instability condition that the birth-death ratio r_0/d is larger than a critical value, i.e.,

where we assume that $q > \phi(\theta) + 1$. In any case we can illustrate the stability-controllability tradeoff by stating that we would expect that the parameters of the immune system are such that the system operates in a *slightly* stable setting such as P_1 (in Figure 3) and not a *very* stable setting such as P_2. The analytic stability calculations of Appendix 2 are only valid for small departures from the uniform state $b(x,t) = \bar{b}$. To escape these limitations, we turn to the computer, approximating the integrals in Eqs. (2)–(8) by sums. (See Appendix 1.) Figure 4 shows

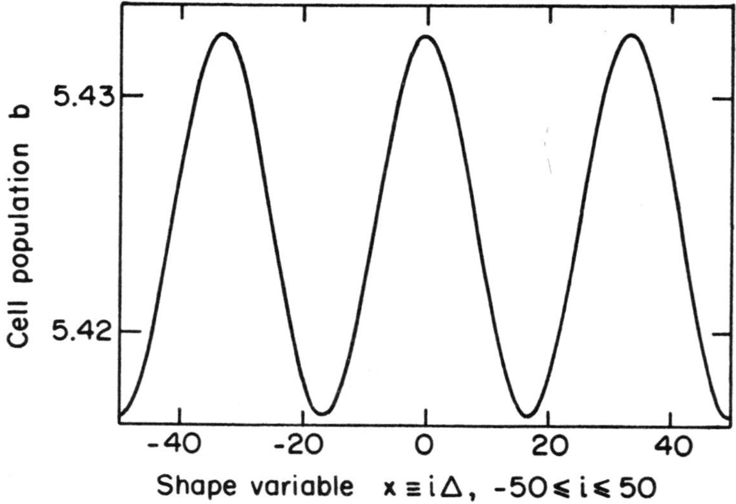

FIGURE 4 The calculated final steady state (uniform to within the errors of the approximate solution method), given an initial state that varies randomly by 10% from the uniform steady state $\bar{b}$. Equations (2)–(8) with periodic boundary conditions and no antigen forcing ($\gamma = 0$). Parameters: $m = d = p = 1$, $q = 9$, $a_M = s_M = \sigma_a = 1$, $\sigma_s = \sqrt{2}$, $r_0 = 8$, $n = 4$, $\lambda = 0$, $\eta = 1.05 \times 10^{-4}$, $N = 50$, $\Delta = 0.16$ [c.f. (A1)].

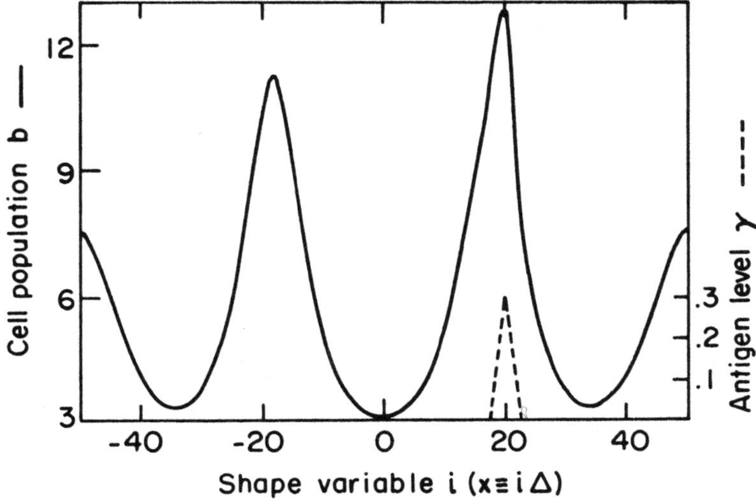

FIGURE 5 A forced stable system at steady state. The antigen forcing function γ of (2) is given by the peak function centered at $N = 20$ (right scale). Note the "echo" of the forcing centered at the complementary shape $N = -20$. In addition there is a third peak, presumably owing to the intrinsic "wavelength" present in the system (see text).

$$\frac{r_0}{d} > \frac{q}{\phi(\theta) + 1} - 1, \tag{19}$$

the typical result after an unforced stable system ($\gamma = 0$) has evolved from random initial conditions; as anticipated, the system has returned to near-uniformity.The remaining departure from steady state is at the level of the error in the approximation scheme. (That there are three peaks is expected from formula (A32a) in Appendix 2.) Figure 5 shows a forced stable system. The forcing given by γ is a peaked function of five units width. (As in Figure 4, and for the same reason, three peaks appear in the response.) Figure 6 differs essentially from Figure 5 only in one respect—conditions are significantly more stable in Figure 6.

As expected, the "immune response" to a given level of antigen is significantly lower under the more stable conditions of Figure 6. Note also that there is a response "width" in shape space that is intrinsic to the system, i.e., the width is not the same as the width of the "antigen" forcing. The intrinsic width is most evident in Figure 5,

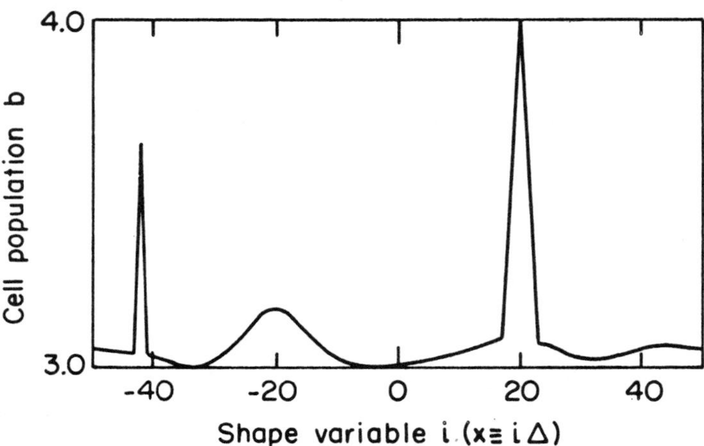

FIGURE 6 A forced system, more stable than Figure 5. Parameters as in Figure 5 except that $r_0 = 6$. Note the smaller magnification in the response, and its narrow focus, compared with Figure 5.

less so in the more stable situation of Figure 6. Figure 7 provides a marked contrast. In it, again conditions are essentially the same except that now $\lambda = -1.8 \times 10^{-3}$ instead of $\lambda = 0$. Recall from our comments on Eq. (9) that a negative value of λ corresponds to an autocatalytic reinforcement that is strongly local in nature. With this negative value of λ, we see from Figure 7 that the entire system response becomes peaked at a single shape. The network is completely destroyed. This shows that even though the immune system in principle is a network, in practice the parameters could be such that network interactions are not at all important.

5. SUMMARY AND DISCUSSION

We have outlined an approach to the immune system that takes the shape of the various immunological determinants as the fundamental (independent) variable. Edelstein and Rosen (1978) and Perelson and Oster (1979) discussed shape space in general terms, but did not propose specific dynamical equations based on shape. See the paper of Percus in this volume for an approach similar to ours.

The major "computational issue" that we have dealt with is the trade-off between the necessity for the immune system to be stable to casual small disturbances, but not so stable as to be virtually inert and unstimulable. We have suggested that this dual requirement will be met by the evolution of a "nearly unstable" immune system. We further postulate that the "near instability" would be attained by means similar to those that lead to instability and pattern formation in several other areas of theoretical biology, namely suitable short-range activation and long-range inhibition. "Range" is measured in shape space. Thus, our postulate has the interpretation that activating influences in the immune system will be more specific than suppressive influences, in the sense that tighter binding to suitable receptors will be required for activation than for suppression. Grossman (1982) has also suggested that activation is relatively specific.

We have illustrated another computation that the immune system presumably carries out, a focussing or defocussing of the antigenic input. Focussing (defocussing) occurs if the range of affinities of molecules evoked in the immune response is narrower (wider) than the range of input affinities. In this first report, we cannot comment in detail upon the relation between our calculations and such issues as affinity maturation (Siskind and Benacerraf, 1969; Wysocki et al. 1986) and immunodominance (Gammon et al, 1987). Suffice it to say that we have seen that the extent of focussing can depend in a rather subtle way, subject to the values of other parameters, on the ratio between the relative specificities required for activation and inhibition (compare Figures 5 and 6.) Alternatively, too specific an activating feedback can entirely negate the network structure (Figure 7). This last point again lends emphasis to the importance of gauging the relative specificities of activation and suppression, as a key factor in determining the character of network interactions, or even in determining whether such interactions play a significant role at all.

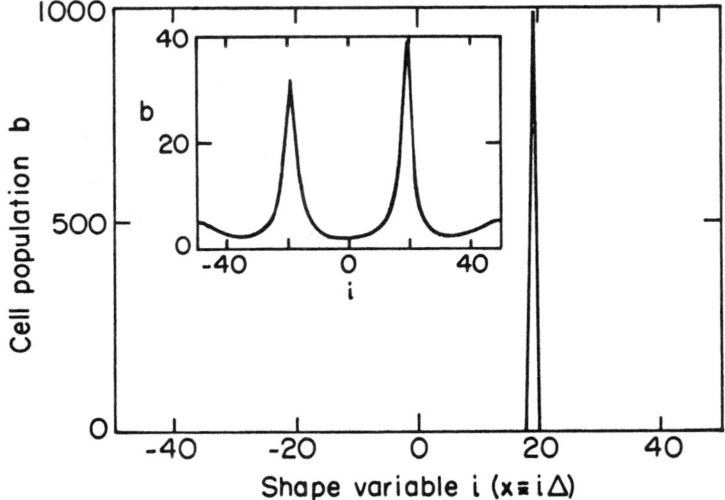

FIGURE 7 As in Figure 6 except that $\lambda = -1.8 \times 10^{-3}$. Inset: $t = 250$. Full graph: $t = 300$; the entire immune response is focussed at the shape most characteristic of the antigenic forcing (and the peak is still growing).

Our simple model shows that the interplay of a number of factors gives rise to the amplification that the immune system can apply to an antigenic stimulus. This suggests that there should be a number of different defects that could result in responses that are inappropriately high or low. Hypergammaglobulinemia as seen in some AIDS patients (Lane, et al, 1983) is perhaps an example of the first condition.

The linear stability calculations of Appendix 2 yield a "critical wavelength" that is a simple estimate of the preferred distance between response peaks in shape space. Figure 7 shows that strong nonlinear effects may drive the system into a state quite different from that predicted by linear stability theory. Nonetheless, experience with nonlinear systems indicates that in many circumstances the critical wavelength often provides at least qualitatively accurate predictions. In the future, we plan to discuss the relation between the critical wavelength and the question of how many response peaks can be present before the immune system becomes saturated. A related question is how adjustments are made so that a permanent long-term memory of a stimulus remains in the immune system after the stimulus is removed. We believe it likely that long-term memory results from changes in parameters that render the immune system permanently unstable *locally* in shape space.

We mentioned, toward the beginning of Section 2, that careful scrutiny must be focussed on the drastic simplification of taking a one-dimensional shape space. In light of the results obtained so far, this assumption seems warranted in a first model. The reason is that the qualitative conclusions that we have deduced seem generalizable to higher dimensional shape spaces. For example, although the quantitative details will certainly change, we believe that with higher dimensional shape spaces, it will remain true that in the presence of relatively more specific activation and less specific inhibition, an instability diagram like that of Figure 3 will still emerge.

Such a diagram, again perhaps with considerable quantitative changes in detail but invariant in general form, is also anticipated when more comprehensive models are considered, with separate equations, say, for B-cells and T-cells, together with soluble antibodies and antigens. Thus, although we certainly hope that better models can yield more qualitative and semi-quantitative understanding of the possible nature of network interactions, we are fairly confident that one insight will remain from the present work—that the immune system has probably evolved to be stable but not too stable.

Our use of the terms "stable" and "unstable" refers to the effect of small transitory disturbances to the uniform "virgin" immune system. For certain parameter domains our model yields unstable behavior; here, the interactions are such that even a tiny local disturbance reverberates so strongly through the system that its final state will contain numerous large peaks of excitation. We stress again our suggestion that the parameters of the immune system are set such that the system is stable, but not overly so.

To our surprise, and in marked contrast with other biological models, we found that instability could not only occur when activation is relatively short range compared to suppression, but also when activation is relatively long range (see Appendix 2). This remarkable behavior stems from the fact that the origin $x = 0$ is not arbitrarily chosen, as it is in most models, but rather represents the single self-complementary shape in our one-dimensional shape universe. It turns out that because of this, there exist (antisymmetric) sinusoidal perturbations that decrease both activation and suppression compared to their basal values in the virgin immune system. Instability can ensue if the decrease in suppression sufficiently outweighs the decrease in activation so that there is a sufficient increase in net activation. (A full explanation of this instability will appear elsewhere.)

It has been claimed that "the 'complete' idiotype network is an absurd immune system" (Langman and Cohn, 1986), apparently since the only perceived alternatives are that either "connectivity is a complex web with multipoint attachments and the network is tied in a Gordian knot" or that "antibody degeneracy is low and. . .the idiotype network fragments into many sets of non-overlapping pairs." Our basic model [Eq. (2)] is formally infinite with some measure of interaction between any pair of shapes. When we pass to the computer, we "approximate" Eq. (2) by a more realistic model that consists of a large but finite number of clones. No unfathomable mysteries appear for either the infinite or finite model. A wide variety of behaviors are possible, depending on the parameter values.

It is clear *a priori* that there is some degree of cross-reactivity between every pair of molecules, so that the formal existence of a network is assured. Even the few simulations that we have presented, however, demonstrate that it is by no means assured that the network has important influences on real immune systems. If such influences exist, unravelling them will, of course, not be easy, and we believe that models such as ours can be helpful in accomplishing this task.

We remark finally that we hope that our schematic theory here can guide "experiments" with far more comprehensive and detailed models such as that discussed by Perelson in the present volume. Such large-scale simulations have the advantage of considerable realism and, hence, of possible semi-quantitative comparison with experiment. But with realism comes a complexity that makes it difficult to obtain an overview of the situation and, hence, to pose strong general questions. Our simpler model and "computational" view of the immune system can help provide such an overview.

ACKNOWLEDGEMENTS

This research was begun during a brief visit of LAS to the Neurosciences Institute, Rockefeller University. Much of it was carried out when LAS enjoyed the hospitality of the Theoretical Biology and Biophysics Group, Los Alamos National Laboratory. Partial support was derived from the U.S. Department of Energy and the U.S.-Israel Binational Science Foundation, Grant 86–107. Thanks to Z. Agur for useful comments on an earlier version of the manuscript, to Y. Barbut and J. Segel for help with the figures, and to C. Weintraub for skilled TEXnical assistance.

APPENDIX 1: APPROXIMATING INTEGRALS BY SUMS AND VICE-VERSA

Let the shape interval $-L \le x \le L$ be divided into $2N$ subintervals of width Δ, where $-L = -N\Delta$, $L = N\Delta$. Let $b_i(t)$ be an alternative notation for $b(x,t)$ when $x = i\Delta$. The average value of the b_i, and an approximation of this average for small Δ [c.f. (3)], are given by

$$\frac{1}{2N}\sum_{i=-N}^{N-1} b_i(t) = \frac{1}{2N\Delta}\sum_{i=-N}^{N-1} b_i(t)\cdot\Delta \approx \frac{1}{2L}\int_{-L}^{L} b(x,t)\,dx. \qquad (A1)$$

[That the two sides of Eq. (A1) are close to one another is an immediate consequence of the definition of the integral.] Approximations completely analogous to Eq. (A1), applied to Eqs. (4) and (5), convert the integral Eq. (2) to a system of differential equations, which were solved numerically to obtain Figures 4–7.

APPENDIX 2: STABILITY CALCULATIONS

Employing Eq. (6), we write the governing Eq. (2) with $\gamma = 0$ as

$$\partial b/\partial t = m - db + R(B,b)bF(A,S), \qquad R(B,b) \equiv d + r(B,b). \qquad (A2)$$

We set $a_M = s_M = 1$ in Eq. (8). This occasions no loss of generality, for general a_M and s_M can be incorporated in F.

In establishing conditions on the boundaries of shape space, we first assumed that $b(x,t) = 0$ for $|x| > L$. This led to strong Mach band phenomena (Ratliff, 1965). We thus switched to periodic boundary conditions

$$b(-L,t) = b(L,t), \qquad (A3)$$

since we felt that there should be no decisive influence of boundary shapes.

Boundary conditions of Eq. (A3) were employed for the computer simulations, but in the stability calculations that follow, we consider an infinite domain ($L = \infty$). We expect that our results will be approximately correct even if Eq. (A3) is imposed, provided that the disturbance wavelength is small compared to L. High accuracy is not necessary, since the purpose of our linear stability analysis is to reveal interesting parameter ranges for simulations of the full nonlinear model.

The constant $\bar{b}$ of the uniform steady state solution (10) must satisfy

$$0 = m - d\bar{b} + R(\bar{b},\bar{b})\bar{b}F(\bar{b},\bar{b}) \qquad (A4)$$

since

$$\int_{-\infty}^{\infty} \begin{pmatrix} a(x,y) \\ s(x,y) \end{pmatrix} dy = \begin{pmatrix} 1 \\ 1 \end{pmatrix}, \tag{A5}$$

and since $B = \bar{b}$.

Let us now examine the stability of such solutions to perturbations of uniform shape. That is, we examine what will happen if the population of every shape is augmented or diminished by the same small amount. Writing

$$b(t) = \bar{b} + b'(t) \tag{A6}$$

we find that

$$db'/dt = m - (\bar{b} + b')d + (\bar{b} + b')R(\bar{b} + b', \bar{b} + b')F(\bar{b} + b', \bar{b} + b'). \tag{A7}$$

Upon linearizing and employing Eq. (A4), we obtain the stability condition

$$-d + RF + \bar{b}R(F_1 + F_2) + \bar{b}(R_1 + R_2)F < 0. \tag{A8}$$

Here

$$R = R(\bar{b}, \bar{b}), \;\; F \equiv F(\bar{b}, \bar{b}), \;\; F_1 = \left.\frac{\partial F(A,S)}{\partial A}\right|_{A=S=\bar{b}}, \tag{A9}$$

$$F_2 = \left.\frac{\partial F(A,S)}{\partial S}\right|_{A=S=\bar{b}}, \;\; R_1 = \left.\frac{\partial R(B,b)}{\partial B}\right|_{B=b=\bar{b}}, \;\; R_2 = \left.\frac{\partial R(B,b)}{\partial b}\right|_{B=b=\bar{b}}. \tag{A10}$$

In accord with the standard practice of linear stability theory, we now consider a perturbation that is exponential in time and sinusoidal in shape. We shall search for perturbations that can destabilize the uniform steady-state. We write

$$b'(x,t) = e^{\lambda t}[\beta e^{ikx} + \beta^* e^{-ikx}], \tag{A11}$$

where an asterisk indicates complex conjugate, and k is the wavenumber of the perturbation. We assume $k \neq 0$, so that the perturbation is sinusoidal and leaves the average value of b unchanged. A typical calculation is

$$\begin{aligned} \int_{-\infty}^{\infty} a(x+y)[\beta e^{iky} + \beta^* e^{-iky}]dy &= \int_{-\infty}^{\infty} a(z)[\beta e^{ik(z-x)} + \beta^* e^{-ik(z-x)}]dz \\ &= \hat{a}(k)[\beta e^{-ikx} + \beta^* e^{ikx}]. \end{aligned} \tag{A12}$$

Here $\hat{a}$ is the Fourier transform

$$\hat{a}(k) = \int_{-\infty}^{\infty} e^{ikz} a(z) dz, \tag{A13}$$

and in calculating the second term of Eq. (A12), we have used the fact that a is an even function. With this, we obtain from Eqs. (A2) and (A6) upon linearization

$$0 = (-\lambda - d + \overline{R}_2 F)(\beta e^{ikx} + \beta^* e^{-ikx}) + R\overline{b}(F_1\hat{a} + F_2\hat{h})(\beta e^{-ikx} + \beta^* e^{ikx}). \tag{A14a}$$

where

$$\overline{R}_2 \equiv R + R_2\overline{b} \tag{A14b}$$

Since the coefficient of $\exp(ikx)$ must vanish separately, we obtain

$$\beta(\overline{R}_2 F - d - \lambda) + \beta^* R\overline{b}(F_1\hat{a} + F_2\hat{h}) = 0. \tag{A15a}$$

The vanishing of the coefficient of $\exp(-ikx)$ implies

$$\beta^*(\overline{R}_2 F - d - \lambda) + \beta R\overline{b}(F_1\hat{a} + F_2\hat{h}) = 0. \tag{A15b}$$

Subtracting the conjugate of Eq. (A14) from Eq. (A15a), we find

$$\beta(\lambda^* - \lambda) = 0.$$

For a nontrivial solution, $\beta \neq 0$ so that λ is real and behavior is monotonic. Since λ is real, Eq. (A15b) is merely the conjugate of Eq. (A15a). If

$$\beta \equiv \beta_r + i\beta_i \tag{A16}$$

upon taking the real and imaginary parts of Eq. (A15), we obtain

$$\beta_r[\overline{R}_2 F - d - \lambda + R\overline{b}(F_1\hat{a} + F_2\hat{h})] = 0, \quad \beta_i[\overline{R}_2 F - d - \lambda - R\overline{b}(F_1\hat{a} + F_2\hat{h})] = 0. \tag{A17}$$

Nontrivial solutions result when

$$\beta_r = 0,\ \beta_i \neq 0,\ \lambda = \lambda^- \quad \text{and} \quad \beta_i = 0,\ \beta_r \neq 0,\ \lambda = \lambda^+ \tag{A18}$$

where

$$\lambda^{\pm} = \overline{R}_2 F - d \pm R\overline{b}(F_1\hat{a} + F_2\hat{h}), \qquad k \neq 0. \tag{A19}$$

To proceed, we note that since a and h are Gaussians, their Fourier transforms are given by

$$\hat{a}(k) = e^{-\ell}, \qquad \hat{h}(k) = e^{-\theta\ell} \tag{A20}$$

where

$$\ell \equiv k^2\sigma_a^2/2, \qquad \theta \equiv \sigma_s^2/\sigma_a^2. \tag{A21a, b}$$

Thus

$$\lambda^{\pm} = \overline{R}_2 F - d \pm R\bar{b}(F_1 e^{-\ell} - |F_2|e^{-\theta\ell}), \tag{A22}$$

where we have used the fact that by Eq. (6b) $F_2 < 0$.

A perturbation of wavenumber k will grow and destabilize the system if $\lambda(k)$ is positive. To find such perturbations, we look for a maximum in $\lambda(\ell)$, the dispersion relation. As a function of ℓ, the extrema of both eigenvalues is given by the solution ℓ_c of $\partial\lambda^{\pm}/\partial\ell = 0$, i.e.,

$$F_1 e^{-\ell_c} = |F_2|\theta e^{-\theta\ell_c}. \tag{A23}$$

Omitting a positive factor, we have at the extremum

$$\frac{\partial^2\lambda^{\pm}}{\partial\ell^2} \sim \pm(F_1 e^{-\ell_c} - \theta^2|F_2|e^{-\theta\ell_c}) = \pm|F_2|\theta e^{-\theta\ell_c}(1-\theta), \tag{A24}$$

where we have employed Eq. (A23). Thus, the extremum will be an internal maximum (i.e., occur for $\lambda \neq 0$) for λ^+ (λ^-) if and only if $\theta > 1$ ($\theta < 1$).

To see whether Eq. (A23) has a solution for positive ℓ_c, we write

$$\frac{F_1}{|F_2|\theta} = e^{\ell_c(1-\theta)}. \tag{A25}$$

If $\theta > 1$, Eq. (A25) has a solution if and only if $F_1 < |F_2|\theta$. That is

$$\text{if } \theta > 1 \text{ and } \frac{F_1}{|F_2|} < \theta, \text{ then } \ell_c = \frac{1}{\theta - 1}\, \ell n \frac{|F_2|\theta}{F_1}. \tag{A26a, b, c}$$

Similarly

$$\text{if } \theta < 1 \text{ and } \frac{F_1}{|F_2|} > \theta, \text{ then } \ell_c = \frac{1}{1-\theta}\, \ell n \frac{F_1}{|F_2|\theta}. \tag{A27a, b, c}$$

Instability sets in at the wavelength

$$\lambda_c = 2\pi/k_c = \pi\sigma_a(2/\ell_c)^{1/2}. \qquad (A28)$$

Under conditions of Eq. (A26a,b), instability commences when parameters are such that $\lambda^+(\ell_c)$ is just barely positive. Employing Eqs. (A23) and (A26c), we can write this condition as

$$\overline{R}_2F - d + R\bar{b}\Phi > 0 \quad \text{where} \quad \Phi \equiv |F_2|(\theta-1)(|F_2|\theta/F_1)^{-\theta/(\theta-1)}. \qquad (A29)$$

For the simple illustrative case [Eq. (15)], Eqs. (A26c) and (A29) reduce to

$$\ell_c = (\theta-1)^{-1}\ell n\ \theta, \quad \phi(\theta) > -1 + dq/(d+r_0); \quad \phi(\theta) \equiv (\theta-1)\theta^{-\theta/(\theta-1)}. \qquad (A30a,b,c)$$

Under conditions of Eqs. (A27a,b), the instability is associated with λ^- and the counterparts of Eqs. (A29) and (A30) are

$$\overline{R}_2F - d + R\bar{b}\Phi > 0 \quad \text{where} \quad \Phi \equiv |F_2|(1-\theta)(F_1/|F_2|\theta)^{-\theta/(1-\theta)}, \qquad (A31)$$

$$\ell_c = (1-\theta)^{-1}\ell n\ \theta^{-1}, \quad \phi(\theta) > -1 + dq/(d+r_0), \quad \phi(\theta) \equiv (1-\theta)\theta^{-\theta/(1-\theta)}. \qquad (A32a,b,c)$$

In addition we see from (A22) that to avoid an instability for very long wavelengths ($k \to 0$) we must impose the condition

$$\overline{R}_2F - d - R\bar{b}(F_1 - |F_2|) < 0. \qquad (A33)$$

REFERENCES

1. Amit, A. G., R. A. Mariuzza, S. E. V. Phillips, and R. J. Poljak (1986), "Three-Dimensional Structure of an Antigen-Antibody Complex at 2.8 Å Resolution," *Science* **233**, 747–753.
2. Edelstein, L., and R. Rosen (1978), "Enzyme-Substrate Recognition," *J. Theoret. Biol.* **73**, 181–204.
3. Gammon, G., N. Shastri, J. Cogswell, S. Wilbur, S. Sadegh-Nasseri, A. Miller, and E. Sercarz (1987), "The Choice of *T* Cell Epitopes Utilized on a Protein Antigen Depends on Multiple Factors," *Immunol. Reviews* **98**, in press.
4. Geysen, H.M., J. A. Tainer, S. J. Rodda, T. J. Mason, H. Alexander, E. D. Getzoff, and R. A. Lerner (1987), "Chemistry of Antibody Binding to a Protein," *Science* **235**, 1184–1196.
5. Grossman, Z. (1982), "Recognition of Self, Balance of Growth and Competition: Horizontal Networks Regulate Immune Responsiveness, *Eur. J. Immunol.* **12**, 747–756.
6. Jerne, N.K. (1974), "Toward a Network Theory of the Immune System," *Ann. Immunol. (Inst. Pasteur)* **125C**, 373–389.
7. Lane, H. C., H. Masur, L. C. Edgar, L. C. et al. (1983), "Abnormalities of B-Cell Activation and Immunoregulation in Patients with the Acquired Immunodeficiency Syndrome," *New Engl. J. Med.* **309**, 453–458.
8. Langman, R. D., and M. Cohn (1986), "The 'Complete' Idiotype Network is an Absurd Immune System," *Immunol. Today* **7**, 100–101.
9. Levin, S.A., and L. A. Segel (1985), "Pattern Generation in Space and Aspect by Interaction and Redistribution," *SIAM Review* **27**, 45–67.
10. Marr, D. (1982), *Vision: A Computational Investigation into the Human Representation and Processing of Visual Information.* (San Francisco: W.H. Freeman).
11. Marrack, P. (1987), "New Insights into Antigen Recognition," *Science* **235**, 1311–1313.
12. Oster, G. (1988), "Lateral Inhibition Models of Developmental Processes," *Nonlinearity in Biology in Medicine*, Eds. A. S. Perelson, B. Goldstein, and M. Dembo (New York: Elsevier), in press.
13. Perelson, A. S., and G. F. Oster (1979), "Theoretical Studies of Clonal Selection: Minimal Antibody Repertoire Size and Reliability of Self-Non-Self Discrimination," *J. Theoret. Biol.* **81**, 645–670.
14. Ratliff, F. (1965), *Mach Bands. Quantitative Studies on Neural Networks in the Retina* (San Francisco: Holden-Day).
15. Rebek, J., Jr. (1987), "Model Studies in Molecular Recognition," *Science* **235**, 1478–1484.
16. Siskind, G. W., and B. Benacerraf (1969), "Cell Selection by Antigen in the Immune Response," *Adv. Immunol.* **10**, 1–50.

17. Wysocki, L., T. Manser, and M. L. Gefter (1986), "Somatic Evolution of Variable Region Structures During an Immune Response," *Proc. Natl. Acad. Sci. USA* **83**, 1847–1851.

J. K. PERCUS
Courant Institute of Mathematical Sciences and Physics Department, New York University, 251 Mercer Street, New York, New York 10012

Polydispersity in Immune Networks

ABSTRACT

A general model of the operation of a mammalian immune system is developed, in which particular attention is paid to the existence of a very large number of related components. By bunching together lymphocytes, accessory cells, and biochemical messengers into functional units, a simplified description is proposed for the proliferation of these units, depending upon their binding interactions. A toy model prepared in this context has a repertoire in close correspondence with expected properties of such a system. Several extensions of the model are described.

1. INTRODUCTION

Polydispersity is a ubiquitous characteristic of natural phenomena.[1] It refers to the presence of sets of objects which are closely related structurally or functionally, whose differences are described by some "internal" degree of freedom which we shall designate by x. The tag x lies in a space whose topology or even dimensionality need not be obvious, but, of course, must be specified. When polymers are synthesized, it is difficult to avoid a distribution of numbers of monomeric units; the number x is, then, an integer. For copolymers, one may instead choose x as a binary string identifying which monomer occupies which site. If three-dimensional structure plays a crucial role in functionality, x might be chosen as the vector of locations of the atoms, as that of submolecular building blocks, perhaps only that of a few spherical harmonic moments of electron density, or even a single real number which, in some fashion, codes for the full structure. Attention to polydispersity becomes useful in practice when the behavior of an entire subset can be represented as that of a single entity, often by identifying a suitable element which mimics the whole set.[2]

The immune system is polydisperse at almost any level of description. It also suffers from the standard problem of missing biological information. The format we shall employ is, therefore, the "black box" approach to complex systems, an impressionistic description whose further details are hopefully to be supplied in the future. Since we have the grandiose objective of modeling the full human immune system, this strategy may be unavoidable, and we will, in fact, carry it to an extreme. At our level of description, it will, for example, be irrelevant whether we have in mind humoral or cellular response. At any rate, we start by referring to the space of recognizable submolecular entities, antigen if nominally external to the system, internal antigen or epitope if internal, as the space $\{x\}$. We will imagine $\{x\}$ as continuous, if necessary as an approximation to a discrete space on a fine mesh. The *first major assumption* in the present discussion is that the system is well-mixed; it is, therefore, described by the set of time-varying, but spatially independent, concentrations $c(x, t)$. This implies both that we neglect inhomogeneity and, by virtue of the large effective system size, we neglect the stochastic nature of molecular,..., cellular number, thereby making inaccessible such phenomena as extinction.

Our *second major assumption*, equally sweeping, is that the reaction of the immune system to a given antigen can be encompassed by that of rigid functional units, lumping together antigen-presenting cell and the ensuing branched structure of lymphocytes and lymphokines.[3] This functional unit—our black box—will be identified by its principal recognition site, e.g., paratope for B-cell recognition, and may again be denoted by x. Our interest will, of course, center on the density of functional units, $\rho(x, t)$. What we want to do is to characterize and identify the density patterns which signal the polydisperse subsets that act in concert, in face of the fine and ramified network of interactions between all molecular species and functional units. This will be achieved by examining a small number of "toy" models. Our conclusion, put briefly, will be that the general repertoire of low zone tolerance, high zone tolerance, normal immune response, as well as an assortment of pathologies, is already well represented at the very primitive level considered, suggesting that a more detailed description will be rewarded by a quite deep understanding of the highly involved phenomenology of the immune system.

2. BASIC MODEL

Let us consider a possible format for the dynamics of the densities $c(x,t)$ and $\rho(x,t)$, which hews closely to commonly held beliefs. To start with, let us define the external antigen concentration as $a(x,t)$. A functional unit specified by y will also expose internal antigens, not equally accessible. If the accessibility of x on y is given by $w(x,y)$, then the total antigen concentration will, of course, be given by

$$c(x,t) = a(x,t) + \int w(x,y)\rho(y,t)dy, \tag{2.1}$$

where dy denotes an appropriate measure on the space $\{x\}$. The stimulation or repression of functional units of type x will now depend upon the binding of submolecular entities to its sites.[4] It is not necessary to separate stimulation and repression. Instead, we simply observe that if x is bound at sites $z_1, z_2, \ldots$ by the local concentrations of $x_1, x_2, \ldots$, then the resulting contribution to the relative growth rate of x may be written in the general form

$$dk(x,t) = k\left(x; z_1, z_2, \ldots; x_1, x_2, \ldots; c(x_1,t), c(x_2,t)\ldots\right) \prod dx_i \prod dz_i, \tag{2.2}$$

which may be positive or negative. Thus, we make no attempt to trace the hierarchy of antibodies $Ab - n$ to $Ab - m - 1$ to a given antigen, but rather work with the full state of the system.

We now make another *major approximation* for the purposes of the present analysis and expand (2.2) only to first order in the concentrations. If the zeroth order is amalgamated into an autonomous growth rate that we will soon include, this means that (2.2) is replaced by

$$k_1(x,t) = \int \Lambda_1(x,y)c(y,t)dy \tag{2.3}$$

for appropriate kernel Λ_1. One can imagine $\Lambda_1(x,y) = \int E(x,y')B(y',y)dy'$ as the composition of the effect $E(x,y')$ of the binding of y' on x on the growth rate of x, and the binding rate $B(y',y)$ of y' with y. The simplification (2.3) is drastic. It allows for neither direct competition nor synergism between different sites on the functional unit. Nonetheless, we will see that the resulting repertoire is quite extensive.

Equations (2.2) and (2.3) carry another implication that is certainly not true, namely that the binding events contribute instantaneously to the growth coefficient. The functional unit, instead, harbors a number of internal delay and switching mechanisms. We shall take into account in a preliminary way the fact that the environment need not change slowly in comparison with the growth by supposing that the instantaneous growth rate consists of two parts, $k_1(x,t)$ determined as in

(2.3), and $R(x,t)$ which approaches a target growth rate $k_2(x,t)$ with time constant $\tau(x)$:

$$\begin{aligned} \tau(x)\dot{R}(x,t) + R(x,t) &= k_2(x,t) \\ k_2(x,t) &= \int \Lambda_2(x,y)c(y,t)dy. \end{aligned} \tag{2.4}$$

To complete the dynamics, we postulate an autonomous birth rate $\alpha(x)$ and death rate $\mu(x)$ for functional units of type x. Combining (2.1), (2.3), and (2.4), and using the abbreviations

$$\begin{aligned} K_\sigma(x,y) &= \int \Lambda_\sigma(x,z)w(z,y)dz \\ a_\sigma(x,t) &= \int \Lambda_\sigma(x,z)a(z,t)dz, \qquad \sigma = 1,2 \end{aligned} \tag{2.5}$$

the dynamics resulting from $\dot{\rho} = \alpha - \mu\rho + \rho\Lambda_1 c + \rho R$ then takes the form

$$\begin{aligned} \dot{\rho}(x,t) =& \alpha(x) - \mu(x)\rho(x,t) + \\ & \rho(x,t)\left[\int K_1(x,y)\rho(y,t)dy + R(x,t) + a_1(x,t)\right] \\ \tau(x)\dot{R}(x,t) + R(x,t) =& \int K_2(x,y)\rho(y,t)dy + a_2(x,t), \end{aligned} \tag{2.6}$$

which is our basic model. In principle, we should append the obvious equation for the degradation of $a(x,t)$(ref. [5]) and a less obvious one for its production, but we will restrict our attention to situations in which $a(x,t)$ is a known function. The dynamics of non-cellular $c(x,t)$ we shall simply ignore.

Equations (2.6) represent an immune network,[6] in which the production of x-functional units is tied in principle to the concentrations of all other units. At a detailed level of description, x should presumably be regarded as discrete, so that the connections really involve graph topology, but, in the continuum blurring which we have in mind, this connectivity is, instead, represented in a more statistical fashion by the kernels $K_\sigma(x,y)$. Generalization to a spectrum of time constants likewise represents a sophistication which is superfluous at the present stage.

3. EQUILIBRIUM PROTOTYPE

Our principal interest in this first phase of the analysis will be in the equilibrium, time-independent, concentration "profiles" that result from (2.6). Let us defer any

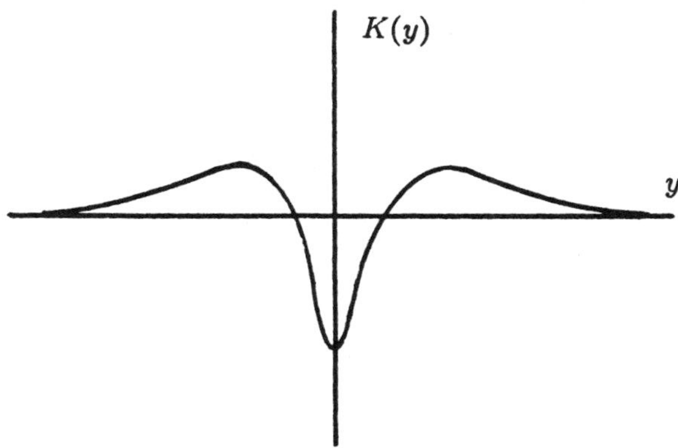

FIGURE 1 Interaction strength vs. distance in species space.

detailed study of stability until later. Then, setting $\dot{\rho} = \dot{R} = 0$ and eliminating R, (2.6) reduces to

$$\frac{\alpha(x)}{\rho(x)} - \mu(x) + \int K(x,z)\rho(z)dz + \overline{a}(x) = 0 \tag{3.1}$$

where

$$K(x,z) = K_1(x,z) + K_2(x,z),$$

and

$$\overline{a}(x) = \int (\Lambda_1(x,z) + \Lambda_2(x,z))\, a(z)dz.$$

To get off the ground, suppose there exists a metric in which, in the absence of external antigen, the system becomes translation-invariant in D-dimensional Euclidean $\{x\}$ space: only differences of variables occur in coefficients. Then, (3.1) becomes simply

$$\frac{\alpha}{\rho(x)} - \mu + \int K(x-z)\rho(z)dz = 0. \tag{3.2}$$

In a rough way, we might expect nearby paratopes to be inhibitory,[7] those further away stimulatory, and much further away neutral, as shown in the one-dimensional caricature of Figure 1. But the states of this system need not be translation-invariant. To find them, we first use the moment expansion trick introduced by Van der Waals[8] for an analogous situation. If the concentration profile varies slowly on the scale of K, we expand

$$\int K(z)\rho(x-z)dz = \int K(z)\left[\rho(x) - z.\nabla\rho(x) + \frac{1}{2}zz : \nabla\nabla\rho(x)\ldots\right]dz. \tag{3.3}$$

If $K(z)$ is taken to be isotropic, an extension of the Hoffmann[9] symmetric connection assumption, this then becomes

$$\int K(z)\rho(x-z)dz = b\rho(x) + c\,\nabla^2\,\rho(x)\ldots \tag{3.4}$$

where

$$b = \int K(z)dz,\; c = \frac{1}{2D}\int z^2 K(z)dz.$$

b and c are different measures of the connectivity of the kernel K. Let us further simplify to the $D = 1$ case, the "toy" model previously referred to, so that

$$\alpha/\rho - \mu + b\rho + c\rho'' = 0. \tag{3.5}$$

Figure 1 clearly suggests that $c > 0$, but makes no firm statement about the sign of b. Multiplying (3.5) by ρ' and integrating, the pseudomechanical $F = ma$ problem has the energy integral

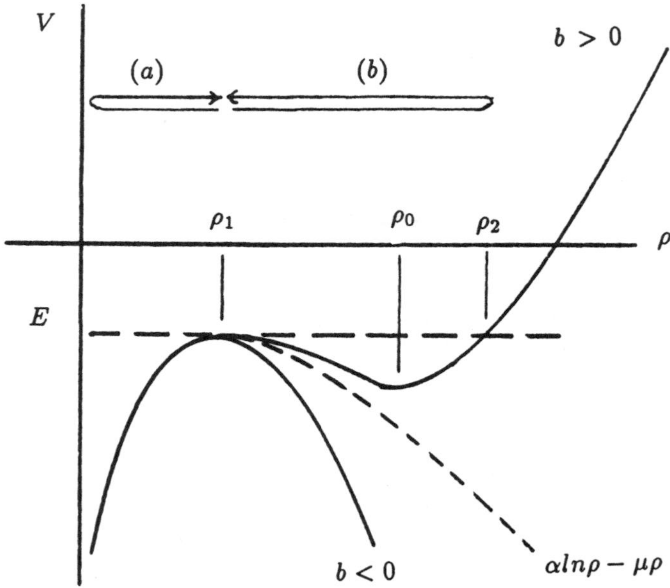

FIGURE 2 Effective potential vs. concentration

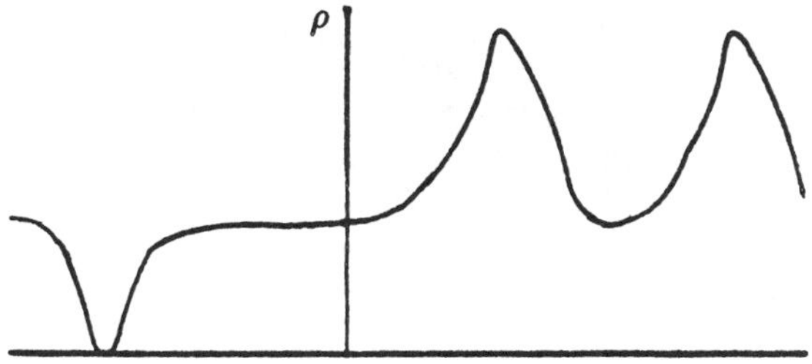

FIGURE 3 Concentration profile of mature model network.

$$\frac{1}{2}c\rho'^2 + V(\rho) = E \tag{3.6}$$

where

$$V(\rho) = \alpha ln\rho - \mu\rho + \frac{1}{2}\rho^2.$$

The "potential" curve $V(\rho)$ shown in Figure 2 allows us to quickly analyze the behavior of our system; note that according to (3.5), $\rho = 0$ acts as a hard wall. Looking ahead, the stationary point ρ_0 for $b > 0$ will never be stable with respect to the dynamics, but ρ_1 may be. There are two possibilities. If ρ_1 is stable, $\rho(x) = \rho_1$ presumably represents the uniformly occupied virgin state of the network. If ρ_1 is unstable, the concentration profile will not be uniform. If $b < 0$, one can, instead, generate a convergent profile (a), where E is tangent to $V(\rho)$, which spends an infinite amount of "time" at $\rho = \rho_1$, dropping to zero in a localized region of $\{x\}$ space. This is a "hole" in the immune repertoire.

If ρ_1 is unstable and $b > 0$, there are two possible trajectories. Again, (a) constitutes a hole, a region of immune tolerance, while (b) generates a spike, a region of immune competence. Without looking explicitly at the dynamics, one can predict with some confidence how these regions are attained on the introduction of an antigenic source $\overline{a}(x)$ in (3.1). Consider first a virgin state $\rho = \rho_1(0)$ which is being driven unstable; if a small dose $\overline{a}(x)$ is now introduced, it is clear from (3.1) and (3.4) that $\delta\rho'' = -\overline{a}/c$, pushing ρ down to curve (a) as instability sets in, a clonal deletion mechanism.[10] If a spike of type (b) has already been established when antigen is introduced, the consequences depend upon both amplitude and spread of $\overline{a}(x)$. For a large spread, $\overline{a}$ is locally a constant, and, since the combination $-\mu + \overline{a}$ appears in (3.1), trajectory (b) will be retained after removal of $\overline{a}$, if $\overline{a}$ is small enough that the maximum ρ_1 continues to exist during its presence. A larger dose, however, eradicates ρ_1, allowing the trajectory to descend to $\rho = 0$, which can be retained as a hole when the high dose is removed. Thus, high-dose tolerance occurs, but similar low-dose behaviour is not obvious, perhaps related to the paucity of observations of the latter.[11]

If the above is to be taken literally, a mature immune network consists of an aggregation of holes and spikes in a $\rho(x) = \rho_1$ background (Figure 3), the former immune tolerant region established by low-dose exposure during excitation form the virgin state, high doses thereafter, or perhaps non-uniformity of $\alpha(x)$ and $\mu(x)$. The latter are presumably immunologically competent, the result of moderate dose, but depend upon the dynamics which we have not yet studied. What we have done, then, in a preliminary way is to identify the polydisperse subsets—the holes and spikes—which are the "macroscopic" objects of the system.

4. STABILITY OF PROTOTYPE

If we neglect time delay, so that $\tau = 0$ in (2.6), elimination of R in the presence of antigen yields the dynamics

$$\dot{\rho}(x,t) = \alpha - \mu\rho(x,t) + \overline{a}(x)\rho(x,t) + \rho(x,t)\int K(x,y)\rho(y,t)dy. \tag{4.1}$$

We can attempt to assess the general course of the dynamics by setting up the entropy (negative Lyapounov[12]) functional

$$S[\rho,t] = \int \left[\alpha ln\rho(x,t) + \left(\overline{a}(x) - \mu\right)\rho(x,t)\right]dx + \frac{1}{2}\int\int \rho(x,t)K(x,y)\rho(y,t)dxdy. \tag{4.2}$$

Hence, *assuming a symmetric connection,*[9] $K(x,y) = K(y,x)$, we have

$$(d/dt)S[\rho,t] = \int \dot{\rho}(x,t)[\alpha\rho(x,t) + \overline{a}(x) - \mu]dx + \int \dot{\rho}(x,t)\int K(x,y)\rho(y,t)dydx,$$

or

$$\frac{dS[\rho,t]}{dt} = \int \dot{\rho}(x,t)^2/\rho(x,t)dx, \tag{4.3}$$

so that $S[\rho,t]$ is non-decreasing in time. If $S[\rho,t]$ is bounded from above, requiring that K be a negative definite kernel, $S[\rho,t]$ must approach a limit as $t \to \infty$, indeed specified by $\dot{\rho}(x,t) = 0$. For example, (3.4) will satisfy this condition if $b < 0$.

The limiting $\overline{\rho}(x)$ may, however, be uninteresting, i.e., only an antigen-modified virgin state. To see this, let us go at once to the dynamics corresponding to an antigen-free toy model (3.5) in which we control the system by putting it into a box $x \in [0, L]$, but mimic homogeneous space by setting Neumann boundary conditions. It is no harder to allow for an arbitrary source term, and so

$$\dot{\rho} = f(\rho) + c\rho\rho'' \qquad (4.4)$$
$$\rho'(0) = \rho'(L) = 0.$$

An equilibrium state must satisfy

$$f(\overline{\rho}) + c\overline{\rho}\,\overline{\rho}'' = 0 \qquad (4.5)$$
$$\overline{\rho}'(0) = \overline{\rho}'(L) = 0;$$

we are concerned with the stability of such a state. To test stability, we make the standard perturbation

$$\rho(x,t) = \overline{\rho}(x) + \delta\rho(x)e^{\lambda t} \qquad (4.6)$$

and ask whether $Re(\lambda) > 0$ can occur. Inserting (4.6) into (4.4), using (4.5), we have to analyze

$$\lambda\delta\rho = \left(f'(\overline{\rho}) + c\overline{\rho}'' + c\overline{\rho}d^2/dx^2\right)\delta\rho \qquad (4.7)$$
$$\delta\rho'(0) = \delta\rho'(L) = 0.$$

The maximal eigenvalue $\lambda_{\max}$ of (4.7) must be real and have the variational characterization

$$\lambda_{\max} = \mathrm{Max}_{\psi} N^{-1}\left[\int_0^L (1/\overline{\rho})(f'(\overline{\rho}) + c\overline{\rho}'')\psi^2 dx - c\int_0^L \psi'^2 dx\right] \qquad (4.8)$$

where

$$N = \int_0^L (1/\overline{\rho})\psi^2 dx,$$

the boundary conditions being an automatic consequence of boundary value variations. But suppose that $\overline{\rho}$ is not uniform. Choosing

$$\tilde{\psi} = \overline{\rho}', \qquad (4.9)$$

differentiation of (4.5) tells us that

$$\left(f'(\overline{\rho}) + c\overline{\rho}''\right)\tilde{\psi} + c\overline{\rho}\tilde{\psi}'' = 0 \qquad (4.10)$$
$$\tilde{\psi}(0) = \tilde{\psi}(L) = 0$$

Substituting into (4.8), we see that $\lambda_{\max} \geq 0$, but also that $\lambda_{\max} = 0$ only if $\tilde{\psi}'(0) = 0 = \tilde{\psi}'(L)$ as well. Since this is not generally the case, we conclude that $\lambda_{\max} > 0$: any non-uniform state is unstable.[13]

In order to invoke the phenomenology of Section 3, we must, therefore, either go beyond the prototype (3.5) or allow for a delay $\tau > 0$. Consider the latter option.

Then (2.6), antigen -free and in moment expansion approximation, becomes in the parameter region of interest

$$\begin{aligned} \dot{\rho} &= \alpha - \mu\rho + \rho(b_1\rho + c_1\rho'' + R) \\ \tau\dot{R} + R &= b_2\rho + c_2\rho'' \\ b_1 + b_2 &= b > 0, \qquad c_1 + c_2 = c > 0. \end{aligned} \tag{4.11}$$

It is very easy—and significant—to assess the uniform equilibrium solutions of (4.1). They are, of course, given by $0 = \alpha - \mu\overline{\rho} + \overline{\rho}(b_1\overline{\rho} + b_2\overline{\rho})$, i.e., $\overline{\rho} = \rho_1$ or ρ_0 and $\overline{R} = b_2\overline{\rho}$, where

$$\begin{aligned} \rho_1 &= \frac{1}{2b}\left[\mu - (\mu^2 - 4\alpha b)^{1/2}\right] \\ \rho_0 &= \frac{1}{2b}\left[\mu + (\mu^2 - 4\alpha b)^{1/2}\right]. \end{aligned} \tag{4.12}$$

Applying a perturbation $\binom{\rho}{R} = \binom{\overline{\rho}}{\overline{R}} + e^{\lambda t}e^{ikx}\binom{\delta\rho}{\delta R}$, (4.11) now reads

$$\lambda\begin{pmatrix}\delta\rho \\ \delta R\end{pmatrix} = \begin{pmatrix} -\mu + \overline{\rho}(2b_1 + b_2 - c_1k^2) & \overline{\rho} \\ (b_2 - c_2k^2)/\tau & -1/\tau \end{pmatrix}\begin{pmatrix}\delta\rho \\ \delta R\end{pmatrix}, \tag{4.13}$$

so that

$$\tau\lambda^2 + \left[1 + \tau(\mu - (2b_1 + b_2 - c_1k^2)\overline{\rho}\right] + \mu - (2b - ck^2)\overline{\rho} = 0 \tag{4.14}$$

Suppose that $\tau = 0$. Then, according to (4.14),

$$\lambda(\rho_1) = -(\mu^2 - 4\alpha b)^{1/2} - ck^2\rho_1. \tag{4.15}$$

Clearly, the uniform $\rho(x) = \rho_1$ is a stable attractor and that is why the associated non-uniform solutions are unstable. Thus, a minimum requirement for a stable non-uniform solution is that $\rho(x) = \rho_1$ be unstable. Now, let $\tau > 0$. Since $-(2b - ck^2)\rho_1 > 0$, the condition $\lambda > 0$ is equivalent to $1 + \tau\left(\mu - (2b_1 + b_2 - c_1k^2)\rho_1\right) < 0$ or

$$b_2 + c_1k^2 < \frac{-2b\left[\tau^{-1} + (\mu^2 - 4\alpha b)^{1/2}\right]}{\left[\mu - (\mu^2 - 4\alpha b)^{1/2}\right]}. \tag{4.16}$$

For example, a band of k's will be excited if $c_1 > 0$, but b_2 is sufficiently negative: large delayed short-range growth inhibition. Furthermore, if maturation of the immune system is accompanied by increase of the intrinsic rate from 0, the required instability threshold for b_2 rises from $-\infty$ to $-2b/\tau\mu$, and it is reasonable to imagine that it is crossed during this period. To be sure, it is not clear that (4.11), with saturation nowhere inserted, will converge dynamically to a non-uniform state, but saturation, e.g., of $b_1\rho$, can readily be appended and the constants c_1 and c_2 tailored for the purpose.

5. OTHER MODELS

It appears, then, that the roster of possibilities predicted by equilibrium analysis gives a very good indication of the possible repertoire of an immune network. Let us discuss briefly a couple of extensions of the translation-invariant toy model of Section 3, as well as a substantial modification. The objective, in a very general way, is to design systems of the form (3.1), with reasonable $K(x,y)$, that exhibit alternative non-uniform states, accessible by environmental control. Since the general equilibrium profile for symmetric K satisfies the variational principle

$$(\delta/\delta\rho(x))\left\{\int[\alpha(x)ln\rho(x)+(\overline{a}(x)-\mu(x))\rho(x)]dx + \frac{1}{2}\int\int\rho(x)K(x,y)\rho(y)dxdy\right\}=0, \tag{5.1}$$

any qualitative idea of the structure of the profile can readily be converted to quantitative form.

From the point of view of the estimated complexity of the immune network, a reasonable $K(x,y)$ might be one with a substantial random character. As a first attempt at constructing such an object, consider a single Fourier coefficient in (3.2),

$$\alpha/\rho(x)-\mu+A\int cos(x-z)\rho(z)dz=0, \tag{5.2}$$

on an interval of length L (an integer multiple of 2π) in one-dimensional space. Assuming that $\rho(x)$ is even, then $\int cos(x-z)\rho(z)dz = \int(cos\,x\,cos\,z + sin\,x\,sin\,z)\rho(z)dz = \rho_+\,cos\,x$, where

$$\rho_+=\int cos\,z\,\rho(z)dz \tag{5.3}$$

so that, from (5.2),

$$\rho(x)=\alpha/(\mu-A\rho_+\,cos\,x). \tag{5.4}$$

Inserting (5.4) into (5.3), we have the consistency relation for ρ_+

$$\rho_+=(\alpha L/A\rho_+)\left[-1+\mu(\mu^2-A^2\rho_+^2)^{-1/2}\right] \tag{5.5}$$

which, in the form of a cubic equation for ρ_+^2, tells us that

$$|\rho_+|=0 \qquad \text{for } A>2\mu^2/L\alpha, \tag{5.6}$$

and

$$\text{either } |\rho_+|=0 \quad \text{or} \quad \overline{\rho}\neq 0 \quad \text{for } A<2\mu^2/L\alpha,$$

again, a high-zone tolerance effect.

A generalization in a different direction is to high dimension D. If $v = D - 1$, and we assume radial symmetry for the concentration profile, (3.4) now yields

$$\alpha/\rho - \mu + b\rho + c\rho'' + (\nu c/x)\rho' = 0. \tag{5.7}$$

Under the scaling $x \to \nu^{1/2}x$, we have

$$\alpha/\rho - \mu + b\rho + (c/x)\rho' + (c/\nu)\rho'' = 0,$$

so that, if the limit $\nu \to \infty$ is taken, the equation becomes

$$\alpha/\rho - \mu + b\rho + (c/x)\rho' = 0. \tag{5.8}$$

The solution of (5.8) is immediate. It takes the form

$$\frac{|\ \rho - \rho_0\ |^{\rho_0}}{|\ \rho - \rho_1\ |^{\rho_1}} = K\, e^{-(b/2c)(\rho_0 - \rho_1)x^2} \tag{5.9}$$

where $\rho_0 > \rho_1$ are in the notation of (4.12). Now, the asymptotic environment $|\ x\ | \to \infty$ produces the uniform value $\rho = \rho_0$, leading always to two types of trajectories, one with localized depleted $\rho \sim \rho_1$, and one with localized increased ρ of any size.

The choice of a translation-invariant connection kernel is not unreasonable in default of any explicit information. But there is more than anecdotal information available on the nature of the molecular species which bind effectively: they have physical structures which are complementary to allow a close fit. What this means is that, to each species' point x, we can associate the species x^c so that the binding strength of x to y goes as the closeness of x to y^c. The occurrence of autobodies,[14] antibodies which bind themselves, shows that one can have species which closely satisfy $x^c = x$. Taking the presumed structure into account, it makes sense to express the growth rate coefficient as $\int K(x, y^c)\rho(y,t)dy$. In fact, defining $\rho^c(y,t) = \rho(y^c,t)$, and supposing that the measure is invariant under complementation, the equilibrium equations can be written in dual form.

$$\begin{aligned} \lambda/\rho - \mu + \int K\rho^c &= 0 \\ \lambda\rho^c - \mu + \int K\rho &= 0. \end{aligned} \tag{5.10}$$

At this stage, the translation-invariance assumption on K can again be made, but now, as shown in Figure 4, small $x - y$ is stimulatory, large $x - y$ repressive, reversing the curve of Figure 1. Carrying out the same moment expansion and inserting the $\tau = 0$ dynamics at no extra cost, we then have

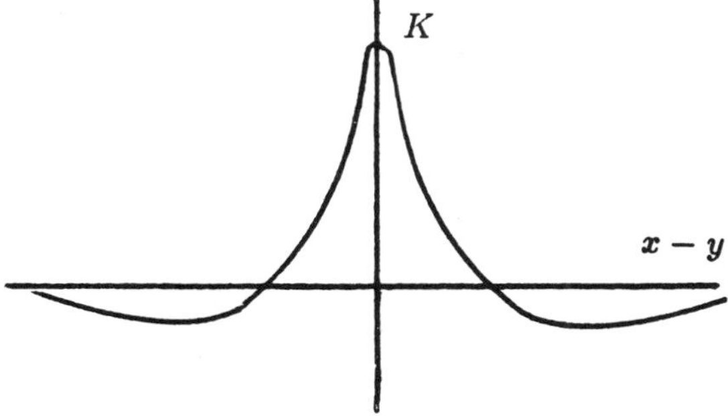

FIGURE 4 Interaction strength for reduced Segel-Perelson model.

$$\begin{aligned}\dot{\rho}(x,t) &= \lambda - \mu\rho(x,t) - \rho(x,t)(b\rho^c(x,t) + (c/2D)\,\nabla^2\,\rho^c(x,t))\\ \dot{\rho}^c(x,t) &= \lambda - \mu\rho^c(x,t) - \rho^c(x,t)(b\rho(x,t) + (c/2D)\,\nabla^2\,\rho(x,t)).\end{aligned} \tag{5.11}$$

This may be completed by using the simplest model for the complement operation: $x^c = -x$,[15] which satisfies the required $x^{cc} = x$. Equation (5.11), as a second-order system in time, can indeed give rise to a stable non-uniform solution, but also, as a second-order system, is a good deal harder to analyze.[15]

CONCLUSIONS

We have examined models of the connections among grouped components or functional units of a spatially homogeneous immunological system. These models give rise to an interconnected pattern of growth rates for the components which was expanded to first order in the component concentrations. Even the most primitive realizations manifest characteristic properties of immune competence and tolerance, leading us to conclude that a much-more-detailed spectrum of repertoires may be available at relatively low cost as we refine our description of the connectivity and of the internal dynamics of the functional units.

ACKNOWLEDGMENTS

This paper is supported in part by DOE contract DE-AC02-7603077.

REFERENCES

1. See, e.g., Gualtieri, J. A., J. M. Kincaid, and G. Morrison (1982), "Phase Equilibria in Polydisperse Fluids," *J. Chem. Phys.* **77**, 521.
2. Alberty, A., and I. Oppenheim (1984), "A Continuous Thermodynamic Approach to Chemical Equilibrium Within an Isomer Group," *J. Chem. Phys.* **81**, 4603.
3. For an overview, see Roitt, I. (1984), *Essential Immunology* (Blackwell). For much more detailed mechanisms, see, e.g., Marchuk, G. I., and R. V. Petrov (1983), "The Mathematical Model of the Anti-Viral Immune Response," *Mathematical Modeling in Immunity and Medicine*, Eds. Marchuk and Belykh (North-Holland), 161.
   Prikrylova, D. (1985), in *Immunology and Epidemiology*, Eds. Hoffmann and Hraba (Springer).
4. Richter, P. H. (1978), "The Network Idea dn the Immune Response," *Theoretical Immunology*, Eds. Bell, Perelson, and Pimbley (Dekker), 539.
5. See, e.g., Marchuk, G. I. (1983), *Mathematical Models in Immunology* (Springer).
6. Jerne, N. K. (1974), "Towards a Network Theory of the Immune System," *Ann. Immuno. (Inst. Pasteur)* **125C**, 373.
7. See, e.g., Kennedy, R. C., J. L. Melnick, and G. R. Dressman (1986), "Anti-Idiotypes and Immunity," *Scientific American*, July, 1986, 48.
8. Van der Waals, J. D. (1894), "The Thermodynamic Theory of Capillarity," *Z. Phy. Chem.* **13**, 657.
9. Hoffmann, G. W., and A. Cooper-Willis (1983), "Symmetry, Complexity and Stability in Immune System Network Theory," *Mathematical Modeling in Immunity and Medicine*, Eds. Marchuk and Belykh (North-Holland), 31.
10. See, e.g., Guillet et al. (1987), "Immunological Self, Nonself Discrimination," *Science* **235**, 865.
11. Dintzis, H., personal communication.
12. See, e.g., Malkin, I. G. (1952), "Theory of Stability of Motion," *U.S. Atomic Energy Commission* **AEC-tr-3352**.
13. See also Belintsev, B. N., M. A. Livshits, M. V. Vol'kenshtein (1979), "Investigation of the Stability of Spatially Inhomogeneous States of Distributed Systems," *Biophysics* **23**, 1073.
14. Kang, C.-Y. and H. Kohler (1986), "A Novel Chimerio Antibody with Circular Network Characteristics: Autobody," *Annals N. Y. Acad. Sci.* **114**, 475.
15. Segel, L. A., and A. S. Perelson, this volume.

FRANCISCO J. VARELA,[*] ANTONIO COUTINHO,[†] BRUNO DUPIRE,[*] and NELSON N. VAZ[‡]
[*]CREA, Ecole Polytechnique, 1 rue Descartes, Paris 75005; [†]Institut Pasteur, 23 rue du Docteur Roux, Paris 75015; and [‡]Department of Immunology, University of Minas Gerais, Belo Horizonte, Brasil

Cognitive Networks: Immune, Neural, and Otherwise

1. COGNITIVE NETWORKS: THE CONTEXT

1.1 GOALS OF THIS PAPER

There is a strong intuitive sense in which immune systems are *cognitive*: they recognize molecular shapes, remember the history of encounters of an individual organism, define the boundaries of a molecular "self," and make inferences about molecular species likely to be encountered. By and large immunology has left these admittedly cognitive terms undefined or at a metaphorical level and has concentrated, instead, on the molecular details of immune components.

The first intention of this paper is to argue that by so doing one is leaving unexamined and shrouded in a fog of mystery the most interesting domain of phenomena the immune system affords to animals: their cognitive abilities. To advance in this direction, one must be willing to embark on an explicit examination of the cognitive *mechanisms* proper to the immune system.

Theoretical Immunology, Part Two, SFI Studies in the Sciences of Complexity
Ed. A. S. Perelson, Addison-Wesley Publishing Company, 1988

This task is timely, for in cognitive science[1] there has been an explosive realization that biological networks afford the model *par excellence* to account with great simplicity for uniquely cognitive capacities such as fast discrimination and memory. By and large, such cognitive networks are taken as analogs of *neural* networks. A second intention of this paper is to further clarify for cognitive scientists how immune systems can provide a distinctly different class of network architecture also capable of sophisticated cognitive performances. Immune nets merit to be examined on their own as a separate large class next to neural networks.

In the remainder of this section, we provide some minimal context for the recent development in cognitive networks which allows us to phrase immune events in a sharper focus. In Section 2, the substance of the paper, we propose what immune networks are, including their most relevant dynamics in the form of a model. Finally, we derive some conclusions in Section 3.

1.2 FROM SYMBOLS TO NETWORKS

The study of biological and artificial cognitive mechanisms was heavily marked by the tradition that considered any form of knowledge as necessarily linked to symbols and rules, in the tradition of logic. This gave rise to the *symbolic*[2] paradigm, where cognition is identified with information processing: rule-based manipulation of symbols. After many years of work with this symbolic paradigm, especially in neuroscience and artificial intelligence, it has become clear that such mechanisms are far too brittle, too inflexible to approach living expertise.

The alternative view revived recently from initial ideas dating back to the 1950's, is usually referred to as *connectionism*. Basically, the idea is to leave symbols aside and to start any analysis (or construction) from simple computing elements, each one carrying some value of activation which is calculated on the basis of the other elements in the network through a dynamical rule. In a typical connectionist model, the weight of the influences between the elements varies. Thus, the network's performance is embodied in a distributed form over the connections, whence the designation.[3] We will argue here that immune networks share with connectionist ideas the distributed dynamical base, but that their mode of change is fundamentally different.

A key idea in a network perspective is that the on-going activity of units, together with constraints from the system's surroundings, constantly produces *emerging* global patterns over the entire network which constitutes its performance. The network itself decides how to tune its component elements in mutual relationships that gives the entire system a capacity (recognition, memory, etc), which is not

[1] We use this term to designate the federation of neuroscience, artificial intelligence, linguistics, and cognitive psychology, which since the later 70's is associated with the scientific study of cognitive processes in animals and machines.

[2] Depending on preferences, the symbolic paradism is also called the cognitivist or computational paradigm. For an introductory discussion, see H. Gardner, 1984.

[3] For a full discussion of connectionism, see Rummelhart and McClelland, 1987.

available to the components in isolation. These emergent properties are the great attractive feature of the network approach, and one that needs to be explored more explicitly for immune networks.

For our purposes here, it is important to remark that, although the connectionist (dynamical network) approach is usually seen as close to the brain and neural networks, this is not necessarily the best understanding of them. Most of the time, the relation with detailed neural systems is suggestive rather than precise. In contrast, what is important is that they are not symbolic machines in the traditional sense, and they operate on the basis of many components with local interactions. This is, of course, also the case for immune networks. It is at this level that immune networks should be studied, offering insights into the way biological networks can achieve cognitive capacities with enormous adaptability, a performance unknown to neural networks.

That is the direction of analysis we will follow here. We now turn to consider in detail the composition and dynamics of immune networks.

2. IMMUNE NETWORKS

2.1 IMMUNE EVENTS AND SHAPE SPACE

The immune system's "knowledge" (whatever that might mean in more precise terms), is about molecular shapes and profiles. Since it has access to nearly every corner in a vertebrate organism, it encounters nearly all available molecular species of the individual. These are either (a) produced by the organism itself, or (b) have penetrated the organism normally through mucosal surfaces, by the food eaten or the air breathed.

The stuff the Immune System (IS) is made of is heterogeneous, in terms of cells and molecules. All structures in the body interact with immunocompetent cells. Many of these are produced by other cells, and lymphocytes themselves produce all sort of molecules regulating gene expression and the activity of other non-immune cells. However, out of this whole array of components, we shall sharply distinguish lymphocytes on the basis of their diversity of variable regions (V-regions). We, thus, take the structural and functional basis of these immune abilities as derived from: (a) the vast diversity of lymphocyte receptors and free antibodies, and (b) the nature of their interactions with their ligands. The scale of this diversity should not be underestimated. In a mouse, currently one estimates 10^4–10^5 the number of protein-encoding genes, while the size of the potential immune repertoire is set at about 10^9–10^{10}. In what follows, we let f_i and b_i denote the concentration of free antibodies and cell-bound antibodies (respectively) of the i-th kind, with $i = 1, \ldots, N \approx 10^9$.

A consequence of the above choice is that it is appropriate to consider all immune events as occurring in a *shape space*, S: a low-dimensional metric space, where each axis stands for a physico-chemical measure characterizing a molecular shape. Typical examples of such parameters would be: electric charge, hydrophobicity, amino acid sequence, and so on (Perelson and Oster, 1979). We will assume that some 5–7 of such measurements suffice to characterize a molecular configuration (belonging to the IS or not) as a point $s \in S$. In particular, in our previous notation, the subscript $i = 1, \ldots, N$ denotes a location within S. Every pair (i, j) of molecular components will relate to each other within this shape space through the affinity m_{ij} of their interactions, which reflects the probability of remaining associated.

2.2 DOMAINS OF INFLUENCE

The weak, non-covalent forces that define immune interactions allow for reversibility and some degeneracy in the detail of reactive surfaces. It follows from all these combined factors that the IS is characterized by an enormous degeneracy and redundancy. Thus, for example, it is not surprising that the requirement for sufficient diversity and repertoire completeness is met by a tadpole with 10^4 or so clonal components as much as by a mouse IS containing 10^8–10^9. Obviously, however, the tadpole IS is less precise on molecular details, since its completeness owes more to degeneracy than to diversity.

This tells, then, that each IS component located at i will have within shape space a distributed *domain of influence*, defined as the distribution of affinity over S. For the finite set of locations occupied by the components of the IS, $j = 1, \ldots, N$, we write the value of the domain of influence of the i-th species as a row of the interaction matrix m_{ij}. Experimentally the values of this matrix are the measured affinity between a given antibody and any other component (free or bound). It is on the basis of their mutual domains of influence that two components will interact to some degree or not. Immune crossreactions keep two species together, neutralized from other interactions for a given period of time. The number of complexed $(i - j)$ pairs at any point in time is, thus, related to the product of their affinity and their molar concentrations; that is a quadratic term of the form: $m_{ij} f_i f_j$.

2.3 DYNAMICS

Clearly, free molecules cannot do much: their action requires some form of binding. Thus, the relevant events in the IS are not the molecules by themselves, but rather their interactions. This is significant, since by centering on interactions of immune components, one is *ipso facto* considering the IS *dynamics*: the active range of changes of immune components as a result of their mutual constraints and reciprocal actions.

It is well known, of course, that immune components are not static. In extreme conditions, a given antibody can vary up to a million fold in concentration. We will propose some specific dynamical rules for the IS in what follows, but before embarking on the analysis of details of immune dynamics, it is important to examine the overall capacities emerging from such an IS network.

2.4 COGNITIVE CAPACITIES

In our view, the IS *asserts a molecular self* during ontogeny, and for the entire lifetime of the individual, it keeps a memory of what this molecular self is. Fundamentally, the IS is an *identity mechanism* in shape space, much as the nervous system is an identity mechanism in the physical three-dimensional space. The IS dynamics is the mechanism that makes the establishment of such a molecular identity possible.

It is as a result of this assertive molecular identity that an individual who had measles in childhood is different from what he would have been had he not been in contact with the virus, or how an IS changes if the person switched from an omnivorous to a vegetarian diet. The IS keeps track of all this history, while defining and maintaining a sensorial-like interface at the molecular level. It must be stressed that the self is in no way a well-defined (neither pre-defined) repertoire, a list of authorized molecules, but rather a set of viable states, of mutually compatible groupings, of dynamical patterns. In effect, a molecule is neither self nor anti-self, as a musical note does not belong more to a composer than to another one. The self is not just a static border in the shape space, delineating friend from foe. Moreover, the self is not a genetic constant. It bears the genetic make-up of the individual and of its past history, while shaping itself along an unforeseen path.

2.5 METADYNAMICS

A unique quality of the IS identity is its adaptability. In fact, any possible new element in shape space, even if newly synthesized (and, hence, with no evolutionary history), can interact with a functioning IS. This quality of "completeness" (Coutinho, 1980) or better: open-endedness (Jerne, 1985) is not the result of learning, but of an intrinsic feature of the way the molecular identity is established while at the same time allowed to change. We refer to this as the *metadynamics* of the IS: the continuous production of new variables, of novel molecular alternatives, most of which never enter into the dynamics of the IS itself, but very rapidly decay and disappear.

The IS metadynamics is based on the unique non-conservatism of V-regions, genetically specified through 4 to 6 distinct gene segments, drawn from a large polygenic pool. Furthermore, the mechanisms leading to the production of mature genes by the rearrangement of those gene segments are error-prone and lack precision in the sites of recombination, giving rise to yet new combinations not to be found in germ-lines. These novelties are further amplified by the activity of an enzyme that randomly adds nucleotides at the sites of recombination, without copying them from a template (Tonegawa, 1983). Hence, it is biologically evident that such mechanisms have been introduced in this system through evolution, but as a result of their presence, the V-regions generated in the organism are not evolutionary selected.

This enormous reservoir of possibilities is the key to the open-ended quality for novelty, both resulting from the system itself, and from unrelated environmental changes.

2.6 IMMUNE DYNAMICS AS NETWORK

The dynamics of the IS was clearly brought to the fore by Jerne's network hypothesis, based on the diversity and degeneracy of variable V-regions (Jerne, 1974). The original postulate is fundamentally correct: normal individuals *do* contain a set of lymphocyte receptors and free antibodies which mutually interact with varying frequencies (Holmberg et al., 1984; Kearny and Meenal, 1986). In this paper, we interpret the admittedly fragmentary evidence, as implying that V-region connectivity is a key property from which the IS asserts a molecular identity, and not a mere epiphenomenon of diversity/degeneracy (see for discussion Urbain, 1986).

It is important, however, to stress the participation in the network of both free and bound receptors. Although the free V-region sub-network is by far dominant in quantity, the sub-network of bound components acts as regulator of the entire network. This is discussed in further detail below, but constitutes a key to the IS's capacity for stability and regulation.

At any given time, the network of free and bound components defines a surface of potential coupling in shape space, which is simply the superposition of the domains of influence of all participating components. Conversely, every point s in shape space will be accessible to the IS to the extent that there are components with some degree of affinity at that point and in a non-negligible concentration. Formally, this can be expressed as

$$\sigma_s = \sum_j m_{sj} f_j. \qquad s \in S,\ j = 1, \ldots, N \tag{1}$$

We shall refer to the coupling surface defined in (1) as the IS's *cognitive domain*, since it represents the limits (or boundaries in S) of the molecular self. For each s, σ_s, also named ***sensivity*** at s, may be viewed as the degree of engagement of s in the network, for it corresponds to the probability of meeting any other species.

2.7 THE COGNITIVE DOMAIN OF THE IS

Although Jerne's idiotypic network is well know among immunologists, it is less often discussed how immune connectivity extends beyond the "web of variable domains" to include all other molecules in the organism which bind to V-regions, that is, that can be coupled with the IS's cognitive domain.

Needless to say, immune components do not individually "know" about anything; they simply bind or do not bind to molecular surfaces. But in keeping with the style of analysis of networks in general, it is the entire ensemble of components which endows the system with a cognitive capacity which is not located anywhere in particular, but embodied in the entire system. More precisely, since molecular complementarities create classes (if A binds to B, and A binds to C, then B and C are equivalent under A), the immune system will necessarily carry molecular complementarities or mimicries to all molecules which the cognitive domain of the IS is open to (can couple with). This on-going internal set of images coincides with what the specific constituents of the molecular self are.

Clearly, one can define "self" from a biochemical or genetic or even *a priori* basis. But from our vantage point, the only valid sense of *immunological self* is the one defined by the dynamics of the network itself. What does not enter into its cognitive domain is ignored (i.e., it is non-sense). This is in clear contrast to the traditional notion that IS sets a boundary between self in contradistinction to a supposed non-self. From our perspective, there is only self and its slight variations. That which is foreign is only so because it is similar to (or only slightly different from) self: the *Unheimlich* of that considered as foreign can only come from this proximity (as Freud pointed out long ago).

This foreignness-as-almost-self is dealt with by the system's metadynamics: the continuous production of novel V-regions with random diverse shapes, which, more often than not, makes little sense for the existing self as they do not interact significantly with existing molecular shapes in the network. For the system to become responsive to a "new" molecular species, it is necessary that what at the beginning are rather isolated new V-regions become selected and incorporated into the network itself by expansion or reinforcement of the cognitive domain into that region. This might never happen; often secondary contacts with non-self molecules evoke "immune responses" which are indistinguishable from "primary responses": no memory was generated in the first contact because the network was not engaged, and, thus, there is nothing to recognize.

It is very likely that many situations will be less clear cut than the above discussion suggests. Some aspects of antigens will fall in varying degrees into the cognitive domain of the network—otherwise, they could not be said to be antigens! Other species may certainly evoke a transient response of disconnected V-regions. Multideterminant antigens, therefore, provide missing connections and generate novel interactions of previously isolated V-regions within the network. Alternatively, the concentration of the novel V-regions hyper-expressed in the immune response will reach significant levels as a self molecule and, consequently, integrate the immune network. In either case, the IS will undergo adaptive compensation through its dynamics and changes in its cognitive domain through metadynamical innovations. Thus, at any point in the life of an organism, the self of an individual is unique and covers a limited domain of shape space.

2.8 LYMPHOCYTE PRODUCTION AND ACTIVATION

Whatever the variability in the production of possible V-regions, it is clear that, ultimately, there is a single V-region in each lymphocyte: they are clonal. Thus, the upper limit for the size of an IS's repertoire is equal to the number of lymphocytes in the individual, and the production of new V-regions requires the production of new cells.

Hence, for the on-going network change, it is key to understand the manner in which a newly produced lymphocyte, advertising a V-region on its surface, can be selected and incorporated in the on-going IS dynamics.

A large number of lymphocytes are produced throughout life in the central lymphoid organs (bone marrow and thymus). But, while B lymphocyte production in bone marrow continues at high rates until death, the generation of T lymphocyte in thymus is considerably slowed in adulthood. Roughly 20% of the total number B cells in an organism is renewed daily from the bone marrow. The large number of these decay rapidly, with mean survival times of a few hours or days, so that the total lymphocyte volume remains roughly constant. Some lymphocytes which migrate to peripheral lymphoid organs are selected for persistence or long life-span, from a few weeks to some months. Up to 10% of the newly formed cells enter this pool. The turnover rates in this compartment are some tenfold lower than in the short-lived pool (Freitas et al., 1986)[4]

Thus, it is clear that there is an enormous production of novelty in newly produced lymphocytes which will never be used, and what is used depends on the configuration of the existing network.

In contrast with many other cells in the organism, lymphocytes are produced as resting cells with no effective actions (effector functions). B cells can produce and secrete antibodies; T cells can help or suppress B cells and even destroy other cells. Yet to do any of this, lymphocytes must be activated from their initial resting state. In our description of the IS, we deal only with activated lymphocytes, and thus, the transition from rest to activity is central to understanding immune phenomena.

In nearly all cases, lymphocyte activation results from V-region interaction with other ligands. These can vary in details; while B cells bind whole, free molecules to their surface receptors, T cells appear to interact with short linear fragments of molecules previously associated with a specific set of self molecules: the product of the MHC genes (Schwartz et al., 1985). There are far too many fragmentary details to cover here. Suffice it to say that, although T cells appear to be activated by some appropriate molecular profiles (Larsson et al., 1984), B lymphocytes fail to do so (Coutinho et al., 1984b). Interactions of B lymphocyte receptors with ligands only induce the resting cell to a stage of higher susceptibility to be activated by a T helper lymphocyte.

Activation of lymphocytes may result in mitosis, with both cells expressing the same V-region. Clearly, the proliferative potential of a lymphocyte is considerable: a single cell can produce thousands of descendents. But the contribution of cell division to clonal amplification in the normal operation of the IS is still unclear. At any rate, activation leads to the acquisition of effector functions: synthesizing antibodies at a rate of several thousand molecules per second (in contrast to one

[4]The population of T lymphocyte is somewhat different and less well studied. Many T cells are produced daily, but fail to be exported to the peripheral IS (about 1/30 of a total of about 30×10^6/day). A good many of these appear to persist in the periphery for long periods of time.

molecule/sec in resting cells). Thus, a unique V-region species on a resting lymphocyte represents some 10^5 molecules in the whole system, while an activated B cell in its 2–3 day life-cycle can produce up to 10^9 molecules of free antibody with the same V-region (Melchers et al, 1974; Jerne, 1984). Thus, network dynamics with regard to B cells is primarily one of activation-dependent changes in the rates of antibody production per cell, rather than cell multiplication. What then regulates B cell activation?

Activation always requires the participation of T cells. The two cooperating lymphocytes must engage in direct cell-to-cell contact for a period of time, before "non-specific" factors can come into play. This contact requires complementarity above a critical threshold, involving V-regions on both receptor types. There are at least three types of activation mechanisms one can envisage. In the first, B and T display complementary V-regions activating each other. It is plausible that this mechanism accounts for a good part of all immune activity in normal individuals (Tite et al., 1986; Bandeira et al., 1987). Such a mechanism is based on V-repertoires, and is independent of the rest of molecular self.

Two other mechanisms are known to exist, both involving V-regions and other self molecules. Resting B cells binding a molecule to surface antibody receptors can present it (or parts thereof) on their surface, and thus, become targets for selective interactions with competent T cells displaying the appropriate complementarities. This is the classical "antigen-bridge" model for T-B collaboration (for discussion, see Coutinho, 1984). There is little information about the frequency of this mechanism in immune dynamics, but believed to be common.

Finally, another type of B-T collaboration has been elucidated. Ligand binding to resting B cells (whether those ligands be somatic self molecules or free antibodies) induces the cell into a state of hyperactivity with regard to possible interactions with T helper cells, primarily because of the hyper-expression of surface molecules required for interaction with T lymphocytes—the MHC gene products (Monroe and Cambier, 1983). In the presence of T helper cells of self-related specificity, such hyperactive B lymphocytes engage in productive collaboration. This type of mechanism does not require that both cooperating cells display complementarities to the same molecule, and most importantly, it does not seem to require a strict receptor specificity from the cooperating T cell altogether. Since B cells cannot know which molecules bind to their receptor, this kind of mechanism strongly favors the activation of B cells that are very connected to the existing network. From recent experiments, we believe this mechanism plays a major role in the activation of normal individuals (Pereira et al., 1986b and to be published; Bandeira et al., 1987).

Thus, B cell activation also falls within the domain of influence of the immune network, and another loop is closed. Once activated, B cells produce large amounts of free antibodies with the same binding preferences as the receptors that had been selected for activation; it follows that these free components will bind and neutralize the very same molecules that led to their activation. Thus, while interactions between circulating antibodies and cell bound receptors are stimulatory, those between circulating antibodies in the immune network are inhibitory for the activation of the B lymphocyte which can produce them (independently of the nature of the activation mechanism summarized above). The clonality of V-region expression finds in these stabilizing dynamical loops its most clear relevance.

Although we have not analyzed the mechanism of T cell activation and T cell repertoires, we have assumed that a network of B cells and antibodies *does* require T cell participation. In the present version of the IS, we simply assume that appropriate T helper cells are available in normal individuals, as supported from recent evidence (Pereira et al., 1985, 1986b). Also, we will not explicitly discuss here the role of T suppression, but assume that similar mechanisms as those discussed for T helper cells will apply (Pereira et al., 1985). Figure 1 summarizes a view of IS dynamics in a schematic way.

2.9 EQUATIONS FOR IS

The foregoing ideas on the IS dynamics, stabilizing factors, and change, can be summarized in the following generic set of equations:

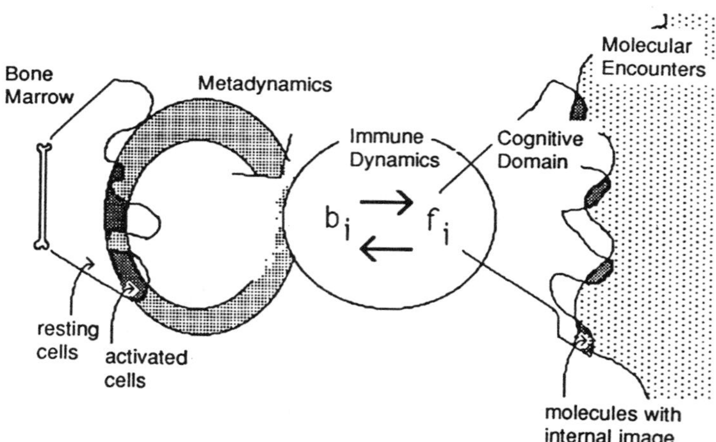

FIGURE 1 Schematic view of IS dynamics.

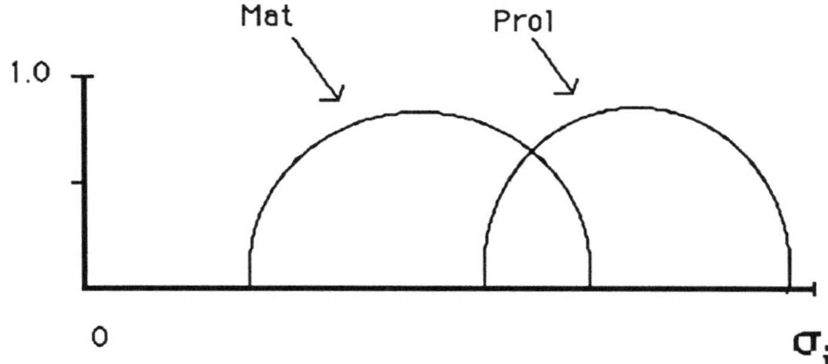

FIGURE 2 Distributed control of the (2a) dynamics under the influence of the antibodies species present, showing up as a proliferation function *Prol*, similar in shape to *Mat*.

$$\frac{df_i}{dt} = -\sigma_i f_i + Mat(\sigma_i) b_i \qquad (2a)$$

$$\frac{db_i}{dt} = -b_i + Prol(\sigma_i)\, b_i + Meta(i) \qquad (2b)$$

$$i \longleftarrow Meta(i), \qquad i = 1, \ldots, N. \qquad (2c)$$

where σ_i denotes the sensitivity at i [see Eq. (1) in 2.6]. (Scaling constants have been omitted for clarity.)

Eq. (2a) describes the main dynamics of the IS through changing concentrations of V-regions: the species located at i in shape space will increase in concentration according to the level of activated B cells capable of producing it, scaled by a maturation function *Mat*, embodying both B cell activation through network contacts and T cell regulation (Figure 2). The f_i's will tend to be decreased in proportion to the network's sensitivity at i, being unable to bind to other ligands. Eq. (2b) represents the distributed control of the previous dynamics Eq. (2a), but itself under the influence of the antibodies species present, showing up as a proliferation function *Prol*, similar in shape to *Mat* (Figure 2).

Spontaneous cell death will reduce b_i's numbers. This pair of linked equations represent both the dynamics of the IS components and the way in which B cell are activated by the state of the network constantly specifying a cognitive domain through which the IS can couple.

The term in Eq. (2c) is not properly an equation, but a metadynamical *algorithm*: it represents the probability that a given species i will be produced by the bone marrow, and thus, a corresponding term with index i will play a role at all in the above dynamics. Formally, it indicates that the variable vector active in Eq. (2a-b) will not be a fixed one, as in a classical differential equation, but will change continuously (Farmer et al., 1986). This probability function is different for each individual according to their specific V-gene endowment, but fixed over life, like a steady generator. Whether an i-th species will be incorporated into the network, proliferate, or mature, is a function of the entire IS's dynamics Eq. (2a-b).

2.10 SOME RESULTS AND OPEN POSSIBILITIES

Our intention in this article is not to make an exhaustive analysis of the model proposed above, but rather to indicate its general characteristics, their interest in immunology, and their relationship with neural networks. A more detailed analysis of the simulation results and their applications will appear elsewhere (Varela, Coutinho, and Dupire, in preparation).

The first item to remark is that the *qualitative* behavior of this family of equations is relatively stable within a broad range of parameters, while at the same time quite varied in the behavior of individual components. This is required if one is to come moderately near the intricacies of biological immune dynamics. The basis of this qualitative behavior is, of course, the non-linearity of the laws we have chosen. Other possible simpler variants do not give satisfactory results. For example, if both *Prol* and *Mat* are constants, Eq. (2) turns out to have unstable behavior, with all the f_i either imploding or exploding. Thus, Eq. (2) embodies a carefully selected and realistic set of local rules of immune networks which seem to be over the minimal threshold of complexity required to begin to be useful in the context of biological observations.

The richness of the dynamics can be appreciated through two examples. First, simulations reveal that Eq. (2) exhibits oscillations for many of the variables. That is precisely the kind of behavior one gets in measuring antibody concentration over a period of several months (Lundqvist et al., 1987). A purely cellular analysis of this fact is bound to fail, since it is an network phenomenon, spontaneously produced by the local rules.

A second interesting emergent property is the fact that, although the system undergoes constant metadynamical change, and can be subjected to a history of interactions with independent antigenic sources, there is a large core of species which stay remarkably regular. The source of this memory of the system resides in the strength of the subsets of variables which act as buffer to one another against change, in complex topological patterns in shape space. In other words, the memory capacity of an immune network is revealed here as a property of topological self-organization in the shape space proper to the non-linear equations.

Histories of antigenic interactions are easily accommodated within this framework, as it should be obvious by now. In fact, it is enough to add to Eq. (2a) a term of the form

$$\sum_j m_{ij} f_j a_j \qquad j = 1, \ldots, M \tag{3}$$

where a_j are the amounts of antigens the IS encounters in a unit of time. Obviously, these encounters will cause a shift in the relative concentrations within the network. But, in contrast to the view most currently adopted, we have left this term for last, to emphasize that in our perspective the IS is not antigen-driven, but self-referential. That is, by far the most interesting dynamics to study is proper to the immune components themselves (Vaz and Varela, 1978; Coutinho et al., 1984a).

That this is biologically sensible is directly seen by considering experiments with "antigen free" mice. These are germ-free animals maintained for a few generations on chemically defined, low-molecular-weight diets and, of course, on filtered air. Therefore, they do not contain or come in contact with antigens that are foreign to their own body. Such mice maintain, nevertheless, levels of immune activity in the spleen that are comparable to those of normal infected mice eating antigenic diets. Both the numbers of activated and "effector" lymphocytes and their turnover, as well as the levels of circulating IgM antibodies are the same in antigen-free and conventional mice (Hooijkaas et al., 1984; Pereira et al., 1986b). This self-centered activity engenders a cognitive domain which *defines* antigenicity, and for which external encounters provide boundary conditions, but not directive inputs.

These results are admittedly only suggestive. The point is to outline a way of doing research on immune networks. Every one of the qualitative results mentioned here needs to be properly studied and quantified. But our approach is meant to provide also a tool for research which constitutes its own validation. For this purpose, the optimal implementation of the model should be on real-time calculation displaying on a computer the updated state of the IS as a cloud of its sensitivities in the context of the self molecules present as constraints. The IS would then become readily visible as an active unity, relating to itself and its environment, and where the user could actively interact by adding or subtracting species, or altering dynamical parameters. In such a more advanced implementation, immune networks become a tool for research capable of explaining known results and suggests future ones.

3. CONCLUSIONS

3.1 NEURAL VS. IMMUNE

The immune and the neural systems have been compared often since they represent the most potent forms of biological cognition known throughout the animal realm. From our perspective, this comparison can be done in detail, given the explicit form of the IS we have put forward. In what follows, we outline what we see as some fundamental *differences* in the mechanism embodied by both cognitive networks.

This comparison is easier to carry out by considering an equivalent set of equations typical of neural network models, say

$$\frac{df_i}{dt} = \sum_j m_{ij} G\{f_j\} - f_i \tag{4a}$$

$$\frac{dm_{ij}}{dt} = F\{f_i, f_j\} \tag{4b}$$

where G is a sigmoid threshold function of the voltages, f_i, the source of the non-linearities in the neural network, and F is a quadratic in terms of both variables as in Hebb's rule.[5] We have written Eq. (4) to emphasize the contrast with Eq. (2) along some main directions:

- First, in immune nets connectivity depends on probabilistic encounters, required by a network made up of components which diffuse in a large volume, in contrast to neural connectivity based on a fixed anatomy over axo-dendritic extensions. This is the basis of the different sources of non-linearities in both systems.
- Secondly, the relevant variables in Eq. (2a) are under the controlling influence of the B cells that produce them. In Eq. (4) instead, it is the connectivity matrix itself which is the controlling factor, as a function of the network's state, as defined by Eq. (4b).
- Third, the algorithmic term Eq. (2c) has no analog in neural networks: this is the source of the rich adaptability immune dynamics exhibit and accounts for a continuously changing topology. This would correspond to a brain where many neurons vanish while others pop up from nowhere.

These points make explicit our initial suggestion that IS seem to provide a radically *different* kind of biological network capable of (at least a few) basic cognitive performances. The full extent of these differences (or their possible equivalence) remains to be investigated in detail.

[5]This family of equations is broadly used in neural modelling; for example, recently by Hopfield and Tank (1985). Many variants do, of course, exist, but they are not our concern here. For a related set of equations and an extensive discussion, see Grossberg (1984).

3.2 ENACTING VS. REPRESENTING

One of the most interesting aspects of the cognitive mechanism present in immune networks is that they operate in the domain of molecular shapes which are not inhabited by man's habitual objects of perception: objects, 3-D space, and other people. This provides an occasion to see how knowledge operates without the anthropomorphic weight of our ordinary perceptual world.

In this sense, the immune networks reveal how biological cognition is a matter of action, or better, of *enaction*: in its very operation the system specifies a domain of relevance (or significance), which becomes a "world" for the animal to act and live with. In fact, the shape space is, without an immune system that lives in it, completely neutral and void of any sense of signification and distinctions of any sort. But as soon as an immune network appears, its own self-assertion creates a clear demarcation of a cognitive domain, and hence, an entire series of discriminations between self and non-sense, proper and inadequate, friendly and foe.

This kind of enactive cognition, so clearly seen in immune networks, has to be contrasted with our usual view of cognition as being a more or less accurate representation of a world already full of signification, and where the system picks up information to solve a given problem, posed in advance (Varela, 1979, 1987). This is the understanding of immune cognition for the antigen-driven clonal selection view of the immune system, where adequate operation means optimal defense of invaders as out-there. This understanding of the IS has also been the base of recent models similar in scope to ours (Farmer et al., 1986).

To be sure, there are some constraints under which the IS must operate, including a clear discrimination of some small number of bacterial shapes which must be kept at bay. But this is not a great cognitive task. The more difficult and creative one is, instead, the operational assertion of a molecular identity which makes vertebrate life possible, which is a positive task and not a defensive one. Immune cognition shows that this world of molecular shapes can be addressed in a vast (infinite?) number of ways, all of them complete, provided that they are consistent enough. The molecular world we inhabit, thus, is not pre-given, and then inhabited *post facto* by our immune systems through some optimal adaptation. It is rather laid down as we walk in it, it is a world brought forth.

This clear demonstration of knowledge as enactive, is another unique contribution of immunology to the yet poorly explored realm of biological cognitive networks and their mechanisms.

ACKNOWLEDGMENTS

FV holds a Chair Scientifique from the Fondation de France; the financial support of the Prince Trust is also gratefully acknowledged. AC acknowledges the supoort of INSERM and DRET.

REFERENCES

1. Bandeira, A., A. Coutinho, C. Martinez, and P. Pereira (1987), "The Origin of Natural Antibodies and the Internal Activity in the Immune System," *Int. Rev. Immunolog*, (in press).
2. Coutinho, A. (1980), "The Self-Nonself Discrimination and the Nature and Acquisition of the Antibody Repertoire," *Ann. Immunol. (Inst. Pasteur)* **132 C**, 131–144.
3. Coutinho, A. (1984), "MHC-Restriction in T Cell/B Cell Interaction," *Ann. Immunol. (Inst. Pasteur)* **135 D**, 71–109.
4. Coutinho, A., L. Forni, D. Homberg, F. Ivars, and N. Vaz (1984a), "From an Antigen Centered, Clonal Perspective of Immune Responses to an Organism-Centered, Network Perspective of Autonomous Activity in a Self-Referential Immune System," *Immunol. Revs.* **79**, 151–169.
5. Coutinho, A., G. Pobor, S. Petterson, T. Leandersson, S. Forsgren, P. Pereria, A. Bandeira, and C. Martinez (1984b), "T Cell-dependent B Cell Activation," *Immunol. Rev.* **78**, 211–224.
6. Farmer, J. D., N. H. Packard, and A. S. Perelson (1986), "The Immune System, Adaptation, and Machine Learning," *Physica* **22 D**, 187–204.
7. Freitas, A., B. Rocha, and A. Coutinho (1986), "Lymphocyte Population Kinetics in the Mouse," *Immunol. Rev.* **91**, 5–37.
8. Gardner, H. (1984), *The Mind's New Science* (New York: Basic Books).
9. Grossberg, S. (1984), *Studies on Mind and Brain* (Boston: D.Reidel).
10. Holmberg, D., S. Forsgren, F. Ivars, and A. Coutinho (1984), "Reactions amongst IgM Antibodies Derived from Normal, Neonatal Mice," *Europ. J. Immunol.* **14**, 435–441.
11. Hooijkaas, H., R. Benner, J. R. Pleasants, and B. Wostmann (1984), "Isotypes and Specificities of Immunoglobulins Produced by Germ-free Mice Fed Chemically Ultrafiltered Antigen-Free Diet," *Europ. J. Immunol.* **14**, 1–27.
12. Hopfield, J., and D. Tank (1985), "Neural Computation of Decisions in Optimization Problems," *Biol. Cybernetics* **52**, 141–152.
13. Jerne, N.K. (1974), "Towards a Network Theory of the Immune System," *Ann. Immunol. (Inst. Pasteur)* **125 C**, 373.
14. Jerne, N.K. (1984), "Idiotypic Networks and Other Preconceived Ideas," *Immunol. Rev.* **79**, 5.
15. Jerne, N.K. (1985), "The Generative Grammar of the Immune System," *EMBO. J.* **4**, 847.
16. Kearney, J., and V. Meenal (1986), "Idiotype-Directed Interaction during Ontogeny Play a Major Role in the Establishment of the Adult B Cell Repertoire," *Immunol. Rev.* **94**, 39–50.
17. Larsson, E.-L., et al. (1984), "Activation and Growth Requirements for Cyclotoxic and Non-Cyclotoxic T Lymphocytes," *Cell. Immunol.* **89**, 223–231.

18. Lundqvist et al. (1987), "Evidence for a Functional Network amongst Natural Antibodies in Normal Mice," (submitted for publication).
19. Melchers, F., L. Laflen, and J. Andersson (1974), in *Control and Proliferation of Animal Cells*, Eds. R. Baserga and B. Clarkson (New York: Cold Spring Harbor Laboratory).
20. Monroe, J., and J. Cambier (1983), "B Cell Activation. III. B Cell Plasma Membrane Depolarization and Hyper Ia Antigen Expression Induced by Receptor Immunoglobulin Cross-linking Are Coupled," *J. Exp. Med.* **158**, 1589.
21. Pereira, P., E. L. Larsson, L. Forni, A. Bandeira, and A. Coutinho (1985), "Natural Effector T Lymphocytes in Normal Mice," *Proc. Natl. Acad. Sci. USA* **82**, 7691.
22. Pereira, P., S. Forsgren, D. Portnoi, A. Bandeira, A. Martinez, and A. Coutinho (1986a), "The Role of Immunoglobulin Receptors in Cognate T-B Cell Collaboration," *Eur. J. Immunol.* **16**, 355–361.
23. Pereira, P., L. Forni, E. L. Larsson, M. Cooper, C. Heusser, and A. Coutinho (1986b), "Autonomous Activation of B and T Cells in Antigen-Free Mice," *Eur. J. Immunol.* **16**, 685–688.
24. Perelson, A., and G. Oster (1979), "Theoretical Studies of Clonal Selection: Minimal Antibody Repertoire Size and Reliability of Self/Non-self Discrimination," *J. Theoret. Biol.* **81**, 645–670.
25. Rummelhart, D., and D. McClelland (1987), *Parallel Distributed Processing Studies on the Microstructure of Cognition* (Massachusetts: MIT Press), 2 vols.
26. Schwartz, R. H., B. S. Fox, E. Fraga, C. Chen, and B. Singh (1985), "The T Lymphocyte Response to Cytochrome c. V. Determination of the Minimum Peptide Size Required for Stimulation of T Cell Clones and Assessment of the Contribution of Each Residue Beyond This Size to Antigenic Potency," *J. Immunol.* **135**, 2598.
27. Tite, J., J. Kaye, K. M. Saizawa, J. Ming, M. E. Katz, L. Smith, and C. Janeway (1986), "Direct Interaction between B and T Lymphocytes bearing Complementarity Receptors," *J. Exp. Med.* **163**, 189.
28. Urbain, J. (1986), "Is the Immune System a Functional Idiotypic Network?," *Ann. Inst. Pasteur/Immun.* **137 C**, 57–100.
29. Tonegawa, S. (1983), "Somatic Generation of Antibody Diversity," *Nature* **302**, 575–581.
30. Varela, F. (1979), *Principles of Biological Autonomy* (New York: Elsevier/ North-Holland).
31. Varela, F. (1987), "Perception and the Origin of Meaning: A Cartography of Modern Ideas," *Understanding Origin*, Ed. F.Varela (Stanford: Stanford Univ. Press), (in press).
32. Vaz, N., and F. Varela (1978), "Self and Non-Sense: An Organism-Centered Approach to Immunology," *Medical Hypothesis* **4**, 231–267.

ALAN S. PERELSON
Theoretical Division, Los Alamos National Laboratory, Los Alamos, NM 87545

Toward a Realistic Model of the Immune System

ABSTRACT

Using idiotypic networks as an example, I propose an approach to developing realistic models of the immune system. I show that systems containing large numbers of different clones, with particular clones being created and destroyed, can be studied by means of computer simulation. By labeling B cell clones and the antibodies that they secrete by a binary string, and using rules of complementarity between binary numbers, I show that one can create complex models with realistic topologies. I point out that the language of graph theory may be useful in characterizing idiotypic networks. I also give an example of a dynamical system of equations, involving antibody, B cell clones and growing antigen that incorporates many of the features of T-independent immune responses. To fully specify these dynamical equations, many important problems in the chemistry of multicomponent mixtures of anti-idiotypic antibodies, B cell receptors and antigen need to be solved. By viewing the immune system as a dynamic entity, whose components may continually be created and destroyed, a new view of the immune system emerges. I suggest that under some circumstances, the immune system may not be operating at a steady-state which is then perturbed by the presence of antigen, but rather may be constantly changing. Thus, like a weather pattern, the immune system may be rather quiescent at some times but raging at other times. Lastly, I suggest that memory can not only be stored by memory cells, but may also be stored dynamically in idiotypic networks.

INTRODUCTION

One of the most notable features of the immune system is its complexity. This complexity is manifested in the specificity and diversity of its components, the high connectivity between components made possible by communication via broadcast (cells secrete soluble molecules which can then be perceived by other cells expressing the appropriate receptor), the ephemeral nature of many of the components (the lifetime of molecules is typically of order minutes to hours; that of cells is typically of order days), and the rapid generation of new system components (approximately one million lymphocytes per second are produced in humans). My aim in this paper is first to review the complexities of the immune system, and then to outline a strategy for developing realistic models of the immune system. I will stress the development of models at the level of individual clones, and describe the development of an idiotypic network model that reflects a modicum of reality in its topology and choice of system elements.

COMPLEXITIES OF THE IMMUNE SYSTEM

The human immune system contains of order 10^{12} cells and probably involves the action of hundreds of classes of molecules. The cells in the immune system comprise a few percent of all cells in the body. They are distributed throughout the tissues and concentrated in the blood, the lymphoid organs such as the thymus, lymph nodes and spleen, and the body tracts that are exposed to the outside environment, such as the gut, respiratory and genitourinary systems. A variety of cell types are involved. Each cell type performs a specific function by means of either direct cellular contact or the secretion of soluble factors. The cells which confer specificity to the system include B and T lymphocytes, and possibly natural killer cells. Macrophages are also involved in the specific response through their role of antigen processing and presentation. Macrophages, monocytes and the granulocytes, such as neutrophils, eosinophils, basophils and mast cells, play a role in phagocytosis and in the production of various immune mediators. All the molecules that are important in the immune response have not yet been identified, but include cell surface receptors and associated glycoproteins such as the antibody receptor, the T cell receptor, CD2, CD3, CD4 and CD8 molecules, the IL-2 receptor, the various Fc receptors, the C3 and the C5a receptor. Also involved are soluble molecules such as serum antibody, the interleukins (IL-1, IL-2, IL-3, IL-4, IL-5), interferon, tumor necrosis factor, lymphotoxin, histamine and the complement components. These lists are far from complete, but they illustrate that a vast array of cell types, and molecules produced by these cells, must be coordinated in order to generate an immune response.

The connectivity between cells in the immune system is not "hard-wired" as it is in the nervous system, but rather relies on transient cell-cell contacts and communication via the secretion of soluble molecules. Communication via soluble molecules can be more flexible than hard-wired connections. As typified by IL-2, such communication can be dynamically regulated at the level of receptor expression and at the level of the concentration of the secreted molecule. (Kevrekidis et al., this volume, Part 1, present a model of T cell activation and proliferation regulated by IL-2). Communication between cells can also involve very specific molecules, such as antibodies. Communication via antibody forms the basis of idiotypic network theory and is dealt with in detail in the next section.

Another source of complexity in the immune system lies in the specificity and diversity of the immune system components. For example, the B cell repertoire as *expressed* in a single mouse, has been estimated to be of order 10^7. However, genetic mechanisms such as somatic mutation, combinatorial association and junctional diversity, can generate a *potential* repertoire of size 10^{15} or greater depending on the extent of somatic mutation (Jerne, 1976). The expressed T cell repertoire is probably of the same order of magnitude as the expressed B cell repertoire. Even though the repertoire is huge, the response to a single epitope is generated by only a small subset of all B and T cell clones. During the course of a response the expressed repertoire may enlarge; new B or T cell types generated in the bone marrow and thymus may contribute to the response and somatic mutation in the periphery may change the specificity of some of the cells engaged in the response (Manser et al., 1985, Kauffman et al., this volume, Part 1). For example, the immune response to the hapten phenyl oxazolone is initially dominated by B cells expressing a single set of germ-line V genes ($V_H - Ox1$ and $V_\kappa - Ox1$). However, as time progresses cells expressing this V-gene combination disappear and are replaced by somatic variants. Later in the response, B cells expressing totally different V region genes dominate the response (cf. Milstein, 1986).

MODELING STRATEGY

How may one realistically model a system, such as the immune system, which not only includes a very large number of elements ($> 10^7$), but also allows the specific elements composing the system to be replaced with different elements over time? In collaboration with J. D. Farmer and N. Packard, I have developed a modeling strategy which is specially designed to handle large, dynamically changing systems (Farmer et al., 1986; Farmer et al., 1987). Rather than simply writing down a set of dynamical equations, our approach involves two components: first the construction of a database of information about the cells and molecules under scrutiny; and second, the definition of an algorithm to construct the relevant dynamical equations based upon the information in the database. As elements in the immune system turnover, the database is updated and the dynamical equations are changed. The number of equations and thus the dimension of the system being studied may also change. It is this property that allows us to easily model systems in which new antibody or cell types are constantly being created and destroyed. This approach will be demonstrated in the next section. However, even if one chooses to ignore dynamic changes in the elements of the system, our modeling strategy provides a means of developing realistic models of the immune system at the level of individual clones.

IMMUNOLOGY IN THE NEXT CENTURY

As immunology progresses into the 21st century, one can envision that all of the various cell types that play a role in the immune system will have been identified. If the current pace of invention and discovery in molecular biology is maintained or accelerates, and if, as planned, the human genome is completely mapped and sequenced, one might expect that most, if not all, of the molecules secreted by cells in the immune system will be known, their genes cloned, and pure recombinant forms made available. The task remaining in immunology will be to understand how the individual cells in the immune system communicate with one another in such a way as to generate coordinate behavior that leads to the elimination of antigen, self-nonself discrimination and phenomena such as memory and learning. I foresee that immunology in the 21st century will be reminiscent of neurophysiology in the 20th century. In neurophysiology, properties of individual neurons are well characterized and yet most of the higher level functions of the nervous system are not understood. In the field of vision, advances in understanding are coming about by distinguishing between the computations made by the visual system, the algorithms used to perform these computations, and the neuronal circuits that implement these computations (cf. Marr, 1982). In immunology a similar conceptual separation between computations, algorithms and the "wet-ware" that implements them may prove valuable. Segel and I (this volume), and Percus (this volume) indicate some first steps toward the founding of a field of *computational immunology* in which one is concerned with discovering the computations and algorithms used by the immune system. To describe the circuitry of the immune system requires a new class of models. The approach mentioned above by Farmer et al. (1986) provides one scheme for the formulation of realistic circuit models. To illustrate the method I consider below the development of realistic idiotypic network models.

A REALISTIC IDIOTYPIC NETWORK MODEL

One of the more complex modeling exercises in immunology has been the attempt to model idiotypic networks. Beginning with the work of Richter (1975, 1978), Hoffmann (1975, 1980), Hiernaux (1977), Seghers (1979) and continuing with the work of Hoffmann (this volume) and De Boer (this volume), models of idiotypic networks have always dealt with prespecified and well-defined populations. Thus, for example, a model might entail idiotypic and anti-idiotypic populations (e.g., Hoffmann's plus-minus model), or they might entail antibodies and/or cells (a distinction between the two is generally not made) at idiotypic level i interacting with idiotypic populations at levels $i-1$ and $i+1$ (e.g., Richter, 1975; 1978). Here, as has become

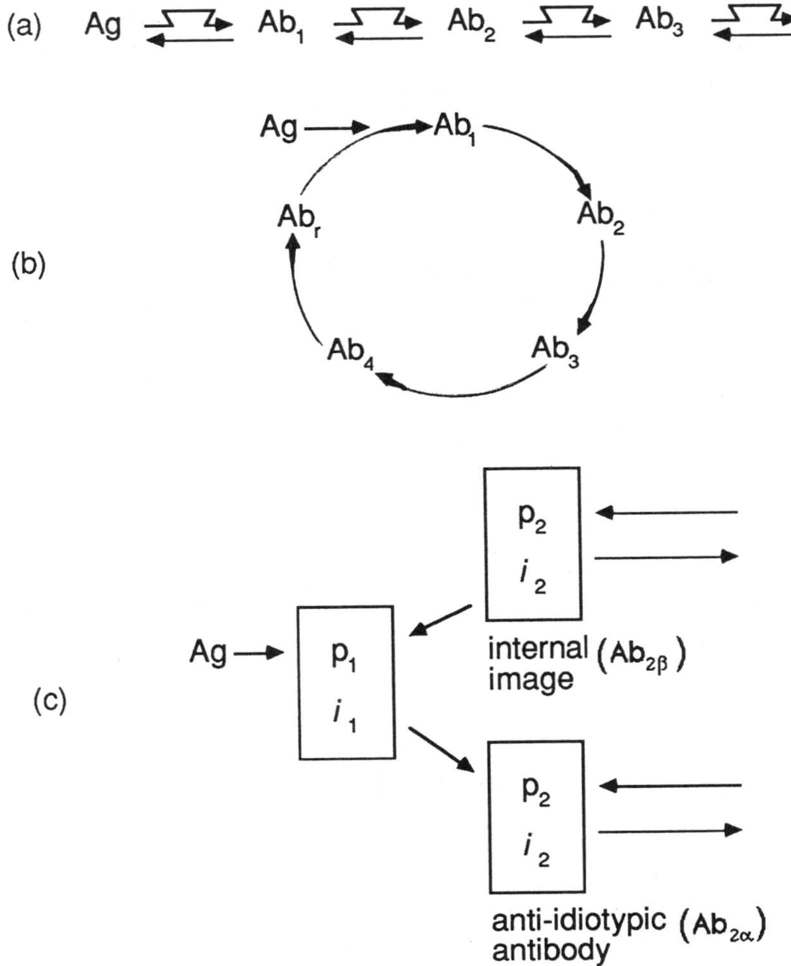

FIGURE 1 (a) Linear idiotopic network. (b) Cyclic idiotypic network. (c) More detailed view of an idiotypic network. Each antibody contains a paratope and an idiotope. An arrow represents recognition. Thus, P_1 recognizes both the antigen (Ag) and the idiotope i_2 on a molecule called the internal image of the antigen.

standard, antigen is considered level 0, antibodies against the antigen are level 1, anti-idiotypic antibodies are level 2, anti-anti-idiotypic antibodies are level 3, and so forth (Figure 1). Since one can continue in this manner forever, concern arose as to whether and where the network would end. Hiernaux (1977) suggested that cycles would arise. In Hoffmann's symmetric network theory $Ab_{i+2} = Ab_i$, so only Ab_1 and Ab_2 need be considered and the problem does not arise. For the Hoffmann, Richter or Hiernaux models it is a straightforward matter to write down a set of differential equations that describe the model and each author has done so. However, are these models realistic? Clearly, Ab_1, Ab_2, etc. are not individual antibodies or clones, but rather refer to sets of molecules and/or clones. Because of this, Jerne (1974) and others recognized that not all idiotypic levels would be distinct. Thus, for example, a clone could simultaneously be in Ab_2 and Ab_4. Other distinctions can also be made. Antibodies A_i can be viewed as having a paratope p_i and an idiotope (or set of idiotopes) i_i. Thus, Jerne et al. (1982) proposed that within the Ab_2 population, two types of molecules could be distinguished: $Ab_{2\alpha}$ molecules with paratopes recognizing Ab_1 idiotopes and $Ab_{2\beta}$ molecules containing idiotopes recognized by the Ab_1 paratope (Figure 1). In addition, there are $Ab_{2\epsilon}$ antibodies, or epibodies (Bona et al., 1982), that simultaneously recognize the antigen and Ab_1 idiotopes. At the level of Ab_3 even more subpopulations can be defined (cf. Figure 7 of Nesterenko, this volume). Greenspan and Roux (this volume) point out the existence of overlaps in idiotypic specificities and thus the consequent inaccuracies of decomposing the idiotypic levels in the manner suggested by Jerne et al. (1982). This all suggests that the models of Figure 1 are greatly oversimplified.

Recent attempts to map idiotypic networks show that they are neither linear nor cyclic, but have much richer topologies (cf. Dwyer et. al., 1986). How then is one to model an idiotypic network? One could try to experimentally map the network, but this is a complex task and it is unlikely that networks of more than 50 or 100 elements could be unraveled. One could assume that in large networks the connections are random and then generate models based on this assumption. De Boer (this volume) uses this methodology. I shall propose a novel approach to the problem.

SHAPE SPACE AND COMPLEMENTARITY RULES

My approach to the construction of network models is a more direct one. I try to model the physical chemical interactions which determine the network connections among antibodies and receptors. One means of proceeding is to use the notion of "shape space" introduced by Perelson and Oster (1979). In this approach one views the paratope (combining region), or any epitope, as having a three-dimensional shape and other features, such as charge, that are important in determining the complementarity between two molecules. One uses these features as coordinates in generalized shape space. A function on generalized shape space determines when two shapes are complementary. Segel and I (this volume) describe a simple one dimensional version of shape space and explicitly construct a function that measures

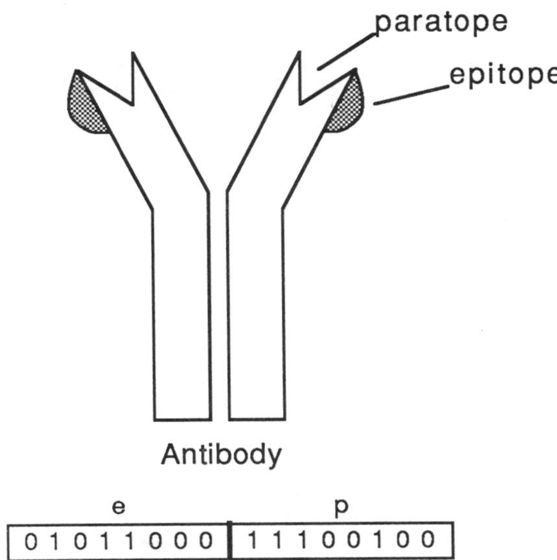

FIGURE 2 Antibody (above) and bit string representation of an antibody.

the degree of complementarity of two shapes. Varela et al. (this volume), Percus (this volume) and Kauffman et al. (this volume, Part 1) also make use of the notion of shape space.

To model a multidimensional shape space, I assign to each antibody combining region a binary string of length N (Figure 2). This binary string represents a point in a shape space which is now a hypercube of dimension N. If $N = 32$, then one can represent $2^{32} \approx 4 \times 10^9$ different molecules in this space. Thus, on a 32-bit computer such as a VAX or SUN workstation, one can represent systems with diversity comparable to that expressed in the mammalian immune system. One way to view this assignment of a binary string to an antibody is to recall that a string of nucleic acids ultimately codes for the antibody V region. This string of nucleic acids is translated into a string of amino acids, which ultimately folds into a three-dimensional protein.

Complementarity between molecules is determined by their three-dimensional structures. At the moment it is not possible to predict the tertiary structure of a molecule given its primary structure. Thus one cannot, in general, determine when two primary sequences will be complementary. In order to proceed with the development of a mathematical model, one needs to assign a rule for the determination of complementarity between shapes. For shapes determined by binary strings (which could be viewed as coded forms of nucleic or amino acid sequences), the simplest rule would be: two shapes are complementary if their binary strings are complementary. In support of this rule, there is evidence suggesting that if a peptide or protein is read from one strand of a double-stranded DNA molecule and a "complementary peptide" is synthesized by reading from the complementary DNA strand, then the peptide and complementary peptide will bind specifically and with high affinity (Bost et al. 1985a, b; Smith et al., 1987; Shai et al., 1987). Further, in the case of the hormone ACTH, antibodies against ACTH and antibodies against the complementary peptide seem to be an idiotypic–anti-idiotypic pair, leading to the speculation that idiotopes and anti-idiotopes may represent complementary sequences in the hypervariable regions of such immunoglobulin pairs (Smith et al., 1987).

One can make the complementarity rule quantitative so that one can assign an affinity between any pair of molecules. A variety of choices are possible. One choice, which has been implemented in a simulation model (Stadnyk, Farmer, Packard and Perelson, in preparation), is to assume that some minimum number of bits m must be complementary for two strings to interact at all. Assign some low affinity, K_L (say, 10^4 M^{-1}) to strings that are complementary at m bits, and some high affinity, K_H (say, 10^{12} M^{-1}) to strings that are complementary at all N bits. For strings that are complementary at i bits, $m \leq i \leq N$, let

$$log K_i = \frac{N-i}{N-m} log K_L + \frac{i-m}{N-m} log K_H \quad . \tag{1}$$

Modifications of this algorithm take into consideration that these bit strings are intended to represent molecules and hence need not be perfectly aligned in order to interact. As described in Farmer et al. (1986), one can slide the bit strings relative to one another and then compute the degree of complementarity. Another algorithm allows for the fact that two randomly chosen strings will on average have half their bits complementary. Thus, in the earlier algorithm m would need to be chosen greater than $N/2$ and one loses some of the scale in affinity differentials between strings. To avoid this loss, one could assign affinities in such a way that only pairs, triplets, ..., of adjacent complementary bits would contribute to the affinity. Stadnyk (1987) examines a rule of this nature where the greater the length of complementary substrings the greater the contribution the substring makes to the overall affinity.

In the original formulation of idiotypic network theory Jerne (1974) envisioned each antibody as being composed of a paratope p and an idiotope or set of idiotopes i. This view can easily be accommodated within the binary string representation, representing each antibody V region by a string which is actually a set of substrings of lengths $n_p, n_{i1}, n_{i2}, \ldots$, say, the first representing the paratope, and subsequent substrings representing various idiotopes on the molecule. In the Farmer, Packard and Perelson model, we represented each antibody by a paratope and a single epitope. Rather than make the epitope a unique determinant, i.e., an idiotope, we hoped that the immune system model would self-regulate in such a way that only unique epitopes would play the role of idiotopes in regulatory interactions.

The evidence reviewed by Greenspan and Roux (this volume) suggests that the paratope and idiotopes of an antibody overlap. Bit string representations of the V region are flexible enough to incorporate overlapping idiotopes and paratopes; however, dealing with overlapping regions is a complexity which I think should be avoided in initial models.

NETWORK TOPOLOGY

An example of an idiotypic network generated by the simplest complementarity rule is given in Figure 3. Here I have chosen a population of ten randomly constructed antibodies. Each antibody is assumed to contain one eight-bit paratope and one eight-bit epitope. Paratopes could be complementary to epitopes or to paratopes. To simplify the diagram I have only displayed paratope-epitope interactions. A

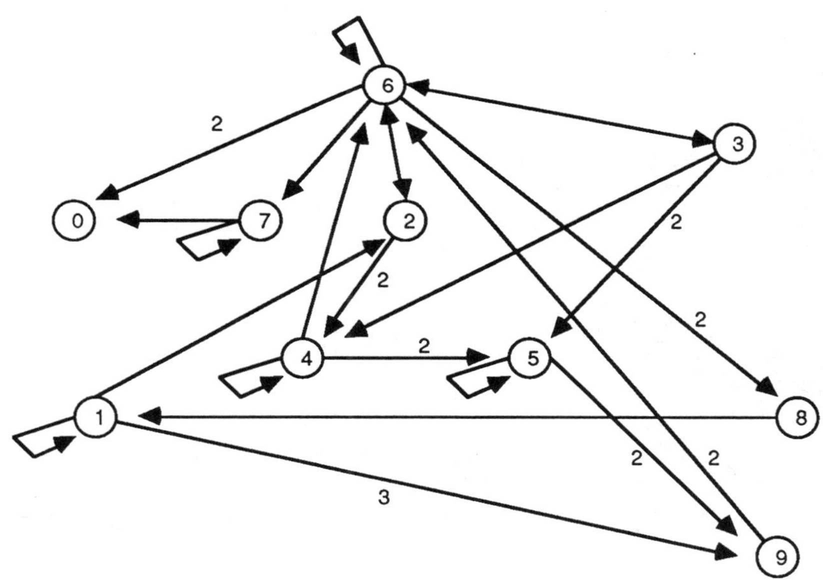

FIGURE 3 An idiotypic network generated using 8-bit paratopes and 8-bit epitopes. Only paratope-epitope interactions are shown. The symbol ④ → ⑤ means that the epitope on antibody 5 is recognized by (i.e., complementary to) the paratope on antibody 4. Three levels of complementarity are noted. The highest level denoted by a 3 indicates that all 8 bits are complementary. There is only one such interaction. The next level, denoted by a 2, indicates that 7 bits are complementary. Most of the interactions are level 1 (the label 1 omitted for simplicity), corresponding to six bits being complementary.

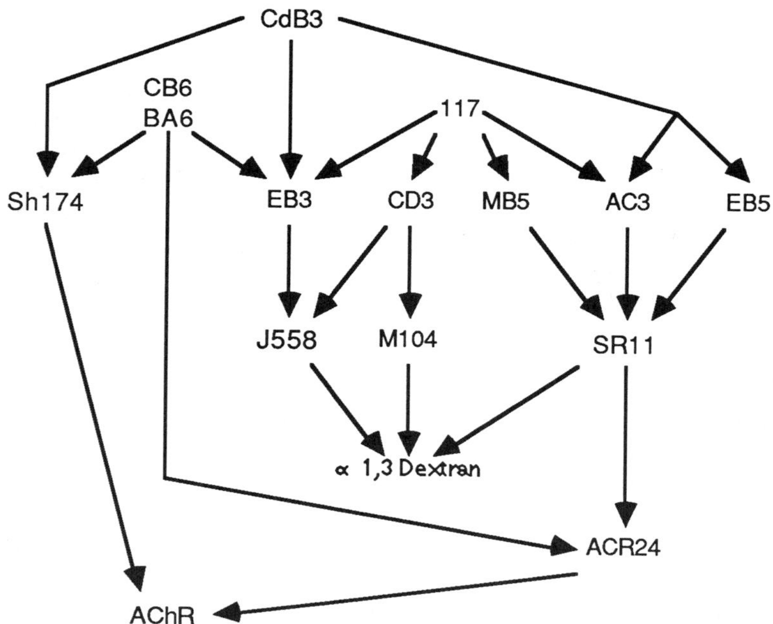

FIGURE 4 Idiotypic network mapped by Dwyer et al. (1986) showing a relationship between the immune responses to the acetylcholine receptor (AChR) and dextran. The vertices in the graph except AChR and dextran are all monoclonal antibodies.

similar diagram could be constructed for the paratope-paratope interactions. In the complementarity rule, the paratopes and epitopes were exactly aligned and it was required that at least six bits be complementary. Thus, the strings could be complementary at 6, 7 or 8 positions, corresponding to three different affinities, say, 10^4, 10^6 and 10^8, respectively. A number of features are notable. First, some antibodies recognize themselves. This is caused by the paratope on the antibody being complementary to the epitope on the same antibody. Interestingly such antibodies have been discovered by H. Kohler and have been called "autobodies" (Chang and Kohler, 1986). Second, some antibodies have much higher connectivity than others. The paratope on antibody 6 recognizes the epitopes on six antibodies, one of them being itself. The paratopes on the other antibodies in the system only recognize the epitopes on between one and three other antibodies. The epitope on antibody 6 is also the most frequently recognized epitope, being recognized by five other antibodies. Some naturally occurring anti-idiotypic antibodies have high connectivity (Holmberg et al., 1984; Holmberg and Coutinho, 1985), and have been suggested to play a role as a "super organizer" (Dwyer et al., 1986). In the simulated network antibody 6 corresponds to such a "high connectivity" antibody. There is nothing special about it; the high connectivity arose by chance.[1] Third, only one interaction is of the highest affinity, six are of intermediate affinity and the remainder are of the lowest affinity. This distribution of affinities seems reminiscent of actual molecular interactions. Most of the interactions of the high connectivity antibody, antibody 6, are of low affinity. This also mimics the experimental situation (Dwyer et al., 1986; Holmberg et al., 1984).

Although Figure 3 was generated by a very simple computer algorithm, it seems to have captured many features of real idiotypic networks. Progress has recently been made in determining the composition of real idiotypic networks (Holmberg et al., 1984; Victor-Kobrin et al., 1985; Kearney and Vakil, 1986; Dwyer et al., 1986). Figure 4 shows a network constructed by Dwyer et al. (1986) that demonstrates a possible connectivity between the immune responses to the acetylcholine receptor (AChR) and α-1,3-dextran (DEX), an epitope found on the surface of certain bacteria. This connectivity is significant because it may provide information about the initiation of the human autoimmune disease myasthenia gravis, which is caused by autoantibodies against the AChR. Comparing Figures 3 and 4, one sees a remarkable similarity between the model network and the experimentally determined network.

A diagram, such as Figure 3, that shows which antibodies have complementary structures needs to be distinguished from a diagram of the actual effective interactions is an idiotypic network. Whether or not an idiotypic and anti-idiotypic antibody productively interact in the immune system will depend both on the concentration of the two antibodies and the affinity of the interaction. The experimentally determined network in Figure 4 was constructed on the basis of reactivity on ELISA and affinities were not measured. Thus, as in the randomly generated network, what was measured was *possible reactivities* between molecules and not actual idiotypic interactions under *in vivo* conditions. Networks such as Figures 3 and 4 might thus be called phantom networks, as opposed to real networks. They exist dimly in the background waiting for the concentrations of their elements to attain high enough values for them to be realized.

The paratopes and epitopes of the antibodies represented in Figure 3 were generated randomly. Using exactly the same algorithm but choosing a different seed for the random number generator gives rise to different networks. Figures 5 and 6 show two other possible networks. The network in Figure 5 is interesting because it decomposes into two subnetworks of five antibodies each, with a single interaction connecting the two subnetworks. Again some antibodies (0, 6, and 8) have higher connectivity than others. Whereas the network in Figure 5 seems rather simple, the network shown in Figure 6 appears very complex. The network cannot be decomposed, and it is not *planar*. A network is said to be planar if it can be drawn in the plane with no lines crossing (Harary, 1969). In Figure 6, many lines cross.

[1] This is not to say that the immune system could not use high connectivity antibodies for special purposes, but it is difficult to see how a control system could be built around a randomly made molecule. If such molecules are encoded in the germ-line then having them play a control function seems more plausible.

This is due to the high connectivity of so many of the antibodies. One would expect that as networks become larger, and contain more antibodies, that the likelihood of their being nonplanar would increase.

IDEAS FROM GRAPH THEORY

When networks are nonplanar, it is difficult to draw the network in such a way that one can gain insight into its structure. We thus need to find methods of characterizing a network from a list of which antibodies interact with which other antibodies. The methodology and language of graph theory can be of assistance. For example, in graph theory a list of interacting components is replaced by a matrix, such as

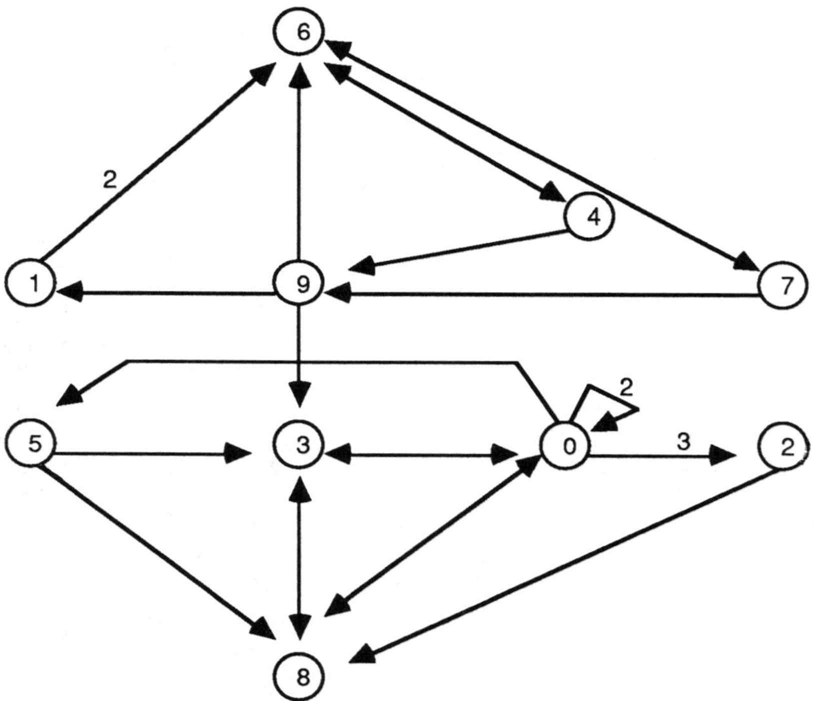

FIGURE 5 As in Figure 3, but different random numbers were chosen for the bit string representation of the antibodies. Note the network is planar and can be decomposed into two subnetworks by removal of the line joining 9 and 3.

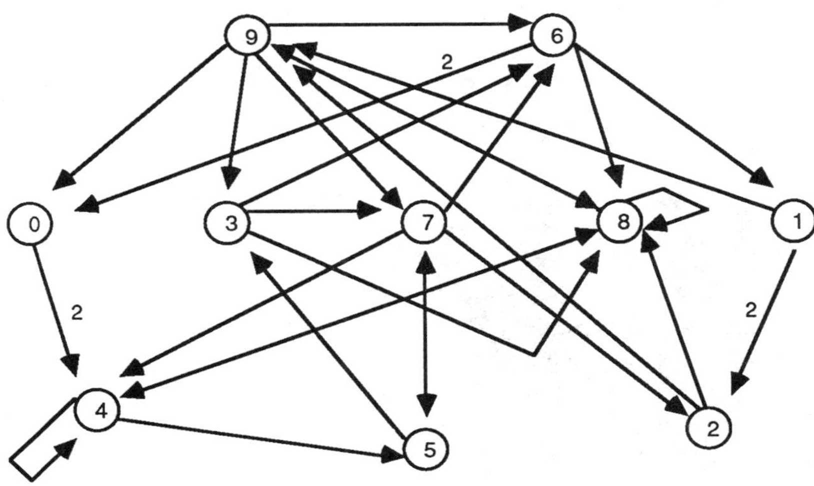

FIGURE 6 Another network generated in the same manner as Figure 3. This network is nonplanar.

the *adjacency matrix*. For a network of n antibodies the adjacency matrix, A, is an $n \times n$ matrix, with each row and column labeled by a node. A 1 is entered into the i, j position of the matrix if antibody i reacts with antibody j, a 0 is entered otherwise (Harary, 1969). (If one wishes one can also keep track of the direction of recognition by putting a 1 in the i, j position only if the paratope on antibody i recognizes the epitope on antibody j. This would then generate a matrix which is not necessarily symmetric). From the adjacency matrix one can determine if any two antibodies are directly connected by idiotypic interactions. Further, one can determine if two antibodies are connected though intermediates. (For example, antibody i may be connected to antibody k which is then connected to antibody j. We then say antibodies i and j are connected by a two-step path). The distance between any two vertices in a graph is the length of the shortest path connecting them. If the network is connected, then the distance between any two antibodies i and j, $i \neq j$, is the least integer m for which the the i, j entry of A^m is nonzero.

In the language of graph theory, we speak of the degree of *connectivity of a graph*. The connectivity of a graph is the minimum number of points (i.e., antibodies) whose removal results in a disconnected or trivial graph. (A graph containing a single point is trivial.) The graph in Figure 5 is 1-connected since removing antibody 9 will disconnect the graph. The *line-connectivity* of a graph is the minimum number of lines (i.e., reactions) whose removal results in a disconnected or trivial graph (Harary, 1969). The graph in Figure 5 is 1-line-connected since removing the reaction between antibody 9 and antibody 3 also disconnects the graph.

Another concept in graph theory that is useful in precisely defining an idiotypic network is the notion of *degree*. The degree of a point (i.e., antibody) is the number of lines incident on it. The degree of an antibody thus corresponds to the number of other antibodies in the network that it reacts with. As one might expect, there are relationships between the degree of the antibodies in a network and the connectivity of the graph (cf. Harary, 1969). De Boer (this volume) and others have used the term connectance to mean the average degree of antibodies in a network.

Other results and definitions in graph theory should be relevant to the characterization of idiotypic networks. What is needed is a way of predicting or characterizing the behavior of an idiotypic network from features of its graph. De Boer obtains some results in this direction; more are needed.

To summarize, representing antibodies by bit strings allows me to use a rule to determine when two antibodies are complementary and to quantify the degree of complementarity. This allows me to assign an affinity to the interaction. The laws of physics and chemistry that determine how proteins fold and interact is the rule that governs real antibody interactions. My rules are *ad hoc* but seem to capture some of the reality of idiotypic networks. Thus, the exact details of the rule may not be important for determining some properties of idiotypic networks. Generating networks in this way allows me to get away from studying simple linear or cyclic idiotypic networks defined by Ab_1 interacting with Ab_2, which interacts with Ab_3, etc.

DYNAMICS OF REALISTIC IMMUNE SYSTEM MODELS

In building realistic models of the immune system, one must not only represent the topology of a network, but also construct dynamical equations which can be used to predict how the various populations of cells and molecules change with respect to time. This is not an easy endeavor, especially given the lack of understanding of the detailed events that lead to cell stimulation, cell proliferation, and secretion of antibody and various lymphokines. Here, through an example, I would like to show how a dynamic model can be created via an algorithm using a database of information about the cells and molecules in the system. Components of the system will be allowed to change due to cell turnover and mutation. This approach was used in Farmer et al. (1986) to create a model for interacting network elements represented by binary strings.

For simplicity, consider a model for the response to a T-independent antigen, or consider a system in which T cell help is sufficiently plentiful that one need not keep track of the T cell population levels. Let B_i and A_i denote the concentrations of B cells of type i and the antibodies of type i that they secrete upon stimulation. The index i is a shorthand for a particular binary string composed of a paratope and an epitope. I assume that B cells of type i can be created *de novo* in the bone marrow or via somatic mutation from some dividing B cell already in the system. Antigen of type k, with concentration G_k may also be present in the system. The index k when used for an antigen is a shorthand for a particular binary string composed of a single epitope. (The generalization to molecules containing multiple epitopes is straightforward and will not be dealt with here). The system could be studied in the absence of antigen to see the development of idiotypic networks, or it could be used to study a clonal selection model by ignoring the possibility of antibody-antibody interactions and focusing only on antibody-antigen interactions. Over a particular time period, we assume that the types of B cells and antigens is fixed. For each molecule of type i (i.e., B cell receptor or antibody), an algorithm is used to generate a list of other molecules j to which molecules of type i are complementary, and the degree of complementarity (i.e., the affinity of interaction). Both paratope-paratope and paratope-epitope interactions are examined and included in the list. Using a rule, such as that given by Eq. (1), an affinity can be assigned to each interaction.

The dynamics of the system can then be modeled by the following set of ordinary differential equations:

$$\frac{dB_i}{dt} = r_B(B_i, B_T)B_i\alpha_i - \mu_B B_i(1-\alpha_i), \tag{2}$$

$$\frac{dA_i}{dt} = sB_i\alpha_i + \epsilon B_i(1-\alpha_i) - \mu_A A_i - \mu_C f_{iC} A_i, \tag{3}$$

$$\frac{dG_k}{dt} = r_G(G_k)G_k - \mu_{kill}G_k - \mu_G G_k \ . \tag{4}$$

B cells can either be stimulated or not. The fraction of stimulated cells of type i is α_i, $0 \leq \alpha_i \leq 1$. Unstimulated cells die at constant rate μ_B. Stimulated B cells do not die, but rather grow at a rate r_B which is a function of both the number of cells of type i, B_i and the total number of B cells in the system, B_T. Antibody of type i is secreted from stimulated B cells of type i at constant rate s, is shed from nonstimulated cells at a low basal rate ϵ (cf. Melchers et al., 1974), and is removed from the system at some constant natural removal rate μ_A. Antibody can also be lost by binding to antigen or to an anti-idiotypic antibody, forming a complex, and then having the complex lost, say by being phagocytosed. The last term in Eq. (3) accounts for this loss; μ_C is the rate of loss of complexes and f_{iC} is the fraction of antibody i in complexes. Because A_i is the total amount of antibody of type i in the system, the simple binding of antibody i to another molecule does not change A_i.

Antigen can either be live or dead. If it is a live antigen, it grows at a rate r_G which can depend on the antigen population level G_k. Chemical, or "dead" antigens, are modeled by choosing $r_G = 0$. Antigen is removed by combining with antibody and then being recognized by effector systems such as the complement system or phagocytic cells. The rate at which the antigen is specifically removed by these mechanisms, μ_{kill}, maybe a function of the concentrations of all antibodies which recognize the antigen and their affinities. Antigen is also nonspecifically removed, say, by the liver, at constant rate μ_G.

To complete the model, the growth, death and stimulation functions need to be specified. The growth of stimulated B cells can be described by an exponential growth law, a logistic growth law or by the class of functions discussed in Segel and Perelson (this volume)

$$r_B = r_{B0} e^{-\lambda B_i} e^{-\eta B_T^n}, \tag{5}$$

where r_{B0} is a constant. If the parameter λ is chosen positive, it can represent inhibitory effects, such as crowding in a lymph node, that reduce growth and limit the ultimate size of a clone. Alternatively, if λ is chosen negative, then self-stimulatory effects of the clone, such as the autocrine secretion of a growth factor, can be modeled. The total number of B cells in an animal rarely increases by more than a factor of two, irrespective of the antigen dose. This dependence on the total population size, $B_T = \sum B_i$ is incorporated into the last term of Eq. (5), η and n being positive constants.

The growth of antigen can also be described by a function of the form of Eq. (5), but for most applications it would suffice to use the logistic law

$$r_G = r_{G0} G_i \left(1 - \frac{G_i}{G_{i,max}}\right), \tag{6}$$

with r_{G0} and $G_{i,max}$, positive constants.

There are various choices for μ_{kill} depending upon the effector mechanism one wishes to model. For example, in complement-dependent killing, a gram-negative bacterium will be lysed if there are one or more IgM antibodies bound to its surface or two or more IgG antibodies within a critical distance of each other on the surface of the bacterium. For concreteness, assume that the antibody is IgM and that there is only one type of antigen present in the system. Then by Poisson statistics the probability of having one or more IgM antibodies bound is

$$p_M = 1 - exp\left(-\sum A_i^b / G\right), \tag{7}$$

where A_i^b is the concentration of antibodies of type i bound to the antigen and G is the antigen concentration. [For IgG-dependent lysis, a related expression can be derived that is quadratic in the IgG concentration (Wiegel and Perelson, 1979; Perelson et al., 1980)]. I assume that μ_{kill} is a constant times p_M, where the constant is the rate of antigen loss given that the complement system is fully engaged. This constant which sets the time-scale for the antigen loss process is independent of the rate of the antigen-antibody interaction which is set by the laws of chemistry. To complete this calculation one needs to analyze the chemistry of antigen-antibody binding. Here I only sketch the details.

To calculate A_i^b assume each bacterium has ρ identical epitopes to which the antibody can bind. Further assume that binding is very rapid compared to the rate of lysis so that chemical equilibrium is attained. For purposes of illustration, I assume antibodies bind monovalently; if antibodies bind multivalently, then this becomes a complex modeling problem (cf. Macken and Perelson, 1985). At equilibrium

$$A_i^b = K_{iG} A_i^f S, \tag{8}$$

where K_{iG} is the affinity of antibody i for an epitope on the antigen, A_i^f is the concentration of antibody i that is free in solution and S, the concentration of unbound epitopes on the surface of bacteria, is given by

$$S = \rho G - \sum_i A_i^b \,. \tag{9}$$

If antibody i can bind other antibodies in the system or B cell receptors, then these interactions need to be modeled in order to determine A_i^f. If antibody-antibody and antibody-receptor interactions can be ignored, so that antibody only binds to antigen then $A_i^f = A_i - A_i^b$. In this simple case, with the assumption that the cell lyses before many antibodies bind so that $S = \rho G$ is a good assumption, one finds that

$$A_i^b = \frac{K_{iG} A_i \rho G}{1 + K_{iG} \rho G} \,. \tag{10}$$

Deriving A_i^b under more general circumstances is a problem that I am currently engaged in.

The chemistry of heterogeneous mixtures of antibodies and antigens also underlies the determination of f_{iC}, the fraction of antibody i in complexes. Antibody can either be free in solution, bound to B cell receptors, bound to other antibodies or bound to antigen. The later two categories comprise immune complexes that I believe would be eliminated by phagocytosis. (Antibody i could also be bound in complexes composed of both cell surface receptors and solution phase antibodies. For example, A_i could bind with one Fab arm to a receptor on B_j and with its other Fab arm to A_j). Because of the rapidity of chemical reactions it should suffice to calculate f_{iC} at equilibrium. The required calculation is similar to that of deriving A_i^b and is currently under consideration. If one chooses to ignore network interactions and only considers antibody-antigen complex formation then classical methods of calculation such as that of Bell (1970, 1971) can be used to determine f_{iC}.

The last function that needs to be specified is α, the fraction of B cells of type i that are stimulated. Determining the conditions for B cell stimulation is a complex problem that is not fully resolved experimentally. Segel and I (this volume) suggest a simple function based on competing stimulatory and inhibitory interactions. More realistic models can be based on the notion that receptor crosslinking is required for cell stimulation (cf. Dintzis and Dintzis, this volume, Part 1; Goldstein, this volume, Part 1; Baird et al., this volume, Part 1). Models have been developed for the crosslinking of bivalent receptors by bivalent or by multivalent antigen (cf. Perelson, 1984; Goldstein and Perelson, 1984; Macken and Perelson, 1985, 1986). Modeling receptor crosslinking by antigen and a heterogeneous population of antibodies is a current research problem.

SUPPLY OF NEW B CELLS

The dynamic Eqs. (2)–(4) are meant to model an epoch over which the number of cell types in the system does not change. This epoch will end when somatic mutation generates a new type of B cell or when the bone marrow supplies a new type of B cell. These new B cells are then added to the list of cells in the system and the range of the index i is adjusted so as to encompass these new cell types. The paratope and epitope of the receptors on these new cell types are compared with the paratopes and epitopes of all other molecules in the system and the appropriate complementarities and their affinities recorded in the data base. This information is then used in computing the various functions in the model that depend on the chemistry of heterogeneous mixtures. A new epoch also begins when a new antigen is injected into the system, when an antigen in the system is successfully eliminated, or when a B cell clone dies out. (To model the elimination of a clone, I set a threshold and if the concentration B_i falls below that threshold, B cell's of type i are eliminated from the system. The threshold is set, say, to correspond to a concentration of one B cell per volume of the immune system. A similar threshold is set for the elimination of antigen).

As a modeling decision I have placed no source term in Eq. (2). Whether or not this is correct depends on ones view of the generation of diversity in the immune system. The bone marrow constantly supplies new B cells. The specific receptor that these new B cells express is determined by a complex genetic process that is designed to generate diversity. If the output of the bone marrow is sufficiently diverse (i.e., if the potential repertoire that can be created by the bone marrow is much larger than the expressed repertoire), then it is highly unlikely that during an epoch the specific types of B cells already in the system are recreated. For this reason I have not placed a source of B cells of type i in Eq.(2). Over long periods of time, the same B cell could be created again. Thus when new B cells are injected into the system at the end of an epoch, I create these B cells at random with no regard for whether or not they already exist in the system.

The absence of a source term in Eq. (2) makes the dynamics of this model different from many others. There is no "virgin" state, as in the model of De Boer (this volume), in which in the absence of interactions the source term matches the natural death term and the concentration of unstimulated cells remains constant. If a B cell is not stimulated it simply dies at rate μ_B. Unstimulated cells thus have an average residence time in the immune system of $1/\mu_B$.

INITIAL CONDITIONS

What are the appropriate initial conditions for Eqs. (2)–(4)? One might choose $G_k(0) = 0$ for all k, so that one could study the formation of a network in the absence of antigen. Also it would seem natural to assume that no antibody is initially present and that the system is populated by a set of B cells with randomly chosen strings representing their receptor paratopes and epitopes. In the absence of antigen and antibody, there are no molecules available to bind and crosslink receptors and stimulate B cells. I see three resolutions of this paradoxical situation. First, one can argue that self antigen is always present and thus setting $G_k = 0$ for all k is unreasonable. In fact, from this it seems to follow that self antigen may be important in the early development of idiotypic networks. Second, one can assume that antibody is generated in the system by the shedding of receptors from B cells or by secretion at some basal rate from unstimulated cells, i.e., $\epsilon \neq 0$. The measured rate of antibody release is very small. Varela et al. (this volume) quote a rate of $1s^{-1}$. If one considers a mouse with 10^8 B cells, each secreting 1 antibody per second, and assuming that the antibody is restricted to a circulatory system of volume 1 cc, then after one day the total serum antibody concentration will be 10^{-8} M. This antibody will be heterogeneous, with the concentration of any particular antibody type depending on the expressed repertoire of the mouse. We know that the repertoire develops sequentially. Thus, if we consider a time at the which the early repertoire is restricted, say to 10^2 clonotypes, then each antibody would be present on average at a concentration of 10^{-10} M. At this concentration, idiotypic stimulation of B cells may be possible, depending of course on the cross-reactivities of the antibodies and the affinities of the antibodies for each other. The finding that early antibodies are highly cross-reactive (Holmberg et al., 1984; Vakil and Kearney, 1986) supports these modeling ideas that require network interactions to obtain idiotypic antibody secretion early in ontogeny. A third resolution of the paradox is also possible. B cell stimulation could occur through direct B cell-B cell interactions (e.g., B cells of complementary types encountering each other and having their receptors crosslink each other). The possibility of such interactions, which depends on the repertoire size, and the cross-reactivity and affinities of the various B cell receptors, can be estimated as in the case considered above. This possibility could be incorporated into the model by an appropriate choice of the stimulation function α. Of course, all three resolutions of the paradox may be operating simultaneously. The model given by Eqs. (2)–(4) is flexible enough to incorporate all three, or to examine the consequences of each one at a time.

DYNAMICS OF THE IMMUNE SYSTEM MAY RESEMBLE WEATHER PATTERNS

The question of whether or not antibody is released by unstimulated cells is of more general interest than the above discussion may indicate. If ϵ is large then as new types of B cells are created in the bone marrow and enter the immune system new antibodies will also "enter" the system. These antibodies, if they attain sufficient concentration, will stimulate anti-idiotypic responses. Thus the immune system will behave as if it is constantly responding to (internal) antigens. Such continual responses will prevent the system from reaching steady-state. Thus, rather than having an idiotypic network in some internally generated time invariant state which is then perturbed when a foreign antigen is encountered (e.g., a virgin state, suppressed state, etc as in many models), we would have an immune system that is in constant motion. I think the dynamics of such an immune system would be more akin to weather patterns. The system never settles down to a steady-state, but rather constantly changes, with local flare ups and storms, and with periods of relative quiescence.

There may be other ways in which weather-pattern-type behavior arises in a network. Even if $\epsilon = 0$ so that antibody shedding is not significant, one could envision that a new B cell, entering a milieu containing many different antibodies, will by chance have its receptors be complementary to some and be stimulated to proliferation and antibody secretion. Thus, the network interactions themselves will cause a perturbation of the network and the appearance of new anti-idiotypic antibodies. The model given by Eqs. (2)–(4) may be able to predict when this perturbation will occur.

The view that the dynamics of the immune system may be constantly changing has not been adapted in any model of the immune system. The simulation model that I have been developing (Farmer et al., 1986; Stadnyk, Farmer, Packard and Perelson, in preparation) shows this behavior under certain circumstances.

IMMUNE MEMORY AND IMMUNE FORGETTING

I would like to end with a speculation on the role of idiotypic networks in immune memory. Memory in the immune system is generally thought to be carried by long-lived cells, called memory cells. Although this may well be the case, memory can also be carried in the form of interacting networks of cells. In Farmer et al. (1986), we suggested that a simple loop of n network elements, (as, for example, in Figure 1, in which Ab_1 recognizes both the antigen and Ab_n) could store a memory of the antigen. After removal of the antigen, idiotypic interactions could maintain an elevated level of Ab_1, so that when the antigen is reencountered the high concentration of Ab_1 antibodies would quickly eliminate it. More realistic models of memory need not rely on a simple loop. In fact, simulations have shown that a loop was a poor example of memory storage—a loop can be destroyed if any of the clones in the loop are eliminated. More complex networks with multiple paths for stimulating Ab_1, that maintain a high level of Ab_1, are better choices for memory storage.

The storage of information in an idiotypic network is a dynamic form of memory. As clones are created and destroyed, the network topology may change, and thus memories may be lost. This is a phenomenon that I call immune forgetting. Certain topologies may be more resilient to change, and may provide longer lasting memories than others. (We know from common vaccination procedures that some memories apparently do last longer than others). However, each network will have a finite capacity for storing information. As in neural networks, the information storage can be viewed as being carried both by the cells and the reactions between them. Because there can be many more reactions than types of clones, (see Figures 3–6), networks may provide an increase in memory storage over that obtained in a model based solely on memory cells.

In summary, I suggest that memory may be dynamic. Thus as new memories are created, say, by stimulated clones growing large and being incorporated into a network, other clones may be eliminated and corresponding portions of the network eliminated. Each network only has a finite capacity for storing information, and this capacity can be exceeded during one's lifetime. Thus, in this view, both immune memory and immue forgetting become natural phenomena.

SUMMARY

In this paper I have outlined an approach to the development of realistic models of the immune system. I have illustrated my approach through an example of an idiotypic network model. In developing the example, I have tried to implement a model in which, as in Jerne's original formulation, antibodies have both paratopes and idiotopes. Further, by using the notion that complementaries between the binary strings representing an epitope and paratope are indicative of complementarities between molecules, I have been able to generate idiotypic network models that have the topology of realistic networks. I have indicated that the theory of linear graphs may provide a vocabulary and means of characterizing idiotypic networks. Because each animal may have a different and dynamically changing idiotypic network, we need results about ensembles of random graphs. Ideally, one would like to obtain results that provide insights into the dynamical behavior of an idiotypic network on the basis of graphical properties of the network. (Results of this type have been obtained in the theory of chemical reaction networks.) I have provided, in somewhat general form, what I believe is a realistic set of dynamical equations for an idiotypic network. To complete the dynamical picture that I began will require a better understanding of the chemistry of mixtures of antibodies and B cell receptors interacting idiotypically. T cells and soluble factors, such as interleukins, will need to be added to such models. Lastly, I have indicated that memory may be stored dynamically and that there may be a realtionship between a network's topology, its dynamics, and its capacity to store information.

Computer simulation provides an attractive means of developing large, realistic immune system models. The network model simulations carried out to date have employed systems of size 100–200 or less. However by using 32-bit binary strings to represent each paratope, one can in principle study systems with 4×10^9 different elements. Although current computer technology does not make simulations of systems with over 10^5 or 10^6 elements economically feasible, I expect this to change. Immune system simulations are ideally suited for parallel computer implementation—each clone can be represented by a separate processor. Using parallel computing techniques, I expect that small immune systems, say, that of a young tadpole with 10^5 - 10^6 lymphocytes, will be feasible to simulate at some low level of detail in the not-too-distant future.

Although in this paper I have stressed applications to idiotypic networks, the methods and the approach are general and apply equally well to non-network models. The idea of following individual clones, labeled by a binary string that represents the unique receptor expressed by that clone, has universal applicability. Within this context, genetic processes that can generate new clones can be modeled, as well as processes that lead to the elimination of specific clones. Thus, even though the components of the system are constantly being replaced, we can obtain a model that captures the continuing nature of the immune system as a whole. As our understanding of the components of the immune system and their modes of communication improves, so too can the degree of realism in our models. Hopefully, by the year 2000, theory will have advanced to the point where all of the major known features will be included in realistic models.

ACKNOWLEDGMENTS

This work was supported by the U. S. Department of Energy. The idea of trying to predict the state of immunology after the year 2000 was inspired by the presidential address given by William Paul at the 1987 annual American Association of Immunologists meeting (Paul, 1987). I thank Catherine A. Macken for reading and providing constructive criticisms of the manuscript.

REFERENCES

1. Bell, G. (1970), "Mathematical Model of Conal Selection and Antibody Production," *J. Theoret. Biol.* **29**, 191–232.
2. Bell, G. (1971), "Mathematical Model of Clonal Selection and Antibody Production. II," *J. Theoret. Biol.* **33**, 339–378.
3. Bona, C., S. Finley, S. Waters and H. G. Kunkel (1982), "Anti-Immunoglobulin Antibodies III. Properties of Sequential Anti-Idiotype Antibodies to Heterologous Anti-γ-Globulins. Detection of Reactivity of Antibodies with Epitopes of Fc Fragments (Homobodies) and with Epitopes and Idiotopes (Epibodies)," *J. Exp. Med.* **156**, 986–999.
4. Bost, K. L., E. M. Smith and J. E. Blalock (1985a), "Similarity between the Corticotropin (ACTH) Receptor and a Peptide Encoded by an RNA that is Complementary to ACTH mRNA," *Proc. Natl. Acad. Sci. USA* **82**, 1372–1375.
5. Bost, K. L., E. M. Smith and J. E. Blalock (1985b), "Regions of Complementarity between the Messenger RNAs for Epidermal Growth Factor, Transferrin, Interleukin-2 and their Respective Receptors," *Biochem. Biophys. Res. Commun.* **128**, 1373.
6. Chang, C.-Y. and H. Kohler (1986), "A Novel Chimeric Antibody with Circular Network Characteristics: Autobody," *Ann. N.Y. Acad. Sci.* **475**, 114–122.
7. Dwyer, D. S., M. Vakil and J. F. Kearney (1986), "Idiotypic Network Connectivity and a Possible Cause of Myasthenia Gravis," *J. Exp. Med.* **164**, 1310–1318.
8. Farmer, J. D., Packard, N. H. and A. S. Perelson (1986), "The Immune System, Adaptation, and Machine Learning," *Physica* **22D**, 187–204.
9. Farmer, J. D., S. A. Kauffman, N. H. Packard and A. S. Perelson (1987), "Adaptive Dynamic Networks as Models for the Immune System and Autocatalytic Sets," *Ann. N.Y. Acad. Sci.* **504**, 118–130.
10. Goldstein, B. and A. S. Perelson (1984), "Equilibrium Theory for the Clustering of Bivalent Cell Surface Receptors by Trivalent Ligands: With Application to Histamine Release from Basophils," *Biophys. J.* **45**, 1109–1123.
11. Harary, F. (1969), *Graph Theory* (Reading, MA: Addison-Wesley).
12. Hiernaux, J. (1977), "Some Remarks on the Stability of the Idiotypic Network. *Immunochem.* **14**, 733–739.
13. Hoffmann, G. W. (1975), "A Theory of Regulation and Self-Nonself Discrimination in an Immune Network," *Eur. J. Immunol.* **5**, 638–647.
14. Hoffmann, G. W. (1980), "On Network Theory and H-2 Restriction," *Contemp. Top. Immunol.* **11**, 185–226.
15. Holmberg, D., and A. Coutinho (1985), "Natural Antibodies and Autoimmunity," *Immunol. Today* **6**, 356–357.

16. Holmberg, D., S. Forgren, F. Ivars and A. Coutinho (1984), "Reactions among IgM Antibodies Derived from Normal, Neonatal Mice," *Eur. J. Immunol.* **14**, 435–441.
17. Jerne, N. K. (1974), "Towards a Network Theory of the Immune System," *Ann. Immunol. (Inst. Pasteur)* **125 C**, 373–389.
18. Jerne, N. K. (1976), "The Immune System: A Web of V-Domains," *Harvey Lectures* **70**, 93–110.
19. Jerne, N. K., J. Roland and P. A. Cazenave (1982), " Recurrent Idiotypes and Internal Images," *EMBO J.* **1**, 243–247.
20. Kearney, J. F. and M. Vakil (1986), "Functional Idiotype Networks during B-Cell Ontogeny," *Ann. Immunol. (Inst. Pasteur)* **137 C**, 77–82.
21. Macken, C. A. and A. S. Perelson (1985), *Branching Processes Applied to Cell Surface Aggregation Phenomena* (New York Springer-Verlag).
22. Macken, C. A. and A. S. Perelson (1986), "Renewal Theory, Geiger Counters, and the Maximum Number of Receptors Bound to a Randomly Haptenated Polymer Chain," *IMA J. Math. Appl. Med. Biol.* **3**, 71–97.
23. Manser, T., L. J. Wysocki, T. Gridley, R. I. Near and M. L. Gefter (1985), "The Molecular Evolution of the Immune Response," *Immunol. Today* **6**, 94–101.
24. Marr, D. (1982), *Vision: A Computational Investigation into the Human Representation and Processing of Visual Information* (San Francisco: W. H. Freeman).
25. Melchers, F., L. Lafleur and J. Andersson (1974), "Immunoglobulin M Synthesis in Resting (G_0) and in Mitogen-Activated B Lymphocytes," *Control of Proliferation in Animal Cells*, Eds. B. Clarkson and R. Baserga (New York: Cold Spring Harbor Laboratory).
26. Milstein, C. (1986), "From Antibody Structure to Immunological Diversification of Immune Response," *Science* **231**, 1261–1268.
27. Paul, W. E. (1987), "Between Two Centuries: Specificity and Regulation in Immunology," *J. Immunol.* **139**, 1–6.
28. Perelson, A. S. (1984), "Some Mathematical Models of Receptor Clustering by Multivalent Ligands," *Cell Surface Dynamics: Concepts and Models*, Eds. A. S. Perelson, C. DeLisi and F. W. Wiegel (New York: Marcel Dekker).
29. Perelson, A. S., B. Goldstein and S. Rocklin (1980), "Optimal Strategies in Immunology. III. The IgM-IgG Switch," *J. Math. Biol.* **10**, 209–256.
30. Perelson, A. S. and Oster, G. F. (1979), "Theoretical Studies of Clonal Selection: Minimal Antibody Repertoire Size and Reliability of Self- Non-Self Discrimination," *J. Theoret. Biol.* **81**, 645–670.
31. Perelson, A. S. and F. W. Wiegel (1979), "A Calculation of the Number of IgG Molecules Required to Fix Complement," *J. Theoret. Biol.* **79**, 317–332.
32. Richter, P. H. (1975), "A Network Theory of the Immune System," *Eur. J. Immunol.* **5**, 350–354.
33. Richter, P. H. (1978), "The Network Idea and the Immune Response," *Theoretical Immunology*, Eds. G. I. Bell, A. S. Perelson and G. H. Pimbley, Jr. (New York: Marcel Dekker).

34. Seghers, M. (1979), "A Quantitative Study of an Idiotypic Cyclic Network," *J. Theoret. Biol.* **80**, 553–576.
35. Shai, Y., M. Flashner and I. M. Chaiken (1987), "Anti-Sense Peptide Recognition of Sense Peptides: Direct Quantitative Characterization with the Riobnuclease S-Peptide System using Analytical High-Performance Affinity Chromatography," *Biochemistry* **26**, 669–675.
36. Smith, L. R., K. L. Bost and J. E. Blalock (1987), "Generation of Idiotypic and Anti-Idiotypic Antibodies by Immunization with Peptides Encoded by Complementary RNA: A Possible Molecular Basis for the Network Theory," *J. Immunol.* **138**, 7–9.
37. Stadnyk, I. (1987), "Schema Recombination in Pattern Recognition Problems," *Proc. 2nd International Conference on Genetic Algorithms and their Applications, MIT, Cambridge, MA.*
38. Vakil, M. and J. F. Kearney (1986), "Functional Characterization of Monoclonal Auto-Anti-Idiotype Antibodies Isolated from the Early B Cell Repertoire of BALB/c Mice," *Eur. J. Immunol.* **16**, 1151–1158.
39. Victor-Kobrin, C., T. Manser, T. M. Moran, T. Imanishi-Kari, M. (1985), "Shared Idiotopes among Antibodies Encoded by Heavy-Chain Variable Region (V_H) Gene Members of the J558 V_H Family as Basis for Cross-Reactive Regulation of Clones with Different Antigen Specificity," *Proc. Natl. Acad. Sci. USA* **82**, 7696–7700.

Index